Beyond 2012

Catastrophe or Awakening?

"*Beyond 2012* is the most comprehensive study of the 2012 phenomena to date. Geoff Stray's detective story leaves no stone unturned as he uncovers the truths and the untruths concerning this fast-approaching date with destiny."

JAY WEIDNER, COAUTHOR OF
THE MYSTERIES OF THE GREAT CROSS OF HENDAYE

"*Beyond 2012* is an invaluable resource. Fair, open-minded, and rational, Geoff Stray's book displays the enormous range of the 2012 phenomenon; a sort of mega-mall of marvelous strangeness, insights, and sudden flashes of truth."

VINCENT BRIDGES, COAUTHOR OF
THE MYSTERIES OF THE GREAT CROSS OF HENDAYE

BEYOND 2012

Catastrophe or Awakening?

A Complete Guide to End-of-Time Predictions

GEOFF STRAY

Bear & Company
Rochester, Vermont

Bear & Company
One Park Street
Rochester, Vermont 05767
www.BearandCompanyBooks.com

Bear & Company is a division of Inner Traditions International

Originally published in the United Kingdom in 2005 by Vital Signs Publishing under the title *Beyond 2012: Catastrophe or Ecstasy—A Complete Guide to End-of-Time Predictions*
Reprinted with minor revisions in 2006 by Vital Signs Publishing
First U.S. edition published in 2009 by Bear & Company

Library of Congress Cataloging-in-Publication Data
Stray, Geoff.
Beyond 2012 : catastrophe or awakening? : a complete guide to end-of-time predictions / Geoff Stray. — 1st U.S. ed.
p. cm.
Includes bibliographical references and index.
Summary: "An illustrated, encyclopedic overview of the prophecies, calendars, and theories that indicate the year 2012 is a threshold of great change for humanity"—Provided by publisher.
ISBN 978-1-59143-097-1
1. Two thousand twelve, A.D. 2. Twenty-first century—Forecasts. 3. Catastrophical, The. I. Title.

CB151.S84 2009
909.83—dc22

2009005414

Printed and bound in the United States by Lake Book Manufacturing

10 9 8 7 6 5 4 3 2 1

Text design and layout by Jonathan Desautels
This book was typeset in Garamond with Ronda used as the display typeface
Book illustrations by Geoff Stray, unless otherwise noted

Every attempt has been made to track all copyright holders. If any errors or omissions have occurred, we will be happy to amend or credit in any future edition.

To contact the author or to contribute ideas and information to the ongoing Dire Gnosis project, please use the e-mail address given on the index page of the Diagnosis 2012 website, at **www.diagnosis2012.co.uk.**

Dedicated to the memory of

Terence McKenna

and of

John G. Stray

CONTENTS

Foreword by John Major Jenkins ix

Acknowledgments xv

Introduction 1

PART 1

Ancient Calendars and Tribal Prophecies

1 The Maya Calendars 9

2 Tribal Prophecies 24

3 Asian Calendars 41

4 I Ching: Ancient Chinese Lunar Calendar 56

5 Abrahamic Religions 70

PART 2

2012 Theories

6 Sunspot Cycles 91

7 Astronomical Claims 100

8 Alternative Archaeology 122

9 The Geomagnetic Field 155

10 Ice Cycles 164

11 0.0.0.0.0 171

12 Fringe Science 178

PART 3

Beyond the Veil

13	Shamanism	203
14	Other Altered States	218
15	UFOs and ETs	233
16	Crop Circles	242
17	Secret Government	259
18	Nostradamus	264
19	The New Age	276

PART 4

Diagnosis

20	Analysis	309
21	Synthesis	336
22	Prognosis	370
	Afterword	400
APPENDIX 1	Two Accounts of Out-of-Body Experiences	403
APPENDIX 2	The Astral Transition	406
APPENDIX 3	Cropgnosis	415
	Notes	421
	Bibliography	462
	About the Author	473
	Index	474

Foreword

INTO THE BEYOND

(Laying the End to Rest)

Writings on 2012 have multiplied exponentially since the end of the last millenium. As we draw closer to 2012, interest focusing on that enigmatic, co-opted, perplexing date will be growing. It is a true and multifaceted vector for all kinds of ideas. The plethora of writings on the subject is a phenomenon in itself, and while discussion of the year 2000 centered largely on Y2K (the "millennium bug"), which proved to be as much a phantasm as many of the much more wacky millennial ideas, there is as yet no discernible locus of the 2012 discussion.

It is for this reason that the Diagnosis 2012 website has been valuable to the ongoing discussion. Since 1998, Geoff Stray has collected and organized a wide spectrum of writings, reports, dreams, sci-fi, and nonfiction studies on the 2012 theme. Although the center of the discussion has yet to be discerned, at least the voices are now assembled under one roof. Stray has been unbiased in what he has allowed into the pavilion of purview. He is a true pioneer, the first 2012-ologist who has sought to collect, survey, contextualize, and comment on the wide spectrum of manifestations related to 2012. Because of his familiarity with all things 2012, he is our best guide into the labyrinths of kaleidoscopic creation and consternation that typify the inner landscape of 2012-land.

While his pavilion has been open to all, Stray has used common sense and discerning analysis to critique the contributions, such that basic errors in theories have been identified, always involving internal inconsistencies in the theory itself, rather than by reference to a set of preconceived doctrines of what 2012 "really" means. I appreciate this quality in Stray's book on 2012 because it allows us to categorize the wide spectrum of writings into fiction and nonfiction, trace the interrelationships, and discern, sometimes, the shared sourcings between different contributions. A bit of order has thus been given to the chaos of creativity that 2012 has spawned.

Geoff Stray clearly leans in a certain direction when it comes to his own views—views that have been distilled, it should be said, from his immersion in the field of ideas—but it is possible to give allegiance to one's own preferential viewpoint while retaining objectivity in assessing all others, and Stray has done this. Another admirable quality evinced by Stray is essential for continuing the conversation on 2012: open-mindedness. We can all strive to practice this virtue, as it is clear even in this incipient phase of the birth of 2012-ology that many views are going to be mutually contradictory, irreconcilable with one another, and thus "unity in multiplicity" may be the ultimate teaching to emerge from our passage through 2012. *Open-mindedness,* rather than meaning the uncritical endorsement of everything, thus takes on the connotation of an inclusive mentality that can embrace contradictory views (even while retaining one's own perspective).

The creation and nurturing of that state of consciousness, that transcendent viewpoint, may initially seem to be simply a prerequisite of furthering serious discussion on 2012. However, such a metaview may indeed coalesce as a supremely important emergent property from our collective dance with 2012. The lesson distilled is simple: higher perspectives are needed to resolve problems and conflicts. And in that speculation we glimpse that something might lie beyond 2012—beyond it in time and space, and beyond it in our own minds. In other words, there can be a "beyond 2012" if we mentally create the space that can embrace that possibility, in effect transcending the apocalyptic nihilism that defines modern civilization and pollutes modern consciousness. Transcendence does not escape or deny that which is transcended, but includes it in a higher vision. We may all come to agree that 2012 is not a final end, which is a notion that seems to be one of the largest and potentially most counterproductive assumptions about it.

This book is called *Beyond 2012*. How are we to go beyond 2012? Is there anything beyond 2012? What is the intended meaning of this phrase? Geoff Stray tells me that "the title intends to impart optimism to the subject, decreasing anxiety that the end of the world is approaching, since it turns out that many of the catastrophe theories are inherently faulted." And so *Beyond 2012* takes the courageous position of clarifying and simplifying the 2012 discussion by winnowing out theories of dubious merit. The 2012 discussion shows that people tend to gravitate toward one of two positions—that 2012 represents a catastrophic end or it represents a new beginning (in some sense). These are difficult questions to resolve, as they touch upon profound meditations on death and immortality, vast cycles of time, and ancient insights into human consciousness that are just recently coming to light. Could it be that our own educational institutions and our own civilization's preconceptions have left us ill-prepared to deal with such profound questions? We, the denizens of the Kali Yuga, the age of spiritual darkness, are struggling to embrace profound higher wisdom, and thus our own transformation does indeed seem to be the prerequisite for any clear understanding of what "the end of time" means. Getting beyond 2012 is about transcending preconceived biases and limitations born from Western civilization's limited worldview.

The 2012 date itself is specifically an artifact of the Maya calendar. It is not a New Age creation of any particular author; it is a reconstructed artifact of the Maya Long Count calendar tradition. Why the ancient calendar makers ended their thirteen-baktun cycle on the date we call December 21, 2012, has been an open question. I believe that date was intended (by its creators) to target a rare alignment between Earth, our Sun, and the center of our Milky Way galaxy—a factual astronomical alignment that takes place only once every 26,000 years. My research follows academic principles of argument and documentation to arrive at this surprising synthesis and reconstruction. After subsequent research, this knowledge of a rare "galactic alignment" also appears to be the centerpiece of other wisdom traditions from around the globe. As such, while the emergence of the 2012 date into contemporary thought is traceable to Maya calendric tradition, the inner wisdom of 2012 is universal. We might even say that 2012 is emerging as an archetype of the end times, and thus resonates with the universal human questions about death and immortality. This avenue of association rightfully opens up 2012 to entire categories of approach that have little to do with my cosmological reconstruction, or even with Maya culture per se. Having said this, we must also honor Maya genius in what is, above all, a profound and advanced cosmological model of human development that uses the Galactic Center as its evolutionary centerpiece.

Readers, by their different natures and preferences, will gravitate toward different categories in this discussion, revealing the relative merit they see in each approach. For some, my archaeophilosophical resurrection of a lost galactic cosmology may seem hopelessly inaccessible, while dream visions catalyzed by ingesting the sacred plant *Salvia divinorum* reveal more interesting data on what 2012 is "really" about. Truth can appear relative—related to ones own position—and that egalitarian principle must be held high, to uphold freedom of expression and belief. But it must also be noted that some of the theories surveyed in *Beyond 2012*—especially in their more recent iterations—gratuitously toss in 2012 to lend superficial support to a theory that otherwise has nothing to do with Maya calendrics or recognizable Maya beliefs. These cut-and-paste cosmologies, such as the new connection between "Planet X" and 2012, indirectly attribute to the Maya ad hoc theories that are poorly supported and that, in fact, espouse beliefs counter to Maya tradition. This is a marketing ploy, plain and simple, to co-opt the awe and apocalyptic punch that 2012 has gained within popular consciousness. It's like name-dropping, and name-dropping doesn't mean you know anything at all about the person named.

Thus some of the theories presented here unfairly latch on to 2012, basically as a marketing ploy, while others are dealing head-on with the real issues of the Maya intention behind 2012. Adhering to the latter approach, I base my reconstruction of the true intention behind the 2012 date on examining the early Maya site that formulated the 2012 calendar. This early Maya site, which experienced its heyday some 2,100 years ago, is called Izapa. In this I am alone among academics and popular writers alike, and my work has therefore generated unique conclusions that are nevertheless backed up by archaeological, iconographical, calendrical, mythological, and astronomical evidence. So we have two extremes of the spectrum: on the one hand, the carefully argued reconstruction of authentic (although up until now forgotten) Maya traditions, and on the other hand, the spurious concoctions of ad hoc ideas that threaten to inject modern distortions into ancient wisdom. Many of the ideas presented in *Beyond 2012* fall somewhere in between, into an intriguing domain of deeper mysteries and higher faculties, of shamanic visions, alternative science, resurfacing traditions, and recovered histories. This domain is worth exploring.

Shall the 2012 discussion be open only to scholars and academics? Definitely not. Professional Mayanists have been singularly unwilling to rationally examine the role that 2012 may play within Maya philosophy and religion. Often, scholars will draw from the silliest hysteria-driven New Age speculations about 2012, and dismiss the whole thing as an unfortunate, but predictable, product of human fears and failings. Although some scholars have recently started to take the 2012 question seriously, for many years independent investigators have been leading the discussion and making breakthroughs. This is actually par for the course in Maya studies, where, historically, the major shifts in understanding have been instigated by self-motivated private investigators like Joseph Goodman, Benjamin Whorf, Yuri Knorosov, and Tatiana Proskouriakoff. Their insights, however, were rejected vehemently and ignored for years before the academic gatekeepers finally embraced them.

The 2012 discussion is formulating itself as a widespread, grass-roots, popular movement—much more so than Y2K/Year 2000. Such broad interest may itself be the antidote to official myopia. If a university think tank emerges to confer on 2012, scholars should refer to *Beyond 2012* as a guide to the multiple theories on 2012. Since the material runs the gamut from inner-dream visions to serious Ph.D.-level studies, academics and social commentators will not be able to toss everything here into an easily dismissed bin of New Age fantasy. I support and encourage serious discussion of 2012, and the current level of the discussion—in quality and cogency—must be raised. *Beyond 2012* does this with admirable comprehension and comprehensiveness, inviting us all to engage with this topic in deeper ways. Again, the keys, exampled by Geoff Stray, are open-mindedness and discernment.

So let us go to the next level in the evolving discussion. Let us have conferences, symposia, think tanks, Yahoo e-mail groups, and café salons. Let us write treatises and poetry, songs and equations. Let us visit Chichén Itzá, Izapa, Egypt, and the inner planes of higher wisdom, and let us reconvene in 2013, when we are beyond 2012, and talk about what accompanies us unscathed through the eye of the needle, the Sun Door at World's End, the Ginnungagap between the ages, the eschatological Symplegades, the still-point threshold crossing, the nexus between death and rebirth.

Vincit omnia veritas: Truth conquers all.

John Major Jenkins,
1 Chicchan, 12.19.11.3.5

John Major Jenkins is a leading independent researcher on ancient Mesoamerican cosmology. He has authored five books on the Maya, including *Maya Cosmogenesis 2012, Tzolkin,* and *Galactic Alignment,* and has given a presentation to the prestigious Institute of Maya Studies in Miami. In March of 1998, he was invited by the Indigenous Council of the Americas to speak at its conference in Merida, Mexico.

ACKNOWLEDGMENTS

Thanks go to the late Terence McKenna for the initial inspiration for this book and for the many spoken word and musical recordings that sustained that inspiration throughout this project. Massive appreciation of Andy Thomas, for applying his proofreading, editing, and publishing skills, and whose encouragement, suggestions, and friendship have been a guiding light over the years.

Special thanks go to John Major Jenkins, for his honesty, integrity, and persistence in uncovering and disseminating the truth, for friendship and support, for the use of his images, and for writing a great introduction to this book. Many thanks also to Vincent Bridges and Darlene, for their help and support and for the use of Darlene's excellent drawings and photographs. Thanks to my ex-lodger, Jan Maloney, for unwavering friendship, for teaching me to use a computer, and for her Scrabble and creative writing skills used in proofreading the text; gratitude for Helen Sewell and Paul Wright, for astrological checking and support. Thanks to Ian Crane for encouragement, and to both him and Lori for their much-needed help during my nomadic period. For help with the Mac, huge thanks go to Jason Porthouse and Allan Brown. To Tania Woodward, Joel Keene, Zyzygyz, Hugh Newman (the New Human), Jay Weidner, Caroline Taylor, John Martineau Clare, Mark and Daphne, thank you all for your support. To my family—John, Eileen, Phyllis, Ange, Keith, Ruth, Nick, Daniel, James, Marina,

Alec, Simon, and that talented rock star Emily: thanks for your love and not openly saying I've lost my marbles. Thanks, too, to all those who have contributed to this project via conversation, e-mails, or their own research and writings, or who have allowed me to use their images, or invited me to speak at their meetings and conferences.

INTRODUCTION

The Maya civilization flourished between 200 AD and 900 AD in Mesoamerica, an area that corresponds to modern southern Mexico, Guatemala, Belize, Honduras, and part of El Salvador. The Maya* had a complex system of calendars, including one called the Long Count, which measured vast stretches of time with uncanny accuracy. It now seems certain (thanks to the work of John Major Jenkins, particularly *Maya Cosmogenesis 2012)* that the Long Count originated in Izapa, which was first inhabited around 1500 BC, but it may have been as late as 355 BC before the Long Count was conceived.

The Long Count (see chapter 1—Tzolkin, Haab, and Long Count—and a detailed description in chapter 19, "The Calleman Solution") includes a cycle that lasts 1,872,000 days, or approximately 5,125 years, which is known as the thirteen-baktun cycle. (Some refer to it as the Great Cycle, but this term has also been used to describe other cycles, so to avoid confusion I call it the thirteen-baktun cycle throughout this book.) For the last hundred years or so, archaeologists have argued over the correlation of Long Count dates to the Gregorian calendar that we use today, and have now mostly agreed on one of two correlations. There

*Note on Maya terminology: I have attempted for the most part to follow the protocol of Maya scholars, in which the word "Mayan" is used only in reference to the language spoken by the Maya, and the word "Maya" is used to describe things pertaining to the Maya, such as Maya calendars.

are a few rogue scholars who continue to disagree, such as Vollemaere, but Mayanist Mike Finley has described Vollemaere's work as being "on the remotest fringes of scholarship." Other correlations, such as the Wells-Fuls correlation, are well reasoned but rely primarily on astronomy, whereas the GMT combines dates recorded in ceramics, dates inscribed on stelae and buildings, cross-referenced Moon and Venus information, dates recorded in the late Classic codices such as the Dresden Codex, de Landa's recorded dates, Aztec dates of the arrival of Cortés, the Chilam Balam and other postconquest books, and the unbroken tzolkin count still being used today in the Highlands of Guatemala.

The two correlations are only two days apart, and they relate the beginning (or zero date) of the current thirteen-baktun cycle to the Gregorian historical date of either August 11, 3114 BC, or August 13, 3114 BC. This means that the cycle ends 1,872,000 days later, on December 21, 2012 AD, or December 23, 2012 AD, respectively. What the archaeologists or Mayanists have yet to agree on is why the Maya chose December 2012 as the end point of the thirteen-baktun cycle.

In 1987, José Argüelles published a book called *The Mayan Factor,* in which he brought the 2012 end point of the thirteen-baktun cycle to the attention of the world at large. I was among the readers of that book, and was particularly fascinated because for the previous five years I had already been studying a calendrical system from another part of the world that had a 2012 termination point!

Terence and Dennis McKenna published a book in 1975, *The Invisible Landscape,* in which they showed that the Chinese oracle the I Ching had originally been used as a lunar calendar in which the sixty-four six-line hexagrams—384 lines in all—represented the 384 days in a thirteen-month lunar year. Following their 1971 journey through the Amazon jungle, in which the brothers had a shared hallucinogenic experience, they developed their insights and devised a method by which to convert the Chinese Book of Changes, the I Ching, into a complex "wave" that mapped the timing of all the changes in the universe, or "the ingression of novelty into space-time." The complex wave was composed of a hierarchy of waves—each one of a magnitude sixty-four times greater than the one below it—that mapped all change, from subatomic event durations up to the duration of the three-dimensional physical universe. The explosions of the atomic bombs at the end of World War II were taken to be sufficiently novel events that they should correspond to a major peak on "the Timewave," and when lined up with these events in the Gregorian calendar, it was found that late in 2012 AD, all the waves and subwaves would peak together, signifying concrescence—that is, an evolutionary pinnacle and dimensional transition.

Incredibly, knowing nothing of the McKennas or the Timewave, Argüelles, in 1984, published a book, *Earth Ascending,* in which he revealed a relationship between the I Ching and the *tzolkin,* the 260-day calendar of the Maya. In 1987, in *The Mayan Factor,* Argüelles went on to show that the tzolkin was a microcosm of the thirteen-baktun cycle (260 *katuns,* in which a single katun consists of 7,200 days). Likewise, when the McKennas wrote the 1975 edition of *The Invisible Landscape,* they knew nothing of Argüelles or the end of the thirteen-baktun cycle of the Maya.

There are too many "coincidences" here to use the word *coincidence*! *Synchronicity,* the word coined by psychologist Carl Jung to describe a meaningful coincidence, is doubly apt: *syn* means same; *khronos* means time.

I started a research project to try to find out what might be scheduled for 2012, and eventually condensed my findings into summary paragraphs in a booklet called *Beyond 2012.* The booklet developed into a website, www.diagnosis2012.co.uk,* attracting input from people from all over the world. It then morphed into this book, in which I have brought all of the research together in an attempt to find a pattern that

*www.diagnosis2012.co.uk—also known as 2012:Dire Gnosis, where Dire means serious or urgent, as well as dreadful, and thus sums up the ambiguous nature of the 2012 phenomenon. Dire Gnosis therefore means, primarily, "important knowledge" as well as implying a diagnostic approach.

underlies all the pieces of the information, misinformation, and disinformation that have been circulating regarding 2012.

Part 1, Ancient Calendars and Tribal Prophecies, looks at the ancient sources of information that are coming to light, suggesting that ancient cultures knew something about an event far in the future and tried to preserve knowledge about it, sending us a message across time. We look at the calendars of the Maya and at the tribal prophecies of Mesoamerica, North and South America, and as far away as Africa and New Zealand. We also take a look at Asian calendars, so-called Western, or Abrahamic, religions, and the I Ching.

In part 2, 2012 Theories, we look at the many ideas that have been prompted by the recent publicity surrounding the calendars and prophecies. Different speculations attempt to prove that our ancestors were trying to warn us about a forthcoming cataclysm, but how convincing are they? Are these just attempts to get some kind of meaning out of the book of Revelation? The theories concern solar cycles, asteroids and comets, rogue planets, plasma bands, the exploding galactic core, and other ideas, and the outcomes are predicted to be everything from burnt-out power stations to total wipeout. One thing is for sure—they can't all be right.

In part 3, Beyond the Veil, we examine contemporary sources of information on 2012, from Nostradamus to modern visionaries and prophets, as well as information gleaned from altered states of consciousness, channelers, UFO contactees, near-death experiencers, psychedelic voyagers, remote viewers, and out-of-body travelers. Even crop circles, those mysterious shapes that appear in grain fields every year, have been connected to 2012.

In part 4, Diagnosis, we take some of the most interesting and convincing items that have emerged during the investigation, apply prehistoric wisdom, and come to a surprising conclusion.

Following part 4 are three appendices that extend the conclusion of this study. Appendix 1 consists of two quotations with comments. Appendix 2 was posted on the Diagnosis 2012 website in 2001.

Appendix 3 is a slightly modified version of a short article posted on the Swirled News crop-circle website (www.swirlednews.com) in 2002.

The conclusions reached in this book should not be seen as "the" answer, since we should always be ready to look at new information and reexamine our position. It is hoped that by looking closely at all the current theories regarding 2012, this book will save people a lot of time trying to reconcile the many different, contradictory, and often confusing collections of information, and thus improve communication, reduce anxiety, and allow research to progress more quickly.

During the course of this study, we shall discuss the experiences of various people who have used hallucinogenic plants and compounds. This does not mean I am recommending that readers do likewise, any more than I recommend that someone try to have a near-death experience. The information that these psychonauts have brought back from inner space is crucial to the discussion, and I am very grateful to all those who have provided it. I am also very grateful to all the theorists who have already grappled with the 2012 enigma, regardless of whether or not I agree with some, or none, of their ideas. The possibility that we may be approaching a very rare event in the history of the planet means this is an essential topic deserving our attention.

If you have any ideas or information to contribute to the ongoing Dire Gnosis project, please contact me through the e-mail address given on the index page of the Diagnosis 2012 website, at www.diagnosis2012.co.uk.

PART 1

Ancient Calendars and Tribal Prophecies

Probably the wisest view is to say: the truth— like the Self—is splintered up over thousands of miles and years; bits are found here and there, then and now, and must be re-collected; bits appear in the Greek naturalists, in Pythagoras, in Plato, Parmenides, in Heraclitus, Neo-Platonism, Zoroastrianism, Taoism, Mani, Gnosticism, orthodox Christianity, Judaism, Brahmanism, Buddhism, Orphism, the other mystery religions. Each religion or philosophy or philosopher contains one or more bits, but the total system interweaves it into falsity, so each as a total system must be rejected, and none is to be accepted at the expense of all the others (e.g., "I am a Christian" or "I follow Mani"). This alone, in itself, is a fascinating thought: here in our spatiotemporal world we have the truth, but it is splintered—exploded like the EIDE—over thousands of years and thousands of miles and (as I say) must be re-collected, as the Self or Soul or EIDOS must be. This is my task.

Philip K. Dick, *In Pursuit of Valis: Selections from the Exegesis*

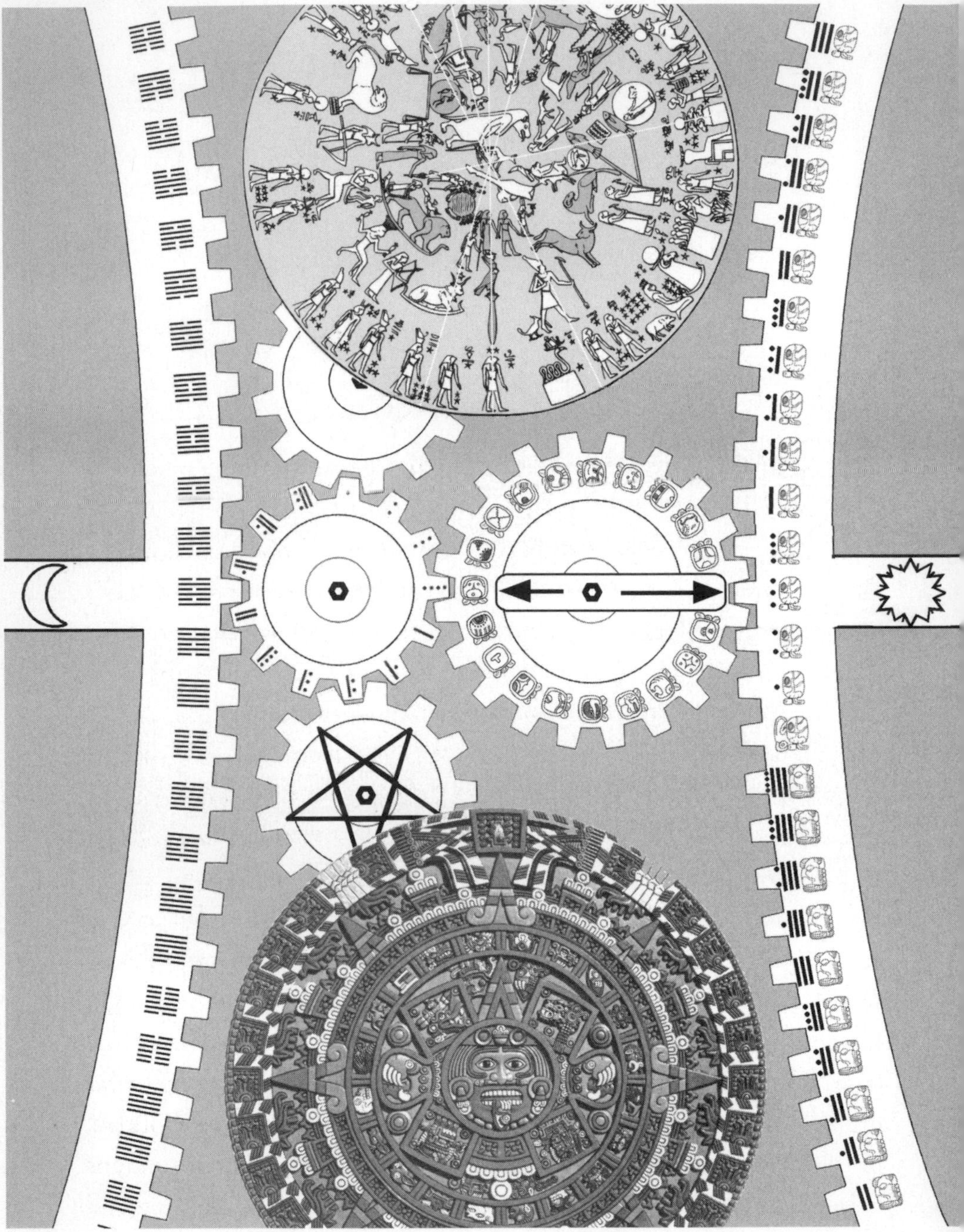

1

THE MAYA CALENDARS

TZOLKIN, HAAB, AND LONG COUNT

The tzolkin, or Sacred Calendar, of the Maya is a 260-day system based on the period of human gestation, according to the descendants of the Maya, who live in the highlands of Guatemala and have used the same unbroken count of days for more than a thousand years. It is composed of twenty day signs, each of which has thirteen variations, and was (and still is) used to determine character traits and time harmonics. It is a little reminiscent of Western astrology, except that the tzolkin does not directly involve the movement of planets relative to stars. Each of the 260 different combinations of day signs and numbers indicates a unique quality for that day, and the tzolkin day that a person is born on is repeated at 260-day intervals as a tzolkin birthday, when the quality of the day that determined that person's character traits is repeated. According to John Major Jenkins's book *Tzolkin,*[1] astronomical cycles that may connect with the tzolkin include such things as the zenith passage interval of the Sun at the latitude of Izapa, Mexico, and the synodic cycle of Mars (one solar orbit by Mars in relation to Earth), which is exactly three tzolkins, or 780 days. It is also an intriguing fact that two of Jupiter's moons have a 260-day cycle. Another consideration is that the morning and evening star phases of Venus last 263 days on average, according to Anthony Aveni,[2] but Jenkins says the period between Venus rising as an evening star and its emergence as a morning star is 258 days, as recorded in an eleventh-century Maya picture book called the Dresden Codex. In addition, three eclipse half-years are equal to two tzolkins.[3] Two hundred sixty days is also the length of time between planting and harvesting maize, the staple diet of the Maya. However, most significant of all is that the Maya daykeepers who live in the Guatemala highlands and still use the tzolkin say that it is based on the period of human gestation.[4]

Bruce Scofield did ten years of research through the 1980s, piecing together the remaining odd scraps of information concerning the meanings of each of the days in the 260-day calendar. He used "historical documents, reports from oral tradition, psychic-archaeological techniques, and common sense."[5] He looked at myths connected with the gods associated with each day sign and, being an astrologer, used a database of about four hundred celebrities and friends with known birthdays and personality traits to look for patterns to "crack the code." A summary of the process and the results can be found in his book *Day-Signs*.[6] It was while reading this book, and the definitions of the twenty day signs, that I clearly recognized my father's character. Before checking to see if the day of his birth actually occurred on the day sign with the characteristics I recognized, I looked through the thirteen variations of it that take place over 260 days. There were two variations that seemed to describe him the best, and when I checked I was amazed to find that he was born on one of these two signs—at odds of 130 to one.

The Maya also used a 365-day calendar called the *haab,* consisting of eighteen "months" of twenty days each (360 days), plus five extra days called *uayeb*. Each of the months had a different name, and the days progressed numerically, like ours, where January 1 is followed by January 2. An example from the haab would be 1 Pop followed by 2 Pop, whereas in the tzolkin, the number and day sign both change from one day to the next. For example, 1 Imix would be followed by 2 Ik. Each day therefore had a tzolkin name and a haab name, and the combination would only happen once every fifty-two "years" (haabs of 365

Fig. 1.1. An improved and updated list of day sign meanings would be Crocodile, Wind, Night, Maize (Lizard), Snake, Death, Deer, Venus or Star (Rabbit), Water, Dog, Monkey, Road or Rain (Grass or Tooth), Corn or Young Maize (Reed), Jaguar, Eagle, Owl, Wisdom (Vulture), Earth (Earthquake, Movement), Flint (Knife, Mirror), Storm (Rain), Lord, Sun (Flower).

days each), or seventy-three tzolkins. This period is called a Calendar Round, which we shall return to shortly.

The Maya measured long time periods by means of a Long Count, in which one 360-day "year" (a *tun*) consists of eighteen twenty-day cycles, or "months," known as *uinals*. Twenty of these tuns add up to a katun; twenty katuns make a baktun (nearly 400 years), and thirteen baktuns add up to a thirteen-baktun cycle of 1,872,000 days (5,200 tuns, or about 5,125 years). The first day, or zero day, when thirteen baktuns had been completed, was recorded as 13.0.0.0.0 (written in Arabic numerals—the numbering system we use today) and is equivalent to 0.0.0.0.0.

Scholars specializing in the Maya have been attempting to correlate the Long Count to our Gregorian calendar since the beginning of the twentieth century. There has been massive variation in the suggested correlations, but as early as 1897, Joseph Goodman suggested that the Maya Creation date was in 3114 BC. In 1905, he refined the correlation to #584280 (the number refers to the Julian day number to which the zero day is correlated). This was unpopular until 1926, when Juan Martinez Hernandez confirmed it (with one day difference—#584281). In 1927, Eric Thompson applied lunar and Venus data to arrive at a four-day adjustment: the #584285 correlation, (expounded upon with more detail in 1935), later confirmed by Floyd Lounsbury, who found evidence to support it by cross-referencing astronomical phenomena with those recorded in the Dresden Codex. In this calculation, the zero day correlates to August 13, 3114 BC, and the end-date to December 23, 2012 AD. In 1950, Thompson did a final reexamination of Goodman's work, combining it with the latest findings, and settled on the #584283 correlation (Goodman-Martinez-Thompson-2, or GMT-2). This puts the start of the thirteen-baktun cycle (day 0.0.0.0.0) on August 11, 3114 BC, and the end-date (known as 13.0.0.0.0) on December 21, 2012 AD (this proposal had first been inferred by Thompson in 1937). This correlation coincides with the unbroken count of days (now known as the True Count, Indigenous Count, or Quiche Count) that has been followed by the Maya people still living in the Guatemala highlands.

Fig. 1.2. If a stela were found bearing the date of the next Creation, 13.0.0.0.0, 4 Ahau, 3 Kankin, corresponding to December 21, 2012 AD, it would look like this. Under the large introductory glyph, the first line reads (l–r): thirteen baktuns, zero katuns. The second line reads: zero tuns, zero uinals. The third line reads: zero kin, plus the tzolkin date 4 Ahau. The bottom line shows the haab date 3 Kankin (and the ruling underworld deity, Night-lord 9).

There is a point to note here, if you ever want to check any dates, or understand the various 2012 theories. To calculate periods of time that traverse the start of our calendar at the theoretical birth of Christ, you have to add an extra year to the calculation because we have no "year zero" in the Gregorian historical year. Since the eighteenth century, astronomers calculating past the BC–AD divide have used their own convention so that they don't have to add the extra year to calculations. Originally they differentiated their system by calling it the "Gregorian astronomical," as opposed to the "Gregorian historical," system. Later they abbreviated it to just using a minus sign. Thus, the thirteen-baktun cycle started in 3113 BC astronomical, or year -3113, both of which are equivalent to 3114 BC historical. The first two examples do not need to have an extra year added to calculations going into AD years; the last one does need the extra year added. The only problem is that some Mayanists (I am still researching who and when) forgot to include the "astronomical" tag on their 3113 BC start-date, or the "historical" tag on their 3114 BC start-date. Others assumed that 3114 BC was an astronomical date, did not add a year, and miscalculated an end-date of 2011. Some researchers made the alternative mistake of taking the 3113 BC date as a historical date, added an extra year, and miscalculated an end-date of 2013.

The 260-unit tzolkin can be used as a map of the thirteen-baktun cycle, which consists of 260 katuns of around twenty years each (see page 14). Each column represents one baktun, the start-dates of which are listed along the top in Long Count notation with the equivalent Gregorian historical year. Katun start-dates in the current baktun are shown in the right-hand margin.

José Argüelles was probably the first to point out (in *The Mayan Factor*) that since the tzolkin is composed of 260 kin, or days, and the thirteen-baktun cycle is composed of 260 katuns, the thirteen-baktun cycle is thus a harmonic of the tzolkin, being exactly 7,200 times larger. This means that the 13 × 20 tzolkin layout can be used to map history, as if it is measuring not individual gestation, but species gestation, and

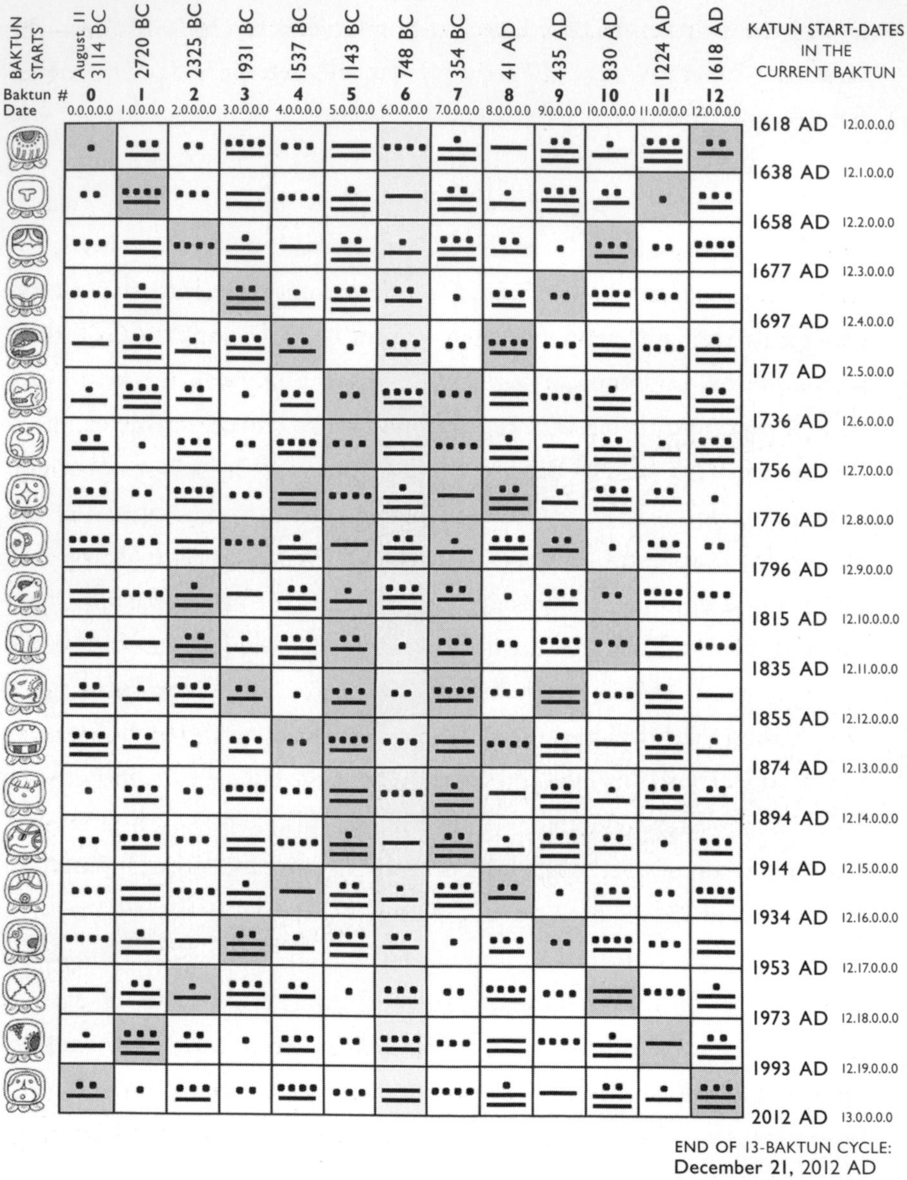

Fig. 1.3. The Sacred Calendar of 260 days, the tzolkin, or Chol Qij. Reading down all the columns, starting at the top left, note that after 13, the numbers go back to 1 again. It takes 260 days for all the combinations of thirteen numbers and twenty day signs to occur. The darker squares were conceived by Tony Shearer (see Shearer, *Beneath the Moon and Under the Sun,* 1975, p. 82), since the sum of the numbers of the four corners is twenty-eight, and the pattern then continues inward, mirrored around a central axis. However, each of the squares on the tzolkin (except the central axis) can be added up to twenty-eight, with their three counterparts.

five of the thirteen-baktun cycles add up to exactly 26,000 tuns—the "Grand Year," or precession of the equinoxes* (see fig. 1.4), a yet higher harmonic of 260.

GALACTIC ALIGNMENT

John Major Jenkins has studied the mythology, architecture, king-accession rites, and mysterious ball games of the Maya (the ball game was played in ball courts all over Mesoamerica and was developed into a symbolic mystery play), and he found encoded there the meaning behind the Maya thirteen-baktun cycle. Jenkins's work is so thorough and his approach so scholarly that he has even convinced some of the archaeological community (usually very entrenched in their views) that he has cracked the meaning of the Long Count and its thirteen-baktun cycle. In his book *Maya Cosmogenesis 2012,* he shows that the thirteen-baktun cycle is a fifth and final cycle in the 26,000-year precession of the equinoxes, except that the Maya measured it from the winter solstice instead of the spring equinox. On December 21, 2012, the solstice Sun will align with the dark rift in the Milky Way (this is a thirty-six-year process, explained below), which the Maya called the Mouth of the Crocodile (or jaguar-toad), the Crocodile Tree being the Milky Way itself.

It would be a good idea to clarify here that the Galactic Center is invisible due to interstellar dust, but has been pinpointed by radio telescopes. When some people speak of the Galactic Center,† they mean the bright central or nuclear bulge; others mean the central part of that bulge

---

*The precession of the equinoxes is the time it takes for the equinoxes to precess back through all twelve signs of the zodiac.

†Throughout the book, the term "Galactic Center" will be used when referring to the astronomical, or radio-telescope-defined center, and the term "visual galactic center" or "apparent Galactic Center" will be used when referring to the crossing point where the 2012 Galactic Alignment will occur (actually from 1980 to 2016). Any energetic effects from the Galactic Center would manifest along the galactic equator anyway, so this is a moot point for many references.

and others mean the radio-telescope-defined center. The ecliptic (the path traveled by the Sun, Moon, and planets, which is the plane of the solar system) crosses the Milky Way close to the fattest, central part of the bulge, just below the dark rift, but the size of the Sun is such that it would touch the dark rift. However, the radio-telescope-defined center of the galaxy is just off the ecliptic, and the Sun will be closest to that in 2219 AD. The Galactic Alignment process described here involves the intersection of the ecliptic with the galactic equator, or, alternatively, the intersection of the winter solstice meridian with the galactic equator.

The dark rift is in the widest part of the Milky Way and corresponds to the direction of the center of the galaxy. As the alignment

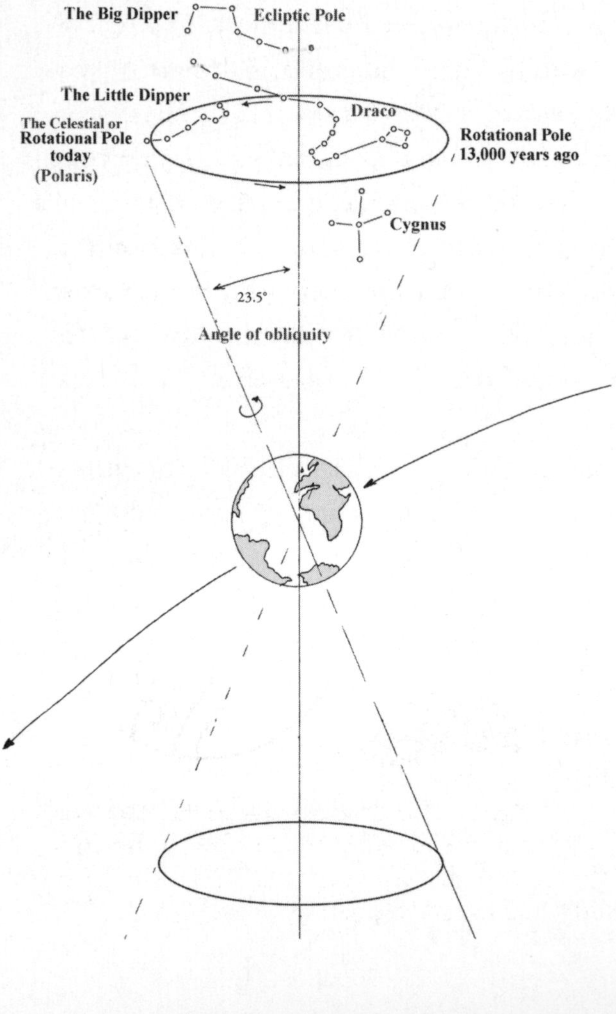

Fig. 1.4. This diagram shows how the rotational pole moves in a circle around the Ecliptic Pole over approximately 26,000 years. This can be measured by noting the slow movement of the constellations of the zodiac behind the rising point of the Sun on solstices or equinoxes. The rising point moves back through the zodiac in the opposite direction to their daily and annual movement, hence the "precession of the equinoxes."

occurs, the solstice meridian will align with the crossing point of the ecliptic (the path of the Sun, Moon, and planets) and the plane of the galaxy, which is called the galactic equator, and pass on to the other side. (The solstice meridian is an arc drawn in the sky that connects the North Celestial Pole to the South Celestial Pole, going through the point in the sky where the Sun reaches its annual most southerly limit.) This alignment, Jenkins says, will signal a "field-effect energy reversal,"[7] allowing us to resonate with the source of the field. The source of the energy field (which includes the electromagnetic spectrum and beyond) is the Galactic Center, and just as water spirals down a plug hole in opposing directions in each of Earth's hemispheres, which are divided by Earth's equator, so our changing orientation to the galactic equator—the plane of the galaxy—will affect us in a similar way. This will be the completion of the "human spiritual embryogenesis"[8] that is measured by the Maya Long Count calendar, culminating in "a pole shift in our collective psyche"[9] and the birth of our higher selves. This very last day of the thirteen-baktun cycle the Maya called Creation Day.

Astronomically speaking, the actual crossing of the winter solstice Sun (or winter solstice meridian) and the galactic equator occurred in 1998, but since the Sun is half a degree wide, this is the halfway point in a process that takes thirty-six years to complete. In other words, it seems likely that any field effects connected with the crossing have already started, and the process will complete around 2016, when the Sun clears the equator. Astronomers have only recently been able to give an accurate estimate of the length of the precession cycle, around 25,800

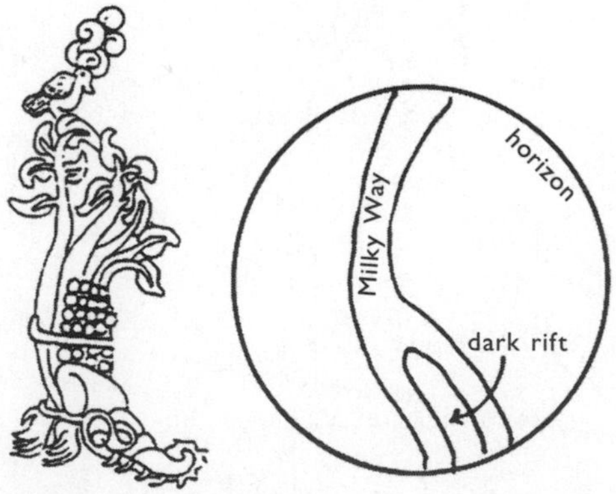

Fig. 1.5. The Crocodile Tree as the Milky Way and the Mouth of the Crocodile as the dark rift. (Compare to fig. 3.2.) From *Maya Cosmogenesis 2012*, by John Major Jenkins.

years,* and even with today's technology, they would be hard-pressed to give a solstice prediction over two thousand years in the future to an accuracy of plus or minus fourteen years like the Maya did. However, some people think that there is no inaccuracy, and that the Maya were using this galactic alignment process to highlight some other event that coincides with it.

THE AZTEC SUNSTONE

In 1790, while installing water pipes in Mexico City's Central Plaza, laborers unearthed a huge carved stone disk (see color plates 1 and 3). Archaeologists tell us it was carved in 1479 AD, in the reign of Axayacatl, just forty years before the Spanish arrived. It is twelve feet in diameter, weighs twenty-four metric tons, and is on display in the National Museum of Anthropology in Mexico City. The disk is known as the Aztec Sunstone, or Calendar Stone, since the central motif is said to show the five gods who represent the five ages of the world, called Suns. The four rectangular panels show the four previous Suns, while

Fig. 1.6. Stela twenty-five from Izapa, showing a scene from the Maya Creation myth, the Popol vuh. Hunahpu shakes Seven Macaw from his perch. Seven Macaw represents the seven stars of the Big Dipper, and precession moved it from its perch (the rotational pole) between 1500 BC and 1000 BC. From *Maya Cosmogenesis 2012,* by John Major Jenkins.

*This is because the rate of precession is slowing down from the classic rate of 25,920 years (the "Platonic year," which was correct around the time of Plato) by around eleven years per century. Precession is caused by the approximate 23-degree tilt of Earth's axis, and the variation in the rate of precession is caused by a variation in the tilt over 41,000 years—a cycle known as "the variation in the obliquity of the ecliptic."

Fig. 1.7. Stela eleven from Izapa, showing the rebirth of First Father, One Hunahpu (father of Hunahpu), in the mouth of the Jaguar-Toad, representing the rising of the winter solstice Sun in the dark rift–galactic alignment. From *Maya Cosmogenesis 2012,* by John Major Jenkins.

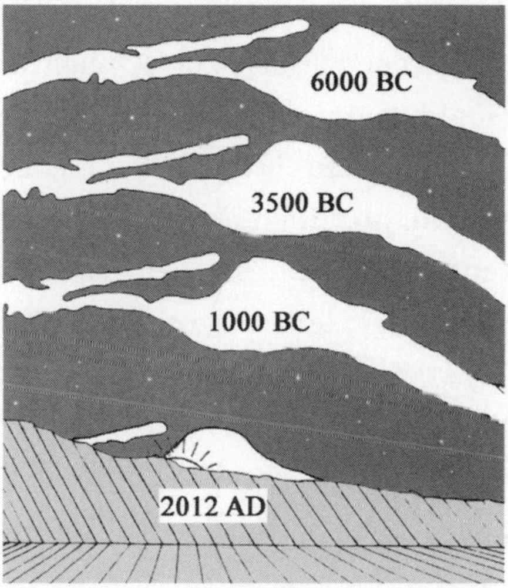

Fig. 1.8. The convergence of December solstice sunrise and galactic equator over 8,000 years, as seen on the horizon from Izapa, Mexico. From *Galactic Alignment,* by John Major Jenkins.

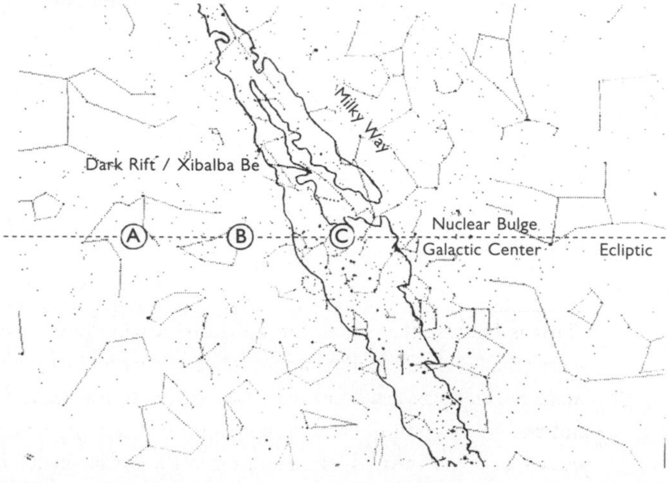

Fig. 1.9. The slow approach of the winter solstice Sun to the Crocodile's mouth, or dark rift, which the Maya also knew as *xibalba be,* the black road, or road to the underworld.
A: 5,000 years ago.
B: 2,000 years ago.
C: 2012. From *Maya Cosmogenesis 2012* by John Major Jenkins.

the central circle shows the face of Tonatiuh (or Huitzilpochtli), the Sun god who rules the present era—the Fifth and final Sun.

The current, or Fifth, Sun is called the Sun of Movement, since it is at the center of the day sign Ollin, which means movement or earthquake. The four previous Suns are shown as the gods or day signs Jaguar, Wind, Rain (or Rain of Fire), and Water, representing the four types of disasters that ended the four previous Suns. The Cuauhtitlan Annals, a Nahuatl manuscript, suggests that the Jaguar Sun ended with an eclipse accompanied by the descent of *tzitzimime*, celestial demons that devoured humankind. John Major Jenkins says the tzitzimime are also predicted by the Aztecs for the end of the Fifth Sun, and sees them as either our own subconscious fears or as "autonomous transdimensional entities."[10] The picture of one of these tzitzimime that Jenkins shows in *Maya Cosmogenesis 2012*[11] (from Codex Magliabechiano) appears to be very similar to the god at the center of the Aztec Sunstone, with claw hands and a flint-knife tongue (see fig. 1.10). Anthropologist and folklorist Professor Gordon Brotherston has shown that the Aztec Sunstone encodes the length of one Sun as 5,200 years, and that the five eras on the stone are thus equal to a 26,000-year cycle of the precession of the equinoxes.[12] Brotherston also shows that the sixteenth-century Quiché Maya epic, the Popol vuh, encodes the same four past eras that are shown on the Sunstone, in the same order as those listed in the Cuautitlan Annals, which originated close to the Aztec capital, Tenochtitlán. The first of the four previous eras ended in a flood (the sign of Water), the second with an eclipse (the descent of flesh-eating tzitzimime, symbolized in the sign of the Jaguar), the third in a rain of volcanic ash (the sign of Rain), and the fourth ended in fierce winds (the sign of Wind). Some Aztec accounts reverse the order of these apocalyptic calamities.

I have taken this a step further with a controversial computer animation I spent several months working on. An examination of the Sunstone showed that it could have been based on an earlier one that had layered, moving parts, which would have made it actually possible to track the tzolkin. It would also have been possible to track the haab, using a partially concealed part of the theoretical original model that I call the Secret Dial, but it could not track the tzolkin and the haab at the same time. However, the tzolkin could be tracked at the same time as the thirteen-baktun cycle of the Maya by using the Secret Dial. This is the most controversial part, since the Aztecs are not normally thought to have had any knowledge of the Long Count calendar (however, they might have based the design on an original with moving parts, belonging to another, earlier civilization). The result is a mechanism that can give every day its own unique day and date position over a period of two thirteen-baktun cycles. This is around 10,250 years (see "The Aztec Pizza," in chapter 16, for more details, and check the following video link and fast-forward to 40:20 and watch for nine minutes: www.video.google.com/videosearch?q=geoff+stray&emb=0&aq=f#).

Fig. 1.10. An Aztec tzitzimitl or sky demon, from the Codex Magliabechiano. From *Maya Cosmogenesis 2012* by John Major Jenkins.

THE CALENDAR ROUND

The civilizations of Mesoamerica all used a 365-day calendar (the haab) alongside their 260-day Sacred Calendar (the tzolkin), and also, in the case of the Maya, their 360-day tun (part of the Long Count calendar). Each day had a tzolkin name and a haab name, and it took exactly fifty-two haabs, or 18,980 days, for the combined day name to be repeated. Every fifty-two (365-day) years—one Calendar Round—there would be an Aztec New Fire ceremony, in which sacrifices were made to stop the world from ending (see color plate 2). The movement of Venus was connected to the Calendar Round, since two Calendar Rounds, or 104

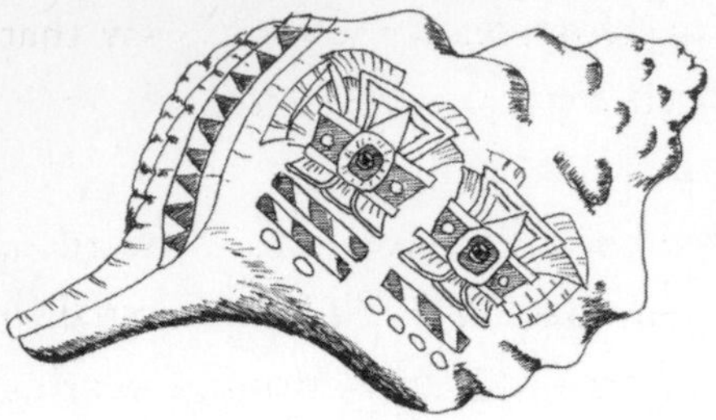

Fig. 1.11. Seashell from Teotihuacán, decorated with two hieroglyphs meaning time cycle, and the numbers twelve and nine. After *Burning Water: Thought and Religion in Ancient Mexico* by Laurette Séjourné.

haabs, and 146 tzolkins was equal to sixty-five Venus cycles of 584 days each. The period of two Calendar Rounds is known as a Venus Round for this reason. In an anonymous manuscript dated 1558, called Leyenda de los Soles (Legend of the Suns),[13] it is recorded that the First Sun lasted 676 years (52 × 13); the Second Sun lasted 364 years (52 × 7); the Third Sun lasted 312 years (52 × 6); and the Fourth Sun lasted 676 years (52 × 13).

It seems that later civilizations, such as the Aztecs, had either not inherited or had forgotten or discarded the Long Count calendar, so the concept of five eras, or world ages, was explained in terms of Calendar Rounds. Evidence that the Aztecs had lost some calendrical knowledge lies in the fact that the earlier Toltecs, from whom they inherited their knowledge of the Calendar Round and New Fire ceremony, were using the Calendar Round to track precession by the movement of the Pleiades, and when they moved south to the Yucatán, they encoded this knowledge in the Pyramid of Kukulcan. A pair of glyphs engraved on a seashell from Teotihuacán,[14] the ancient city taken over by the Toltecs, show date glyphs that include bar-and-dot number symbols, as used by the Maya. The Aztecs reverted to a much cruder, dots-only number system.

Tony Shearer, in his 1971 book *Lord of the Dawn,* says that when Quetzalcoatl, the Feathered Serpent, god of the Aztecs, departed, he left a prophecy of "thirteen Heavens and nine Hells." Each Heaven and Hell cycle was fifty-two years ("years" meaning 365-day haabs) in length, and he would return at the end of the thirteen Heavens; then the nine Hells would start. This thirteen-Heaven cycle seems to correspond to the periods given in the Leyenda de los Soles, implying that the Fifth

Sun lasts 468 (52 × 9) years (haabs). However, some sources say that the fifth era consists of the thirteen Heavens plus the nine Hells—676 + 468 = 1,144 years.[15]

Since Quetzalcoatl was depicted as white skinned and bearded, and was expected to return in the year 1 Reed, and Cortés landed in Mexico in the year 1 Reed (1519 in the Gregorian calendar) bearing what appeared to the Maya as the sign of Quetzalcoatl (the cross of the four directions) on the sails of his fleet, the Aztecs welcomed the Spanish tyrant as the returning Quetzalcoatl. The Spanish then slaughtered them all, burned their books, and destroyed their cities, in the name of their own god.

If nine Calendar Rounds are added to the time of the arrival of Cortés, we arrive at 1987, and this is the thinking behind the Harmonic Convergence (conceived and promoted by José Argüelles), which took place on August 16 and 17, 1987.* To try and unify this Aztec end point with the thirteen-baktun cycle of the Maya and other calendars and philosophies, Argüelles postulated a period from twenty-five to twenty-six years, from 1986–87 to 2012, as a "synchronization phase." In chapter 2, Tribal Prophecies, several elders mention a "period of purification" that seems to correspond to the synchronization phase of Argüelles. Were any of these ancient revelations influenced by more-recent sources of speculation? Perish the thought!

Since the thirteen-baktun cycle is not divisible by the Calendar Round, or even the haab, they could not both begin and end at the same point. However, there is a Tikal Calendar Round that will end on April 2, 2012, and there is a Teotitlan Calendar Round that ended on March 1, 1987.[16]

---

*However, there was a substantial error made in calculating the Harmonic Convergence, and it should have been on January 7, 1987. This is an error of 221 days and is explained fully in note 21 of chapter 19.

2

TRIBAL PROPHECIES

MAYA DAYKEEPER: HUNBATZ MEN

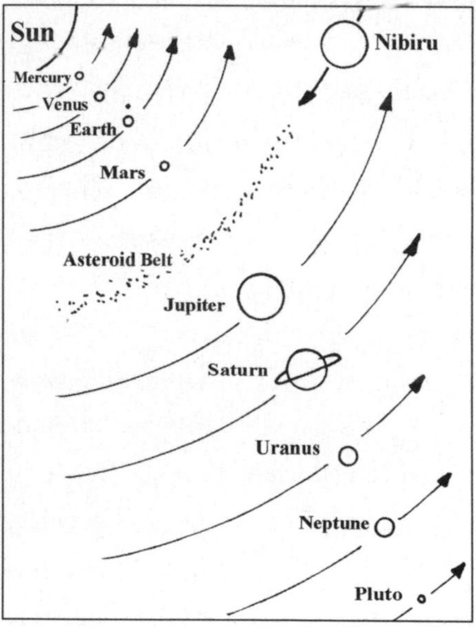

Fig. 2.1. The trajectory of Nibiru as it enters the solar system. After *The Twelfth Planet* by Zecharia Sitchin.

In his book *Secrets of Maya Science/Religion,* Hunbatz Men, a modern Maya daykeeper, says that we should visit sacred sites to correct the outcome of a flaw in human DNA. This will "strengthen our energy bodies" for the coming transition at the end of the thirteen-baktun cycle.

In March 2002, Hunbatz Men held an eight-day course called Initiation in the Cosmic Mayan Healing for Humankind. During the course, he announced that the Maya knew about a massive planet on an elliptical orbit, which was known to the Sumerians as Nibiru (for more information on this, see part 2, "2012 Theories"). The Maya called it Tzoltze ek'. There may have been a connection with "the 26,000-year Cosmic Mayan Calendar," which has a "relationship with the Pleiades and the Mother Earth" and "the major changes taking place on our Sacred Planet."[1]

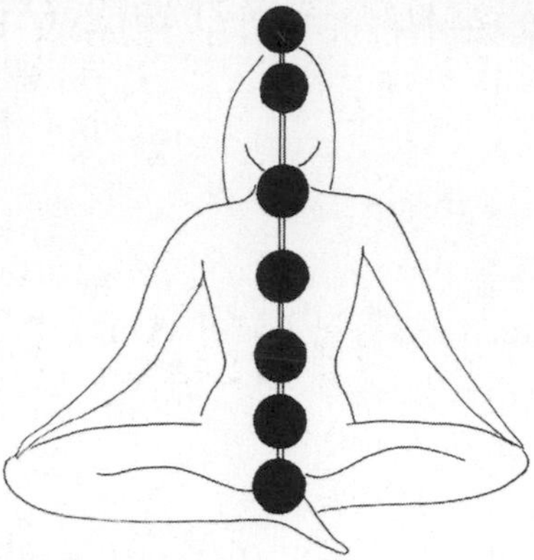

Fig. 2.2. The positions of the seven chakras, or power zones.

He says that the seventh moon of Nibiru will activate kundalini, the serpentine energy that in turn activates the seven chakras, or power zones, along the spine. (*Secrets of Maya Science/Religion* provides evidence that this was not an exclusively Indian and Tibetan teaching.) Tzoltze ek' will also activate our genetic memories and all the pyramids on the Earth and underwater, so we are advised to visit them when it appears in the skies. Nibiru has a period of 6,500 years, not 3,600, he says, and visits us four times every 26,000-year cycle (the "great calendar of the Pleiades" that ends on December 21, 2012). The planet is inhabited by beings who operate on a higher "mental frequency," according to Men.

MAYA-AZTEC SHAMAN: QUETZA-SHA

The Maya-Aztec shaman Quetza-Sha explained in an interview[2] that on July 11, 1991, during the total solar eclipse over Mexico, a "sacred dimensional doorway of the equilibrium opened," and this was the "return of Quetzalcoatl." This was the start of an information-download process, which Quetza-Sha describes using Philip K. Dick's enigmatic terminology*: "It will leave us plasmated with the millenary knowledge of the new time," and this information is to prepare "for the dimensional birth of a new race that comes from the Pleiades."

*In *Valis,* Philip K. Dick defined *plasmate:* "I term the immortal one a plasmate, because it is a form of energy; it is living information. It replicates itself—not through information or in information—but as information." Dick continued: "The plasmate can crossbond with a human, creating what I call a homoplasmate. This annexes the mortal human permanently to the plasmate. We know this as 'birth from above' or 'birth from the Spirit.'"

The creation of the new race is due to occur in 2012. "The year 2012 is not the destruction of the planet, but is the transformation of the spirit, sexual energy, the energy of your heart and your mind in all the dimensions of our existence in the solar system," says Quetza-Sha. The process will continue through 2024 with a fifth-dimensional transformation, to 2029, when humans themselves physically transform into fifth-dimensional beings.

MAYA ELDER: DON ALEJANDRO

At the Latinola website,[3] there is a quotation regarding Maya elder Don Alejandro Cirilo Pérez Oxlaj, head of the National Council of Elders of Guatemala, which says:

> What we do know is that it wouldn't hurt to listen to the words of Don Alejandro who said that on Dec. 20, 2012, Mother Earth will pass inside the center of a magnetic axis and that it may be darkened with a great cloud for 60 to 70 hours and that because of environmental degradation, she may not be strong enough to survive the effects. It will enter another age, but when it does, there will be great and serious events. Earthquakes, maremotos (tsunamis), floods, volcanic eruptions and great illness on the planet Earth. Few survivors will be left.

The authors Patricia Gonzales and Roberto Rodriguez conclude by saying, "Don Alejandro has been sent as a messenger from a council of elders to warn the world that we must change the way we live and take care of Earth." We shall return to Don Alejandro later in the book.

GUATEMALAN AJQ'IJ: CARLOS BARRIOS

Author and healer Steven McFadden has written an article, "Steep Uphill Climb to 2012,"[4] on the talk by Guatemalan *ajq'ij* (a Queche-Maya term for "daykeeper") and priest-anthropologist Carlos Barrios in

Santa Fe in autumn 2002. Barrios says that the twenty-five-year period between 1987 and 2012 is the period between the Fourth and Fifth Worlds.[5] He also says 2012 will not be the end of the world, but it will be transformed. He says the "alignment with the heart of the galaxy in 2012 will open a channel for cosmic energy to flow through the Earth, cleansing it and all that dwells upon it, raising all to a higher level of vibration." However, in another article,[6] Carlos's brother Gerardo Barrios says the Thirteen Heavens cycle began in 1991, but will "gestate" for nine years until 2001, when it will be "born." The period is actually ten years, but it is possible the nine-year period is measured from July 26, 1992, based on the "Dreamspell Time Shift," which was, in turn, related to the 1991 solar eclipse.

INCA PROPHECY: WILLARU HUAYTA

Fig. 2.3. Comparative sizes of (l–r) Earth, Jupiter, and Nibiru.

Willaru Huayta, a Peruvian spiritual messenger and Quechuan native elder, says that 2013 will mark the end of the Incan calendar, and in that year a "huge magnetic asteroid" three times larger than Jupiter will pass close to Earth, "activating the purification of the land," causing cataclysms that will kill off most of humankind. A core of people will survive to become the "seed-people" of the "Sixth Generation." To become a part of that core, we must heal our spiritual sickness by conquering our egos.[7]

INCA PROPHECY: Q'ERO PRIESTHOOD

In her 1999 book, *Keepers of the Ancient Knowledge,* Joan Parisi Wilcox, who has been trained in Peru to the highest current level of Q'ero priesthood (the fourth level), reveals the Q'ero prophecy of spiritual evolution and a golden age. The prophecy is incorporated into a sixteenth-century Catholic philosophy of three ages. These are the Age of the Father, the Age of the Son, and the Age of the Holy Spirit (see color plate 4). The latter, which the Q'ero call Taripay Pacha, began in the period between August 1, 1990, and August 1, 1993, when the world underwent a cosmic transmutation, called a *pachakuti*. Wilcox quotes William Sullivan's translation of the word *pachakuti* from his 1996 book, *The Secret of the Incas,* in which he translates it as "an overturning of space-time." Taripay Pacha is the period extending approximately from 1993 to 2012, during which humankind will spiritually evolve—it is the "Age of Meeting Ourselves Again," when time will end. It is divided into two phases: a seven-year phase to the year 2000, then a twelve-year phase until 2012, when something happens to time, or our perception of it, and the Taripay Pacha becomes a golden age. The three worlds of the Andean cosmos will unite—the *hanaq pacha,* or upper world; the *ukhu pacha,* or lower world; and the *kay pacha,* or mundane physical world, will merge into one (we shall return to this subject in chapter 21). The process is assisted by a technique of Andean shamanism called the "engaging of the energy body."

HOPI ELDERS

The Hopis of Arizona believe that we are living in the fourth of seven eras, the Fourth World, and that there will be a "great purification" just before the start of the Fifth World (and, like the Aztec version, they say that the previous era ended in a flood). Frank Waters's *Book of the Hopi* provides further details:

The First World ended by fire; the Second World ended when the Earth "teetered off balance" and rolled over twice, causing floods and an ice age; the Third World ended in a flood. The Hopis call the transition from one era to the next Emergence, and it is symbolized by a labyrinth identical to ones shown on early Cretan coins. It is also known as the Mother Earth symbol, or Mother and Child. The whole process is seen as a birth and is reenacted during an annual ceremony called Wúwuchim, in which "initiates undergo spiritual rebirth."[8] This ceremony is held in November and includes a New Fire ceremony, similar to that which the Aztecs hold in November once every fifty-two years. The Emergence is symbolized in an underground ceremonial chamber, a *kiva,* with a hole in the floor and a hole in the roof. The hole in the floor is the emergence point (*sipápu* or *sipápuni*) into the Third World, and the ladder to the hole in the roof is the emergence point into the Fourth World.[9]

Fig. 2.4. The Hopi symbol of Emergence, which is exactly the same as the Cretan labyrinth.

Exactly as John Major Jenkins has found with the Aztec New Fire ceremony, the Hopi New Fire ceremony culminates at midnight, when the Pleiades are overhead. This shows a direct connection to the way in which the Toltecs tracked precession and encoded the results into the Pyramid of Kukulcan (see chapter 8). At this point in the ceremony,

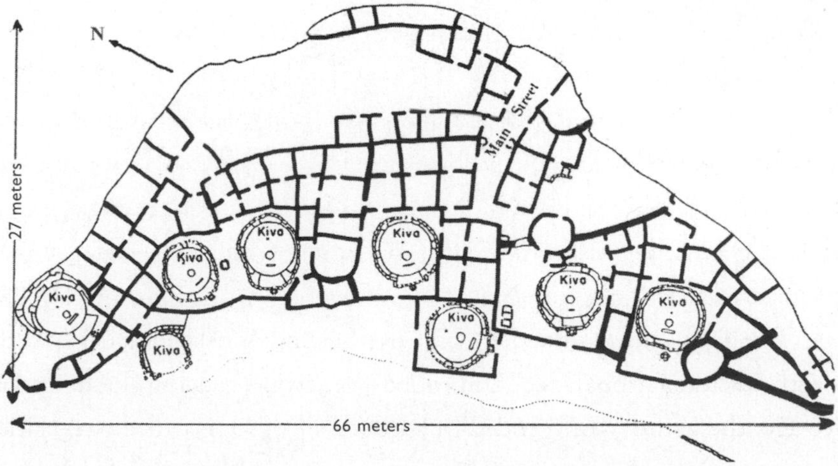

Fig. 2.5. A plan of Mesa Verde, a ruined town of the Anasazi, an extinct tribe of Pueblo Native Americans that lived over a thousand years ago in what is now the American Southwest. Note the many kivas.

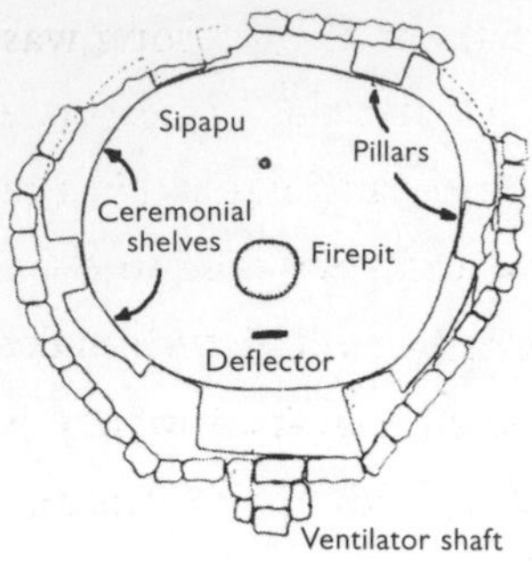

Fig. 2.6. Plan view of a kiva.

"a lone white-robed figure wearing a large white star enters the kiva and announces, 'I am the Beginning and the End.'" Shortly after this, the fire is extinguished, while priests, godfathers, and initiates throw off their clothes and the stars they have been wearing on their foreheads. Chaos ensues, as everyone tries to get up the ladder and out before the World ends.

The Hopis believe that the human body has a series of psychophysical centers, and this system is uncannily similar to that of the Hindu and Buddhist chakras, except that the Hopis list only the upper five centers, whereas the Hindu system lists two additional lower ones. The first Hopi center is at the top of the head and is like a door that is open at birth and death. Originally, humankind communicated with the Creator through it, but during the First World it closed up (this is comparable to the biblical Fall); the second center is at the brain; the third at the throat; the fourth at the heart; the fifth at the naval, or solar plexus. The living body of Earth is based on the same design, with vibratory centers along an axis.[10] In each era, or World, human beings became governed by the next center down, each more material and less spiritual than the one above. The Fourth World, in which we are currently living, is the most material, and at the next transition, development reverses and goes upward, being governed in turn by each higher center.[11] Waters does not state whether each higher center will dominate for a whole era or whether it will be possible to rise through the centers in one era, but the culmination will be a reopening of the door at the crown.

BLUE STAR KACHINA

Kachinas are spirits, of which there are over three hundred, and they are represented in Hopi ceremonies by dancers wearing costumes and masks, who are then "taken over" by the kachinas. In 1914, a song was first sung at the Wúwuchim ceremony predicting that the "Emergence into the Fifth World" would come when the Saquasohuh danced in the plaza. The Saquasohuh is the Blue Star Kachina, and according to *Book of the Hopi* "represents a blue star, far off and invisible, which will make its appearance soon."[12] The other source on the Blue Star prophecy is said to originate from a Native American elder named White Feather, of the Bear Clan, who in 1958 hitchhiked a ride with a Methodist minister and told him of nine prophecies, starting with the coming of white men with guns (which occurred in the sixteenth century). After that, the ancient prophecies were said to have foretold covered wagons, longhorn cattle, railroad tracks, power and telephone lines, concrete roads, and oil spills. The eighth prophecy is the hippie era, which manifested in the 1960s. The only remaining prophecy to be fulfilled is the ninth and last one: "You will hear of a dwelling-place in the heavens, above the Earth, that shall fall with a great crash. It will appear as a blue star. Very soon after this, the ceremonies of my people will cease."[13]

Some have suggested that this referred to the Russian Mir space station that crashed to Earth in March 2001. Others say that the blue star was the comet Hale-Bopp, in 1997. Moira Timms suggested that it was a supernova that appeared on February 23, 1987, and was known as SN1987A. It was the brightest supernova in four centuries, and it showered Earth with subatomic particles and a whole spectrum of radiated waves.[14] In 1963, Frank Waters wrote that "the Emergence to the future Fifth World has begun."[15] Since then other Hopi elders have added that the final stage of this transition period is known as the Great Day of Purification,[16] but all the subsequent Blue Star prophecies seem to be based on the two above, from *Book of the Hopi* and from White Feather (see color plate 5).

In Steven McFadden's book *Profiles in Wisdom,* Hopi elder Oh Shinnah Fastwolf says the Great Purification will be over by 2011, and also says that "it will be marked by the appearance of a new star." However, it seems that Oh Shinnah variously claims to be an

Apache healer or a Cherokee grandmother as well as a Hopi elder. The Fastwolf family is said to have claimed she is an Irish folksinger from Chicago.[17]

Frank Waters first published *Book of the Hopi* in 1963, and to research it he lived on the Hopi reservation for three years (probably 1959–1961) and was subsequently seen as an authority on Hopi matters. In the book, Waters says that the Hopis believe that the Maya, Toltecs, and Aztecs were "aberrant Hopis clans"[18] who remained in Mesoamerica, while the Hopis migrated north. In 1975 he brought out his book *Mexico Mystique,* in which he showed the similarities in the cosmology, mythology, and ceremonies of the Hopis and the Aztecs and Maya. He equates the catastrophic end of the Fifth Sun of the Aztecs with the end of the Fourth World of the Hopis.[19] Also in *Mexico Mystique,* Waters says that the thirteen-baktun cycle began on August 12, 3113 BC, and ends on December 24, 2011 AD. He got this end-date of 2011 from *The Maya,* by Michael Coe (first published in 1966, but Coe's later works give a revised end-date of December 23, 2012.)[20] This is obviously why some people have received the impression that the Hopi prophecies end in 2011.

Alternative historian Murry Hope says the Hopis "share with the ancient Egyptians a knowledge of Sirius, which they know as 'Blue Star Kachina.'"[21] However, according to the glossary in the back of *Book of the Hopi,* the word for Sirius, the Dog Star, is Ponóchona, while the word for Blue Star Kachina is Saquasohuh. Self-styled futurist Gordon Michael Scallion (who claims he had visions and dreams of the Blue Star before he knew about the Hopi prophecy), in his work *Notes from the Cosmos,* provides a possible—or impossible—explanation linking Sirius and the Blue Star:

> This solar system shall become a binary sun system, when the Blue Star returns . . . During the day it will appear as a silvery light, one hundred times brighter than the morning star . . . During the evening, it will appear as a moon . . . The Blue Star is a companion to

> Sirius B . . . The vibrations given off by the Blue Star will enable the soul to have an easier time in communication with its host . . . A new light body is being created.

We shall have more to say about Mr. Scallion later in the book.

DOGON CONNECTION

Readers of Robert Temple's *The Sirius Mystery* may recall that the Dogon tribe of Mali, West Africa, have known for centuries that Sirius, the Dog Star, is a binary star system and that Sirius B is a white dwarf—facts they say were given to their ancestors by visitors from Sirius. They based their Sigui calendar on the orbit time of Sirius B around Sirius A, originally a period of forty-nine to fifty years, as in the case of the orbit, but the Sigui calendar was later reformed to a sixty-year period.[22] In the 1976 first edition, Temple told us that the Dogon knew of a third star, and that he thought it would turn out to be a red dwarf, if and when it was discovered. What you may not have heard, if you haven't read the 1998 edition of *The Sirius Mystery,* is that astronomers have recently discovered Sirius C. In 1995, twenty years after the publication of the first edition, French astronomers Daniel Benest and J. L. Duvent announced, in the *Journal of Astronomy and Astrophysics,* their discovery of a small red dwarf—Sirius C. This has now been confirmed, vindicating Temple's work.

The Dogon say that the "spaceship" of the visitors—the Nommo—

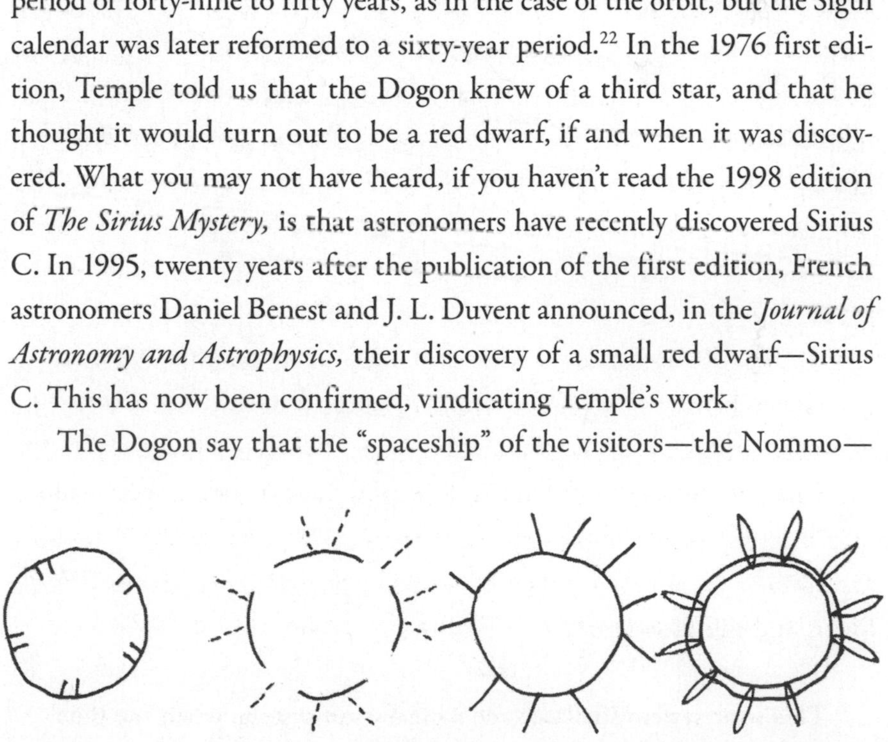

Fig. 2.7. The "emergence" of the Star of the Nommo, as drawn by the Dogon people. After *The Sirius Mystery: New Scientific Evidence for Alien Contact 5,000 Years Ago,* by Robert Temple.

looked "like a new star," and that they will return one day, when "a certain star will reappear," and there will be a "resurrection of the Nommo." They also say that Sirius A and B "were once where the Sun now is." The star will be invisible before it "emerges," and it is drawn with the rays inside the circle. It will only be "formed when the Nommos 'ark' descends, for it is also, symbolically, the resurrected Nommos' 'eye' symbolically."[23] Temple thinks this is Phoebe, Saturn's tenth moon, not to be confused with Phobos, one of Mars's moons that Scallion and others have predicted will soon break loose.

PUEBLO NATIVE AMERICANS: FIFTH WORLD IN 2012

The Pueblos are Native American peoples who live in stone or clay houses in central New Mexico and northeast Arizona. The term includes the Navajo, Hano, Hopi, Zuni, Acoma, and Laguna peoples, who are all said to have descended from the Anasazi, who lived over a thousand years ago.

The son of a Pueblo shaman, Speaking Wind from northern New Mexico has said that the Fifth World will start in December 2012. This conforms to Hopi prophecies, even though some have put the transition at 2011.

> As I said, the final cleansing of Earth began in June 1998. In September 1998, the five brothers (planets) aligned themselves to usher in the cleansing energies of Earth. The chaos everyone has been anticipating, or the tearing away of the illusion and the lie, will begin between January and April 1999 and will continue to escalate until the nine brothers (planets) align themselves on the 5th May 2000. From that date, until the last day of the Fourth World, December 22, 2012, a date taken from our star calendars, everything living will undergo a purge. If mankind will not willingly let go of the illusion, and the lie, it will be stripped away. And we have only begun to feel the effect, since June 1998. The 23rd day of December 2012 is the first day of the Fifth World.[24]

CHEROKEE CALENDAR

In her book *The Cherokee Sacred Calendar,* Raven Hail, a member of the Cherokee Nation of Oklahoma, describes a calendar system that seems to be a cross between the Aztec and Maya Sacred Calendars. It is a 260-day wheel with twenty day signs, which have names very similar to the Aztec version. However, they use a numerical notation identical to the Maya notation. The ephemeris for finding your Sacred Calendar birthday is comparable to that given by Bruce Scofield in his book *Day Signs,* about the Aztec system, and both of these correspond to the Maya True Count. Hail says that a Cherokee World, or Sun, was approximately 5,200 years. (In the Long Count of the Maya, the thirteen-baktun cycle lasts 5,200 tuns, or 360-day years, which is about 5,125 solar years; the Aztecs didn't have a Long Count, but only a concept of Suns, of which the current is the Fifth, but with varying accounts of how long the Suns were.) Previous Worlds, or Suns, have ended in cataclysm, Hail says, and the current one, which is the Sun of Heron (equating to the Sun of Movement in the Aztec system), will end in earthquakes, but nobody knows when.

However, the ephemeris, like Scofield's, lists the day signs for the first day of each thirteen-day period from January 11, 1900. But where

Fig. 2.8. The Cherokee calendar, as shown in *The Cherokee Sacred Calendar—A Handbook of the Ancient Native American Tradition,* by Raven Hail.

Scofield's ephemeris ends on December 20, 2000, Hail's goes on to December 18, 2012, and then lists the day that falls three days later as Four Flower—December 21, 2012, where the ephemeris ends. The extra ephemeris section alone is quite useful, but there is something else that we need to take into account. On the page about this book at Amazon.com, the book receives some very bad reviews, including one by someone called Tahlehquah, who says, "As an official of the Cherokee Nation, let me advise the readers that the 'teachings' of 'Raven Hail' are not endorsed by the Nation, nor in fact, do they follow any Cherokee historical or cultural beliefs that I have ever heard."[25]

CHEROKEE CALENDAR WHEELS

Dan Troxell, who is "of Chickamauga Cherokee blood," has a website called Mysteries of Trox.[26] On the site, he tells of the Cherokee calendar and how it was inspired by "the 52 scales around the mouth of the rattlesnake," and that it measures "Time Untime."

He describes a system of interlocking wheels and says sometimes the wheels were made from stone and sometimes they were formed on the ground using pebbles. These are often mistaken for Medicine Wheels. They represent the movement of "the stars, the heavens and the universe." On the rings are "glyphs and colors," the alignments of which give various meanings.

There are six rings—three visible (the "Rings of the Heavens") and three invisible (the "Rings of the Universe"). One of the three visible wheels has twenty-two sections, each representing fifty-two years. The twenty-two sections are divided into two parts: the first thirteen are years of Light and the following nine are years of Darkness. The whole cycle is thus 1,144 years, and is called a World, consisting of 676 years of Light (13 × 52) and 468 years of Darkness (9 × 52). There is a transitional period from Light to Darkness and vice versa, called a Crossing. These Crossing points vary in length, from zero to twenty-five years, and are a collective struggle comparable to the birth and death transitions of the individual.

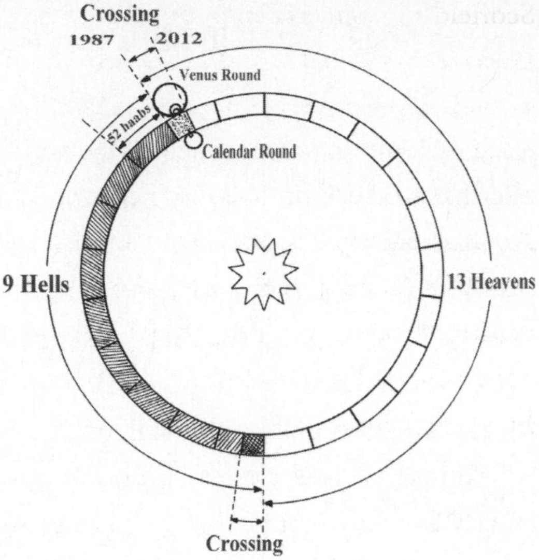

Fig. 2.9. A diagram based on a description of the Cherokee calendar by Dan Troxell.

Here again, we have confirmation of a twenty-five-year transitional period starting in 1987, since Troxell says that "the Cherokee calendar ends in 2012."

Incredibly, Troxell says the Cherokee name for the Great Spirit Creator is Yahowah. This seems to support the claim of the Mormons that one of the Ten Lost Tribes of Israel once lived in America. In fact, Tony Shearer (whom I mentioned at the end of chapter 1) thought that the twenty-two golden plates found by Joseph Smith, founder of the Church of Latter-Day Saints, had come from Palenque, and that the prophet Mormon was Lord Pacal of Palenque. It was also Tony Shearer who first publicized the concept of the ending of the nine Hell cycles on August 16–17, 1987 (the Harmonic Convergence). Dan Troxell's description is exactly like a diagram called the Echo Wheel, by Nick Armstrong, on his website 2012 Unlimited.[27] From the text accompanying the diagram, it is obvious that the inspiration for Armstrong's version is the Shearer information; actually, it is almost identical to Shearer's diagram, which is also called the Echo Wheel (Shearer, *Beneath the Moon and Under the Sun,* 99). I can't help wondering if Dan Troxell was inspired by the same source, or even by Armstrong.

THE SENECA INDIANS

Seneca society's Wolf Clan Teaching Lodge "teaches the wisdom, prophecy, and philosophy of Earth history," and Grandmother Twylah Nitsch[28] is a Seneca elder of the Wolf Clan, who was chosen "many decades ago" to preserve the Wolf Clan teachings of her maternal grandfather, Moses Shongo, the last of the traditional Seneca medicine men: "The Seneca Grandfather Moses Shongo (died ca. 1925) foresaw a 25-year period of purification, lasting until the year 2012 or so, during which the Earth will purge itself."[29]

So here we have a native prophecy concerning not only 2012, but a twenty-five-year period of purification preceding 2012, and that means that the year 1987 was flagged long before 1969 or 1970, when the late Tony Shearer told José Argüelles that August 16–17, 1987, would be the end of the ninth Hell cycle (in the so-called prophecy of Thirteen Heavens and Nine Hells). This is the time that Argüelles called the Harmonic Convergence. This also has a bearing on the Kalachakra prophecy (see chapter 3).

ZULU PROPHECY: THE RETURN OF MU-SHO-SHO-NO-NO

Credo Mutwa, an eighty-year-old Zulu *sanusi* (shaman) and elder from South Africa and author of *Song of the Stars: The Lore of a Zulu Shaman,* gave a talk at the Living Lakes Conference in California on October 2, 1999, that included the following:

> Let me tell you two last things please. One, it is this, that I am told by the great storytellers of our tribes that fresh water is not native to our Earth, that at one time, many thousands of years ago, a terrible star, or the kind called Mu-sho-sho-no-no, the star with a very long tail, descended very close upon our skies. It came so close that the Earth turned upside-down and what had become the sky became

> down, and what was the heavens became up. The whole world was turned upside down. The Sun rose in the south and set in the north. Then came drops of burning black stuff, like molten tar, which burned every living thing on Earth that could not escape. After that came a terrible deluge of water accompanied by winds so great that they blew whole mountaintops away. And after that came huge chunks of ice bigger than any mountain, and the whole world was covered with ice for many generations. After that the surviving people saw an amazing sight. They saw rivers and streams of water that they could drink, and they saw that some of the fishes that escaped from the sea were now living in these rivers. That is the great story of our forefathers. And we are told that this thing is going to happen again very soon. Because the great star, which is the lava of our sun, is going to return on the day of the year of the red bull, which is the year 2012.[30]

We have heard about the Dogon of Mali, West Africa, with their astronomical knowledge and calendar connection to the Blue Star. There is a stone circle in South Africa, in a private game reserve called Timbavati, which means "the falling down of a star" in the Zulu language. It still has an equinox alignment intact. In 1997, a stone circle in the southern Egyptian desert was found to have an alignment to the summer solstice sunrise, and dated 6000 BC, is the "oldest astronomically aligned structure yet discovered anywhere on the planet."[31] There is also a stone observatory at Nomoratunga, in Kenya, connected with the Borana calendar, based on rising stars and lunar phases. There are more stone circles in Ethiopia, Morocco, The Gambia, Senegal, and Togo. The point is that the Long Count has been traced to Izapa. Izapa has Olmec connections. The Olmec carved mysterious giant stone heads with African features. Could there be a calendrical connection between the Long Count and Africa? The 520-day eclipse calendar of the Berbers of Tenerife, one of the Canary Islands, also suggests this.

MAORI PROPHECY: VEIL DISSOLVES IN 2012

The indigenous people of New Zealand, the Maoris, have a creation legend in which Rangi (Sky) and Papa (Earth) were partners and were "closely clasped together." They had several children, including a wind god, sea god, and war god, who lived between these two parents, squashed and without light. One day these children decided to push their parents apart so they could have room to move. They all tried and failed until one of the gods, Tane, pushed them apart, separating them—and then there was light.

There is a second, more secretive, legend that follows from this, which says that there will come a time when the "children of Tane" (humankind) are so busy, so distracted with fighting, greed, and lust, so separated from their original parents, that they—Rangi and Papa—will take that opportunity, while no one is looking, to quickly come back together, destroying everything in the process.

Recent conversations between a young Maori[32] and his *kaumatua* (tribal elders) have revealed that they see the prophecy being fulfilled in the year 2012. When one elder said *"Ka hinga te arai,"* the rest nodded and agreed. Since *hinga* means "to fall" and *arai* means "curtain," the phrase was at first understood to mean "the curtain will fall," as in "final curtain," or the end of the world. However, the elder was one who retained knowledge of the *tuturu Maori,* or "original Maori" language, and it turned out that the original meaning of the word *hinga* was "to dissolve, to be removed, to fade"; and the original meaning of the word *arai* was "veil, thin, separation." Since when people die, the Maoris say they have gone *ki muri i te arai,* "behind the veil" or "passed over," the word *arai* actually means "a separator" between the physical plane and the spiritual plane.

So in 2012, when "ka hinga te arai," the Maoris imply that there will be a "dissolving of the veil," or merging of the physical and spiritual planes.

3

ASIAN CALENDARS

VEDIC TIME CYCLES

In the Hindu system of time cycles, inherited from the Vedic civilization of the Indus Valley, now believed to date to 7000 BC and preserved in manuscripts called the Vedas, the longest cycles last trillions of years, but at the shorter end of the spectrum we have the four yugas:

Krita/Satya Yuga (Golden Age): 1,728,000 years = 2,160 × 800
Treta Yuga (Silver Age): 1,296,000 years = 2,160 × 600
Dvapara Yuga (Copper/Bronze Age): 864,000 years = 2,160 × 400
Kali Yuga (Iron Age): 432,000 years = 2,160 × 200

Each yuga is split into three parts: dawn, day, and dusk.

Krita/Satya Yuga: dawn = 144,000 years; day = 1,440,000 years; dusk = 144,000 years
Treta Yuga: dawn = 108,000 years; day = 1,080,000 years; dusk = 108,000 years
Dvapara Yuga: dawn = 72,000 years; day = 720,000 years; dusk = 72,000 years
Kali Yuga: dawn = 36,000 years; day = 360,000 years; dusk = 36,000 years

If these numbers look familiar, that is because they are precessional numbers, in which a Platonic year (an early term for the precession of the equinoxes, named after the Greek philosopher Plato) is equal to 25,920 years (12 × 2,160, or 360 × 72), or one complete precession of the equinoxes (or solstices); a Platonic month is equal to 2,160 years, or the precession of the equinox through one sign of the zodiac; a Platonic day is equal to 72 years, or the precession of the equinox through one degree; two Platonic months are equal to 4,320 years; half a Platonic month equals 1,080 years; and two Platonic days equal 144 years. These Platonic/precession numbers also occur in the Long Count system of the Maya: 360 days = 1 tun; 3 tuns = 1,080 days; 7,200 days = 1 katun; 144,000 days = 1 baktun; 7,200 tzolkins in the thirteen-baktun cycle.

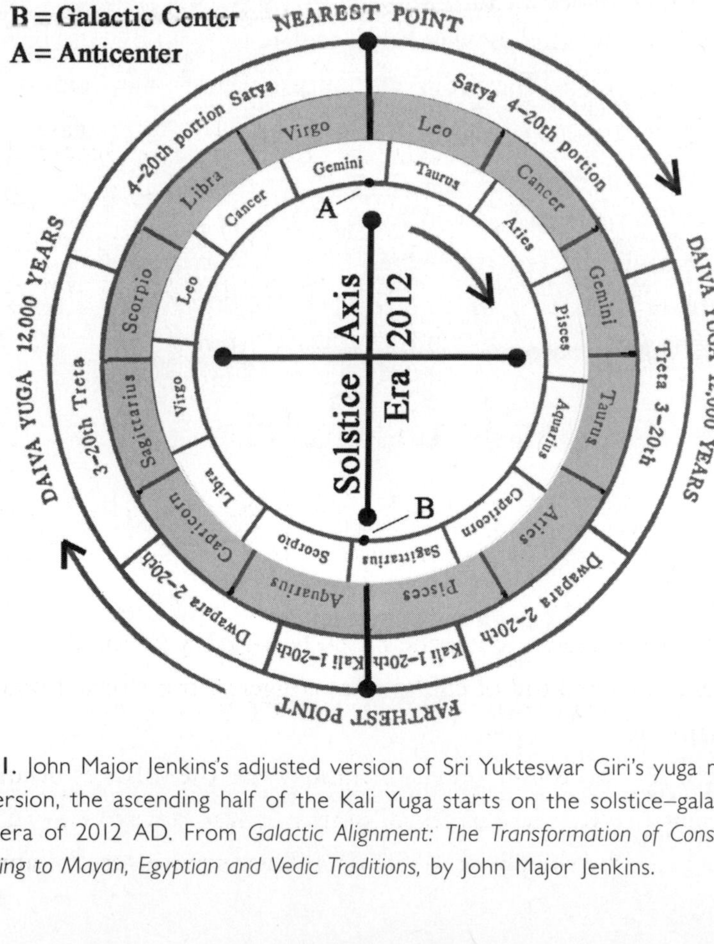

Fig. 3.1. John Major Jenkins's adjusted version of Sri Yukteswar Giri's yuga model. In this version, the ascending half of the Kali Yuga starts on the solstice–galaxy alignment era of 2012 AD. From *Galactic Alignment: The Transformation of Consciousness According to Mayan, Egyptian and Vedic Traditions,* by John Major Jenkins.

The philosopher René Guénon and others say we are now nearing the end of the Kali Yuga, although some sources say it only started in 3102 BC. John Major Jenkins, in his book *Galactic Alignment,* goes into the Vedic yuga system in detail and studies the work of the Hindu saint Yukteswar (born 1855), who discovered an oversight in the Vedas and went on to explain it. In one of the oldest Vedic writings, attributed to the god-man Manu, the four yugas are said to add up to 24,000 years, but when they are enumerated they come to only 12,000 years. The Krita/Satya Yuga lasts 4,800 years; the Treta Yuga lasts 3,600 years; the Dvapara Yuga lasts 2,400 years; and the Kali Yuga lasts 1,200 years.

Yukteswar explained the enigma in a diagram that equates the 12,000 years to a descending half of the 24,000-year cycle and another 12,000 years to an ascending half. Jenkins's study of Yukteswar's analysis shows that the Indian astrologer was talking about the precession cycle, but basing it on a 24,000-year cycle, rather than the (approximate) 25,800-year cycle recognized by astronomy. The "descent into darkness"—the Kali Yuga—started when the summer solstice (June) Sun was aligned with the apparent, or visual, Galactic Center, around 10,800 BC, and the "ascent back into light" takes place now, when the winter solstice (December) Sun aligns with the apparent Galactic Center (galactic equator) around 1998–2012.

GALACTIC ALIGNMENT AND THE VEDAS

In *Galactic Alignment,* Jenkins has shown that not only was knowledge of the precession of the equinoxes encoded into mythology worldwide, but the ancients also knew that there are certain times during the cycle of precession when the Earth-Sun-galaxy relationship allows the influx of some kind of energy that triggers a transformational leap in humankind.

There are two points on the ecliptic where the solstices occur, and the ancients called these solsticial points "gates." These celestial gates were described by the philosopher Macrobius as the gates of Capricorn

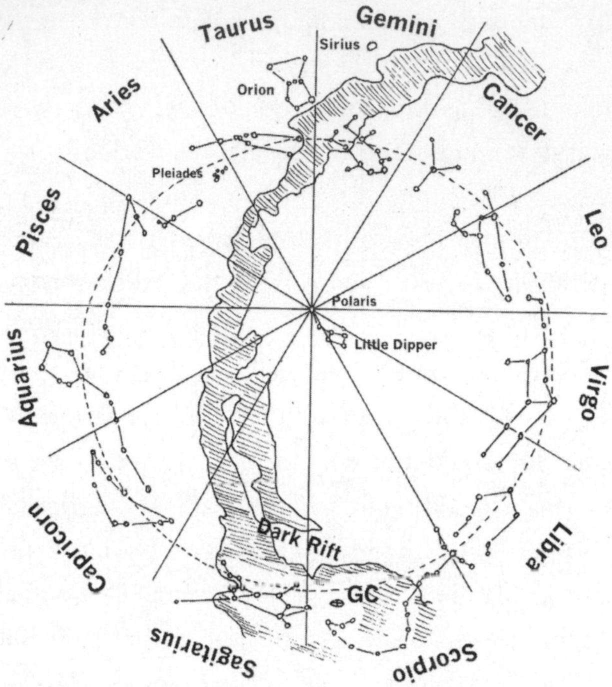

Fig. 3.2. The celestial gates are currently situated where the ecliptic (dotted line) crosses the Milky Way.

and Cancer, and were seen as doorways through which the soul descends to be reborn on Earth and ascends following physical death.

In Joscelyn Godwin's book *Mystery Religions in the Ancient World,* the gate of reincarnation is called the Silver Gate (Cancer), while the gate of ascension is called the Golden Gate (Capricorn), but he says that the Silver Gate is the "Way of Reincarnation," while the Golden Gate is the "Way of the Gods" and leads "beyond the Circle of Necessity, i.e., to release from the round of birth and death. These are the two routes from which the soul can exit from the world at death, and the Capricorn gate is the one through which the gods descend to Earth."[1]

It is clear from Giorgio de Santillana and Hertha von Dechend's book *Hamlet's Mill*[2] (see chapter 8) and from Jenkins's *Galactic Alignment* that these gates don't actually lie in Cancer and Capricorn but in the neighboring constellations of Gemini and Sagittarius. This is

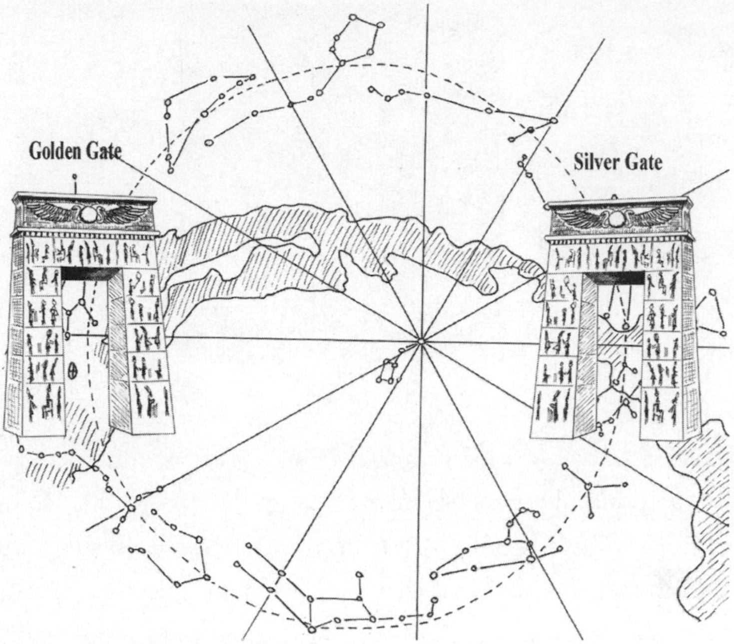

Fig. 3.3. The Golden Gate and the Silver Gate currently lie on the galactic equator.

because precession has moved—precessed—the gates back into the neighboring constellations. The Golden Gate lies at apparent Galactic Center, in Sagittarius, and the Silver Gate lies at Galactic Anticenter (the point in the galactic plane that lies directly opposite Galactic Center) in Gemini (there will be more on this in chapters 7 and 8). Following the work of Oliver Reiser, Jenkins makes a convincing case that the alignment of the solstice Sun with the galactic equator (close to Galactic Center), which is flagged by the end of the thirteen-baktun cycle in 2012, marks the window of time when the divine influences descend.

> Precession changes our angular orientation to the larger magnetic field of the galaxy in which we are embedded. During regularly occurring eras in the precessional cycle, as indicated by the solstice-galaxy alignments (probably the equinox-galaxy alignments, too), the Earth's protective magnetic shield becomes unstable and oscillates.

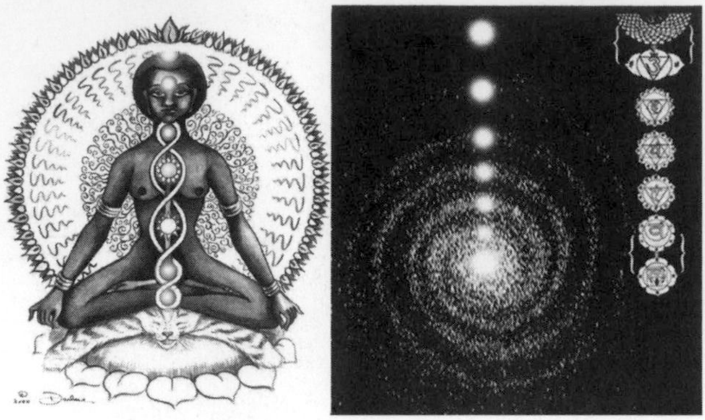

Fig. 3.4. The chakra system in the microcosm and the macrocosm (image on left: Darlene). From *Galactic Alignment: The Transformation of Consciousness According to Mayan, Egyptian, and Vedic Traditions,* by John Major Jenkins.

> Without a complete field reversal being required, this oscillation allows greater amounts of mutational rays to strike the surface of the Earth. While this may result in mutations and a greater chance for "evolution," of greater significance is the possible transformative effect on human consciousness during alignment eras, when human beings are exposed to higher doses of high-frequency radiation.[3]

In other words, when the solsticial points (and possibly the equinoctial points) come into alignment with the galactic plane (the galactic equator), the celestial gates open.

Jenkins refers to Sri Yukteswar's study of the Vedic yuga system and shows how that system is also signaling the galactic alignment. He concludes that there is a galactic chakra system, with the base chakra at apparent Galactic Center and the crown chakra in the direction of the galactic anticenter. Earth is at the level of the fifth (throat) chakra, evolving toward the sixth (*ajna,* or third eye) chakra: "The as yet unspoken message in this survey is that the galactic alignment opens a channel for the Kundalini shakti to flow through the Earth, cleanse (it) us, and excite it (us) into a higher level of being."[4]

However, Jenkins also accepts that the alternative concept, in which Galactic Center is seen as the crown chakra and Galactic Anticenter is seen as the direction of the root, or base, chakra, may be equally valid. Douglas Baker suggests Sirius is the galactic third eye.

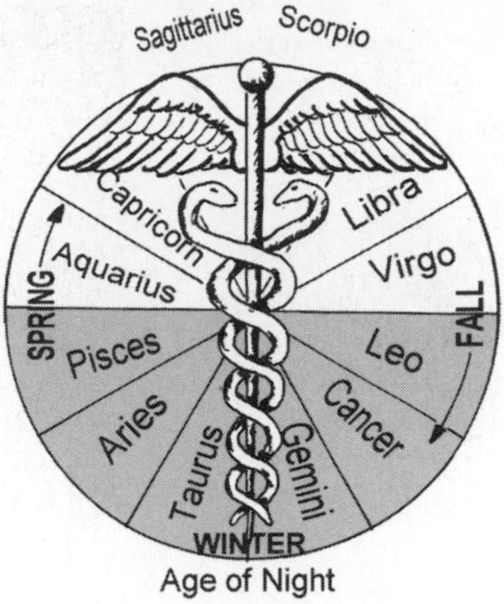

Fig. 3.5. Some interpretations of the galactic chakra system would have the base chakra at Galactic Anticenter near the Gemini–Taurus boundary, and the crown chakra at Galactic Center, near the Sagittarius–Scorpio boundary. After the book *Prophecies to Take You Into the Twenty-First Century,* by Moira Timms.

THE CHINESE CALENDAR

The Chinese calendar consists of sixty-year cycles, which in turn contain five twelve-year cycles. Each year is named after one of twelve animals and one of five elements, with two consecutive years being named after each element: Rat, Ox, Tiger, Rabbit, Dragon, Snake, Horse, Sheep, Monkey, Rooster, Dog, and Pig. These are sometimes called the twelve terrestrial branches. The elements are Wood, Fire, Earth, Metal, and Water.

Each element has a destructive manifestation followed by a constructive manifestation, making the ten celestial stems. The current Chinese sixty-year cycle started on February 22, 1984, with the year Wood Rat, and almost halfway through the sixty years, on January 23, 2012, the Chinese year Water Dragon will begin.

THE TIBETAN CALENDAR

The Tibetan calendar is based on the Chinese system, and started in the year 1027, when the Kalachakra teachings were introduced to Tibet. However, this was the year Fire Rabbit, three years after the start of the Chinese sixty-year cycle, so the current Tibetan sixty-year cycle, or *rab-byung,* therefore started on February 28, 1987. This is the seventeenth cycle of sixty years.

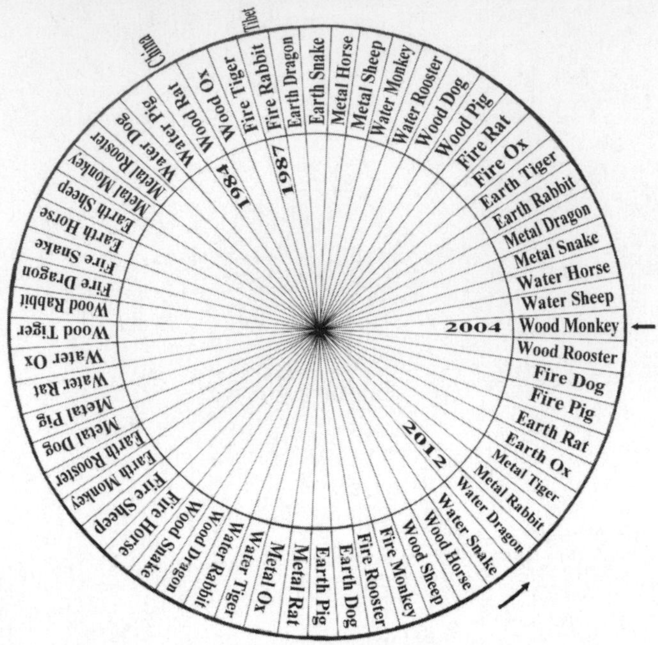

Fig. 3.6. The sixty-year combination cycle of "stems and branches," showing the Chinese and Tibetan starting points of the current cycle and the relative position of 2012.

The twelve Tibetan animals vary slightly from the Chinese and start with Rabbit rather than Rat, due to the three-year difference; the elements are thus listed starting with Fire rather than Wood.*

*Note that the Mesoamerican calendars also incorporate animals into their systems, but as day signs rather than year signs. Seven of them are the same as the Tibetan animals; Snake follows Dragon in both systems, and 2012 is the year of the Dragon in Chinese and Tibetan calendars. Tibet uses the Chinese calendar system of five elements combined with twelve animals to give a sixty-year cycle, the rab-byung, and in both Chinese and Tibetan systems 1986 was the year of the Fire Tiger. Most sources agree that the year Fire Tiger started on February 9, 1986, and ended on January 28, 1987, in China. In the Chinese version, the sixty-year cycle begins with a Wood Rat year, but the Tibetan calendar started in 1027, and this was the fourth year of the Chinese sixty-year cycle, the year Fire Rabbit (or Fire Hare). So the Tibetan sixty-year cycle starts three years after the Chinese sixty-year cycle. "Chinese and Tibetan new years, however, do not always coincide. This is because each of these calendar systems has its own mathematical formulas [formulae] for adding leap-months and for determining the start and length of each month." (The Berzin Archives: www.berzinarchives.com/astro_sc.html.) In 1986, the year Fire Tiger started on February 9 in Tibetan and Chinese systems, but the following year, Fire Rabbit began on January 28, 1987, in China and February 28, 1987, in Tibet. While the previous sixty-year Chinese cycle ended on February 1, 1984, and the current, twenty-seventh (some say seventy-eighth) Chinese cycle began on February 2, 1984, in the Tibetan version, the sixty-year cycle that ended in 1986–87 was the sixteenth sixty-year cycle since the Tibetan calendar started.

THE TAMIL CALENDAR

The Tamil calendar of southern India also comprises a sixty-year cycle, which, like the Tibetan one, restarted in 1987 (but in mid-April rather than February). It is a derivative of the old Hindu solar calendar and is based on the sidereal year. The sixty-year cycles may be connected to the sixty-year Jupiter–Saturn conjunction cycle, but the last conjunction was in May 2000, not close to the end/start of any of the above calendars. Each year in the Tamil calendar has a name, and 2012 is called *nandana,* meaning "delight."

Jenkins has discovered[5] that in Islamic and Indian astrology, the lunar nodes, which precess over 18.61 years,[6] are represented as the head and tail of a dragon. The lunar nodes are the points where the Sun and Moon's orbits intersect. When an eclipse occurred, the dragon was thought to be "eating" the Sun or Moon. According to an Islamic text, when the south lunar node (tail of the dragon) is at exactly three degrees Sagittarius, it is exalted. This is close to Galactic Center, and Jenkins points out that although the precise alignment of the galactic equator with the solstice Sun was in 1998, maybe eclipses could "represent the quicker 'second hand' of the precessional clock."[7]

There will be a total lunar eclipse on winter solstice 2010. So if galactic alignment involving the Sun at the galactic equator was the "hour hand" at winter solstice 1998, and the lunar eclipse on winter solstice 2010 is the "minute hand," when the Moon becomes involved, then maybe on winter solstice 2012 the "second hand" involves the alignment of another body to Galactic Center.

One classic degree of precession equals seventy-two years; winter solstice 1926 heralded the year 1927 (see below), and occurred seventy-two years before winter solstice 1998; a sixth of that is twelve years (1998 to 2010); a sixth of that is two years (2010 to 2012). So the question is: What celestial body will appear in conjunction with Galactic Center on winter solstice 2012? Could it be asteroid 2002AU4 or 1999NW2, both due on that day? Perhaps a comet or meteor shower, or an electro-magnetic pulse or EMP wave appearing as a blue star from Galactic Center? Could it be Nibiru, Planet X, Mercury, Venus, Mars, a supernova in Sagittarius—or something else?

KALACHAKRA: TIBETAN BUDDHIST PROPHECY

Gautama Buddha gave the Kalachakra teachings just before his death (*kāla* means "time" and *chakra* means "wheel," so Kalachakra means "Wheel of Time"). The first king of Shambhala (the first in a line of seven kings) was a witness to the teachings and he took them back to Shambhala, a mythical (and here I mean mythical in its academic sense of a mysterious story with concealed levels of meaning, rather than its popular usage as a simple fiction) kingdom in a hidden valley of the Himalayas. The eighth king founded a line of twenty-five kings of which he was the first (25 + 7 = 32 kings total), and the prophecy says that the last of these will establish a golden age. Each king reigns for a hundred years, so the golden age would then come 3,200 (32 × 100) years after the death of the Buddha. The reported dates of the Buddha's death vary, from 2422 BC to 546 BC, so the coming of the golden age is expected to be sometime between 779 AD and 2655 AD.[8]

Since the Tibetan calendar began in 1027 AD (1026 in some versions), when the Kalachakra teaching was taken to Tibet, most sources agree that the twenty-first king (twenty-eighth overall if you count the original seven kings) assumed power in 1927, and will rule until 2027. So the twenty-fifth and last (thirty-second overall) king would thus be due to arrive in the year 2327 AD. However, some Tibetan lamas believe the twenty-fifth king, Rudra Cakrin, is ruling now, which would still fit in with the possible dates for the death of the Buddha. In fact, the Nyingmapa, or "ancient ones"—the oldest sect in Tibetan Buddhism—say that Padmasambhava (a.k.a. Guru Rinpoche) brought Buddhism from India to Tibet around the eighth century AD. So perhaps the thirteenth (twentieth overall) king started ruling around the eighth century, three hundred years before the Kalachakra teachings were taken to Tibet, and it was the twenty-fifth king (thirty-second overall) who took the throne in 1927. If so, then his reign ends in 2027, and sometime before then he will fight and win the last battle against the "king of barbarians," inaugurating a golden age, which is supposed to last at least a thousand years.

THE KALACHAKRA-2012 CONNECTION

In his 1975 book *The Transformative Vision,* José Argüelles claimed that the sixteenth sixty-year Kalachakra cycle that began in 1927 and was to end in 1987 was the "last 60-year cycle of the Tibetan calendar," and he had already been told in 1969, by the late Tony Shearer, that the Ninth and last Aztec Hell cycle would end in 1987. This seemed to be an amazing coincidence. Shearer's understanding of the "present major cycles of Heavens and Hells" developed from a conversation with Mexican artist José Miguel Covarrubias, who had obtained the information from "a sorceress in the jungle hotlands of Tehuantepec."[9] Shearer determined how the cycles corresponded to the Gregorian calendar, and that information is published in his 1971 book *Lord of the Dawn* and in *Beneath the Moon and Under the Sun.* He found that the end of the Ninth Hell would be August 16, 1987, which, according to a calendar correlation

Fig. 3.7. The emblem of Kalachakra, or the Ten of Power, combining ten syllables that express the essence of the teachings. The outer teaching includes the prophecy, the inner teaching deals with techniques of manipulating energy flow through the psychic centers in order to transform the body into "a vehicle of liberation," and the other teachings include the use of sacred symbols such as this one, which represents the Kalachakra deity, who is shown in a passionate embrace with his consort—the unity of male and female, compassion and wisdom, necessary for enlightenment.

that is only one day away from the True Count, was 13 Ahau—the last day (as he thought, but in actual fact, the Maya saw it as being an endless cycle with no "last day") in the 260-day tzolkin (in the True Count—the unbroken count still used by the descendants of the ancient Maya—the previous day, August 15, 1987, was 13 Ahau). This seemed to be another amazing coincidence. Argüelles announced in his book *The Mayan Factor* that August 16–17, 1987, would be the Harmonic Convergence, inaugurating a twenty-five-year period of "phase shift," culminating in "galactic synchronization" in 2012.

However, Argüelles and Shearer made several errors in their calculations. They said that Cortés landed on Good Friday, April 21, 1519, and that this was the day 1 Reed in the year 1 Reed—the prophesied day for the return of Quetzalcoatl, the mythic savior god-man of the Aztecs (known to the Maya as Kukulcan). Exactly nine Calendar Rounds later (a Calendar Round is 18,980 days, or 52 × 365-day haabs, so nine of them would be 170,820 days), the Ninth Hell, they said, would end on the Harmonic Convergence, August 16, 1987. Using the "terminal year-bearer system," wherein the year is named after the 360th day, 1519 was the year 1 Reed or Ce Acatl, but April 21 (Julian date, or JD) was actually a Thursday—meaning Good Friday was April 22. That is only one day of error, but when we then investigate to find out when the day 1 Reed occurred, we find it was twelve days later, on May 4 (JD). When we then check to see when nine Hells after Good Friday 1519 (170,820 days) would be, we get January 7, 1987 (GD, or Gregorian date), 221 days before the Harmonic Convergence. Nine Calendar Rounds after the day 1 Reed would be January 20, 1987 (GD), which is 208 days before the Harmonic Convergence. Note that Shearer's calculations are based exclusively on Aztec dates and despite the similarities between Aztec and Maya cultures and calendars, the Harmonic Convergence and much other material in *The Mayan Factor* is more Aztec than Maya. The Aztecs, who lived about five hundred years after the Maya classic era, had lost a lot of calendrical knowledge.

Considerable study has allowed for a reappraisal of Shearer's and

Argüelles's statements, so that they can be restated correctly. Since Shearer's information originated in Tehuantepec, and the Tehuantepec River is just south of Oaxaca, and the Teotitlan area is just north of Oaxaca, we can rewrite Argüelles's key paragraph using the facts:

The landing of Cortés at Veracruz occurred in the spring of 1519, which corresponded to the year Ce Acatl (1 Reed), dedicated to the principal god-hero of the Aztecs and their predecessors, the Toltecs, Quetzalcoatl. Furthermore, according to the single source we have, an unknown so-called sorceress from Tehuantepec, the ancient seers (probably Aztecs) had prophesied that this particular year would mark the end of a Heaven period consisting of thirteen fifty-two-year cycles and the beginning of a major Hell period consisting of nine fifty-two-year cycles, or 468 years. However, various districts of Mesoamerica inhabited by Aztec, Maya, Olmec, Zapotec, and Mixtec groups followed different variations of the fifty-two-year cycle, and there is only one that ended in 1519. This is known as the Teotitlan Calendar Round, and the day 2 Reed (2 Acatl or 2 Ben), or June 13, 1519 (JD), marked the beginning of nine fifty-two-year cycles that would terminate on March 1, 1987 (GD), the end of nine Teotitlan Hells. On this day, a twenty-five-haab cycle of purification would begin, ending on February 23, 2012, which is just two days after the Tibetan New Year, or Losar.

The Tibetan calendar started in 1027 with the arrival of the Kalachakra teachings, and the sixteenth sixty-year rab-byung cycle of the calendar ended on February 28, 1987, just two days before the end of the nine Teotitlan Hell cycles. Amazingly, although the Tibetan New Year is governed by the lunar cycle, if we add twenty-five 365-day haabs to the end of the sixteenth sixty-year rab-byung cycle in 1987, we arrive at Losar in the year 2012, which occurs on February 21, two days before the end of the Teotitlan-timed purification period.

There is another significant fact about 1927 that Argüelles missed. The twenty-first king of Shambhala assumed power in 1927, but some lamas believe this is actually the twenty-fifth (thirty-second overall) and last king, who is ruling now (32 × 100 years after the Buddha's death). This last king is the tenth and final incarnation of Vishnu, who will return to vanquish evil, put an end to the age of discord, and inaugurate the golden age. Known by the Tibetan name Rudra Cakrin, his Hindu name is Kalki. In the golden age, which will last at least a thousand years, the prophecy says there will be no more work or fighting, and our lifespans will increase to eight hundred years.

HINDU PROPHECY: KALKI AND THE CRITICAL MASS OF ENLIGHTENED ONES

In *The Way to Shambhala,* Bernbaum says that the golden age will come when the Sun, Moon, and Jupiter all meet in the same quadrant as the Tishya constellation[10] (which is part of Cancer). The next time this happens will be on July 26, 2014, according to my astronomical software,[11] but here is a quote from the RISA (Religion in South Asia) website, revealing that Kalki is already here and the golden age is coming in 2012:

> About 15 million people around the world think of Kalki Bhagavan as the Supreme Being who will usher in a new age. This golden age was inaugurated on September 25, 1995, and will be fully realized in the year 2012. The Vishnu Purana and several other sacred texts consider Kalki to be the last incarnation/avatar of Lord Vishnu. He has descended nine times already [twenty-three times in some versions], and the last one for this cycle will be at the end of Kali Yuga, the present degenerate age. Some Hindu thinkers put the beginning of Kali Yuga around 3102 BC, and since it is supposed to last 432,000 years (according to one reckoning), many Hindus are not expecting Kalki anytime soon. Nevertheless some individuals have been identified as Kalki by their devotees.[12]

On the Skyboom website, a page about this "Avatar of the Golden Age" further explains the 2012 connection: "Others would say that this is not possible, that the old calendars say that the Golden Age won't 'be here' for thousands of years. Currently there is heated debate in India, as some experts have claimed to have found errors in the ancient calendars. This puts the date of the Golden Age very close to the end of the Mayan calendar."[13]

The Golden City website explains that Kalki only needs to enlighten 60,000 people to start a chain reaction, since he is a mukti avatar, "a divine Master through whom the Divine can transfer Enlightenment to any number of people."

> There are now very powerful spiritual energies engulfing the Earth coming from the center of the Universe. These energies are meant to make mankind enlightened, but somehow the process is not starting off as it was supposed to. So Kalki can give and transform these energies to anyone now, but only until 2012. If 60,000 have become enlightened before this, he can help man getting enlightened. If later, the negative powers will take over and we have missed this very rare chance. The future will not at all be nice if man will not transform fully in the near future.[14]

In 1977, another teacher, Sri Goswami Kriyananda, the Guru and Spiritual Preceptor of the Temple of Kriya Yoga in Chicago (who is not from India, despite the name), told a student, author Ray Grasse, that the period from 2010 to 2012 signified a "potentially important window for open contact with non-human races."[15]

4
I CHING
The Ancient Chinese Lunar Calendar

TIMEWAVE ZERO

The I Ching, or Book of Changes, consists of an arrangement of sixty-four hexagrams (six-line figures) consisting of two types of lines, yin and yang, and has been used for millennia as an oracle. Coins or sticks are thrown, the combination of which signifies one or two hexagrams, the second hexagram representing the predicted change in situation from the first one. However, as I mentioned in the introduction to this book, in *The Invisible Landscape,* Terence and Dennis McKenna make a case that the I Ching had originally been used as a lunar calendar, in which the sixty-four six-line hexagrams—384 lines in all—represented the 384 days in a thirteen-month lunar year.

In 1971, the brothers went on an expedition into the Colombian Amazonas, via San Augustine and Bogotá to Florencia. From Florencia they flew to Puerto Leguizamno, an isolated embarkation point on the Rio Putumayo, which is a tributary of the Amazon. They then traveled by boat down the Putumayo for two days and turned onto the Rio Cara-Parana for the last part of the river journey to the mission at San Raphael. From there, they walked 110 kilometers through the forest, to the tiny mission settlement of La Chorerra, where they shared a hallucinogenic experience furnishing information on the pattern behind the unfolding of events in the universe. Terence returned home convinced that concealed in the I Ching lay a mathematical encoding of this pattern, and he proceeded to examine it closely.

The theory that emerged from the I Ching study is known as Novelty Theory, since "I Ching" means "Book of Changes," and the graph that was generated from it maps the "ingression of novelty into spacetime." Philosopher and mathematician Alfred North Whitehead had developed a definition of novelty that was elaborated by Terence McKenna into "the Eschaton" or, simply, "the Timewave." This Timewave displays the ebb and flow of novelty—the wave ascends toward habit or descends into novelty, revealing qualities inherent in time itself, that are repeated fractally at all levels of manifestation. In 1987, mathematician and software developer Peter Meyer produced an MS-DOS (Microsoft Disk Operating System—a personal computer operating system used in the 1980s that was superseded by Windows) that is able to display any section of the graph, from ninty-two minutes up to seven billion years. Later versions of the software, which was originally known as Timewave Zero, incorporated other number sets, culminating in the current version, known as Fractal Time, or just the Timewave (this is the Fractal Time website: www.hermetic.ch/frt/frt.htm). However, before discussing these other versions of the wave, we need to understand the steps that culminated in the Timewave.

Terence noticed that the two main numbers of the I Ching, six (lines in a hexagram) and sixty-four (hexagrams in the sequence), when used as multipliers around a 384-day basic cycle, caused the following resonances:

64 × 6 = 384 days—3 lunations or 1 lunar year
384 days × 64 = 67.29 solar years—6 minor sunspot cycles
67.29 solar years × 64 = 4,306 solar years—2 zodiacal ages
4,306 solar years × 6 = 25,836 solar years—1 complete precession of the equinoxes

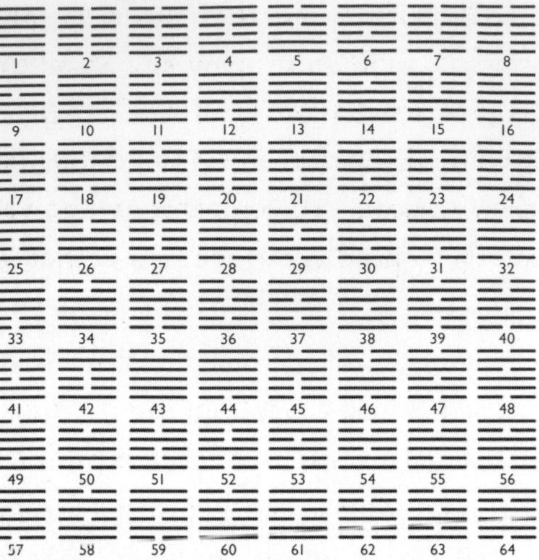

Fig. 4.1. The King Wen sequence of the I Ching

An inspection of an early arrangement of the hexagrams, known as the King Wen sequence, showed beyond any doubt that the hexagrams were deliberately and carefully arranged in this order. For example, the second hexagram of each pair either is a complete reversal of the first (where each line changes to its opposite) or is an inversion of the first (the whole hexagram is turned upside down). He then proceeded to apply a series of operations to the sequence to mathematically extract all the information and convert it into a timewave.

"First operation": the amount of difference between sequential hexagrams (the number of lines that changed to their opposite) was enumerated, then graphed, and the graph was reversed and recombined with itself to produce a simple wave. This was then analyzed for skew in two directions, producing two sets of sixty-four numbers that were then combined into one set. The simple wave was next analyzed for divergence, congruence, and overlap of the two components, and the resulting set of sixty-four numbers was combined with the skew set.

"Second operation": the simple wave was recombined with itself at three levels, where six simple waves were laid end-to-end, representing the six lines of a hexagram; two larger versions were overlaid on the six, to represent two trigrams (half hexagrams of three lines each, of which there are a total of eight that combine to make the sixty-four hexagrams), and finally, one still-larger simple wave was laid on top to represent a whole hexagram. This complex wave has 384 positions

(6 × 64) available for analysis, and the sixty-four final numbers from the "first operation" can be recombined using the complex wave into 384 quantification values, or the number set. When these 384 values are graphed, we arrive at a figure called the Eschaton. To get this far, the McKennas used what was then an advanced computer at a university, the printouts from which are the appendices of the first edition of *The Invisible Landscape.*

This Eschaton wave would be repeated at a series of levels, each sixty-four times greater than the one below, and combined into a "modular wave hierarchy." This later became known as the fractal wave, and is composed of twenty-six levels that govern all existing change, from subquantum events up to the duration of the three-dimensional universe. The term *fractal* had not even been invented at this point, however, and Peter Meyer's Timewave Zero computer software, incorporating all the fractal levels, and first developed in 1987, was still about thirteen years away.

The next step was to correlate the Eschaton wave and history, when history is expressed as a graph of novel events. The explosion of the atomic bomb at Hiroshima on August 5, 1945 (U.S. time), was taken to be a sufficiently "novel" event that it should relate to a major peak on the wave. When we consider the acceleration of technology in the last three hundred years and look at the Eschaton wave as a 4,306-year cycle, it seems obvious where the Hiroshima bomb explosion should lie on the wave. When this is done, there is just one 67.28-year cycle left until novelty hits the baseline, on November 17, 2012.* This is the point where all the waves and subwaves peak together, signifying the "maximum ingression of novelty into spacetime"—an evolutionary omega point (end point, named after the last letter of the Greek alphabet) and dimensional transition known as "concrescence" (originally a biological term meaning "the growing together of related parts, tissues, or cells," adapted by Alfred North Whitehead to refer to the point of maximum novelty).

---

*November 18, if you take the non-U.S. time of August 6, 1945, as the date of the bomb.

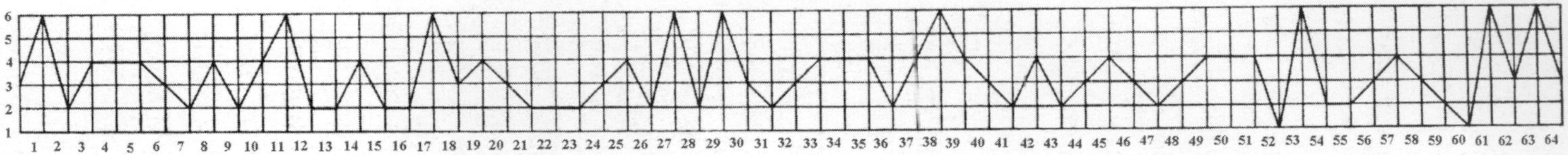

Fig. 4.2. The graph of the numerated difference between each hexagram and the subsequent one. After Terence and Dennis McKenna's *The Invisible Landscape: Mind, Hallucinogens and the I Ching.*

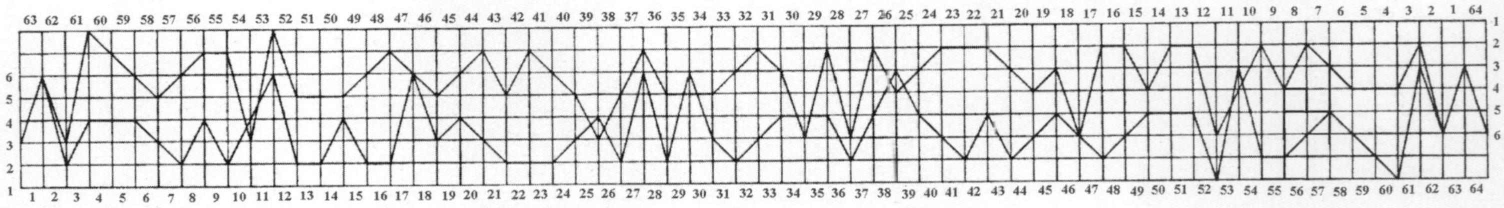

Fig. 4.3. The simple wave. After Terence and Dennis McKenna's *The Invisible Landscape: Mind, Hallucinogens and the I Ching.*

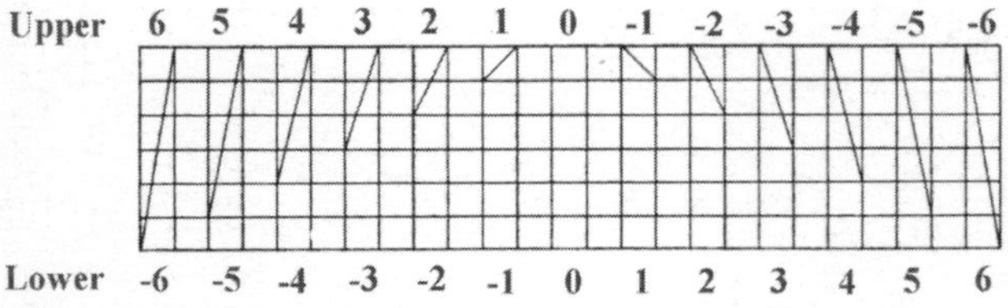

Fig. 4.4. Line segment skew values. After Terence and Dennis McKenna's *The Invisible Landscape: Mind, Hallucinogens and the I Ching.*

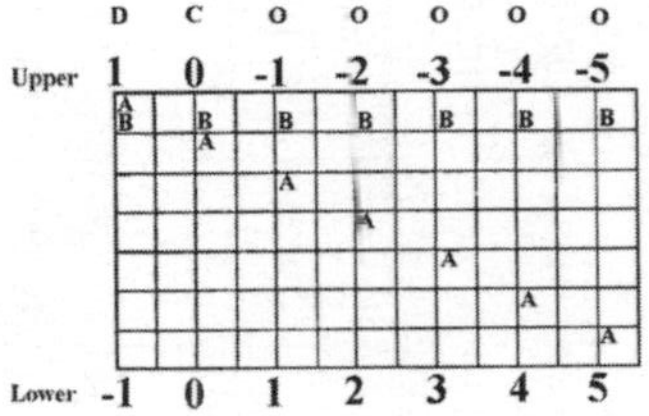

Fig. 4.5. Divergence, congruence, and overlap values. After Terence and Dennis McKenna's *The Invisible Landscape: Mind, Hallucinogens and the I Ching.*

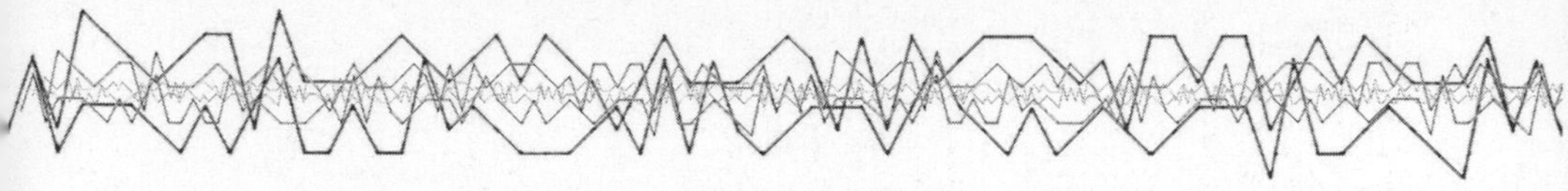

Fig. 4.6. The complex wave. After Terence and Dennis McKenna's *The Invisible Landscape: Mind, Hallucinogens, and the I Ching.*

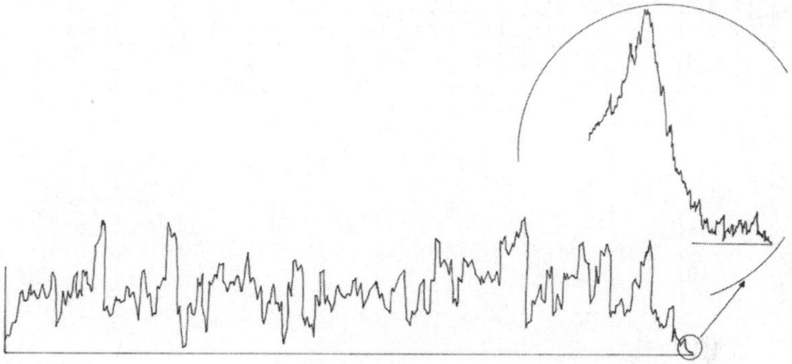

Fig. 4.7. The Eschaton wave with magnified section revealing the fractal nature of the wave—the same shape is repeated at a smaller level. (Baseline = maximum novelty)

When Peter Meyer's software allowed the full fractal wave to be examined, the major novelty spike moved away from August 5, 1945, due to the cumulative effect of all waves and subwaves, but by then Terence McKenna had heard about the thirteen-baktun cycle of the Maya and its end point in December 2012 and had brought the termination point forward to coincide with it, and this was incorporated into the software. In 1996, the mathematics of timewave theory came under criticism by a British mathematician named Matthew Watkins, but it was later recalculated by another mathematician and nuclear-fusion specialist, John Sheliak, using vector analysis. Sheliak confirmed that there was an error, but that it was correctable. The resulting fractal wave (sometimes called Timewave One) was again locked into history by correlating a major novelty spike to the Hiroshima explosion, thus restoring meaning to the fractal wave. However, when the end point is moved forward to coincide with the thirteen-baktun end point on December 21, 2012, the novelty spike in 1945 moves forward to September 8 (not back to July 16, 1945, the date of the Trinity

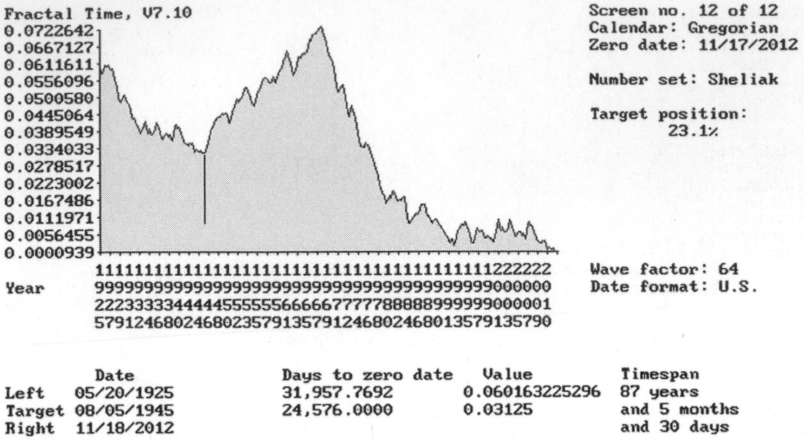

Fig. 4.8. The timewave section covering about 87.5 years before 2012. The vertical line signifies the Hiroshima event in August 1945. Screen shot from Peter Meyer's Fractal Time software. Note maximum novelty is downward, not upward. Maximum novelty value = 0.0000000.

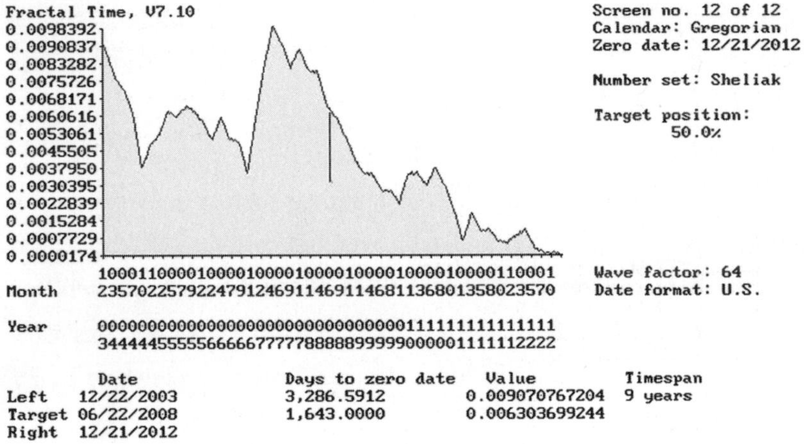

Fig. 4.9. The Sheliak wave 2003–2012. The diagram shows novelty peaks for the nine years leading up to 2012, with the end point set to the Maya end-date of December 21, 2012. Screen shot from Peter Meyer's Fractal Time software.

test, when the first nuclear bomb was exploded, just prior to Hiroshima, as implied by Sheliak's commentary).[1] September 8, 1945, is an unremarkable date—in other words, there is no significant novel event that occurred on it, as we would expect there to be, since the corrected wave's major novelty spike points it out, when the end-date is set to December 21, 2012. This

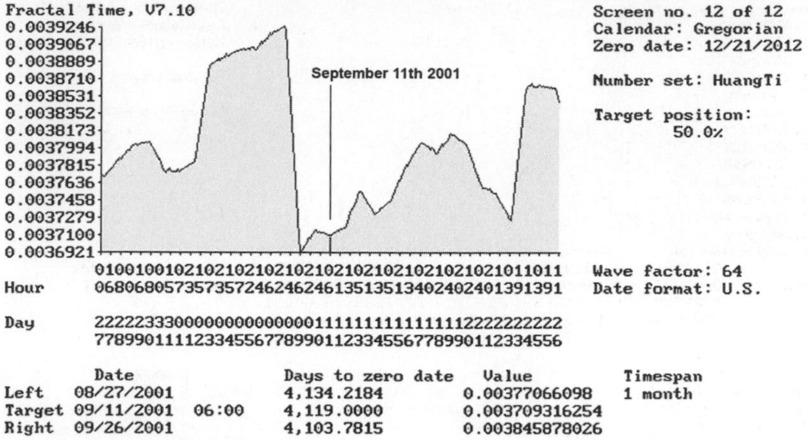

Fig. 4.10. The Huang Ti wave showing a target date (where the vertical line is) of September 11, 2001. The end-date is set to December 21, 2012. When the end-date is moved to December 23, the novelty spike will move to the right, coinciding with the target date vertical line. Screen shot from Peter Meyer's Fractal Time software.

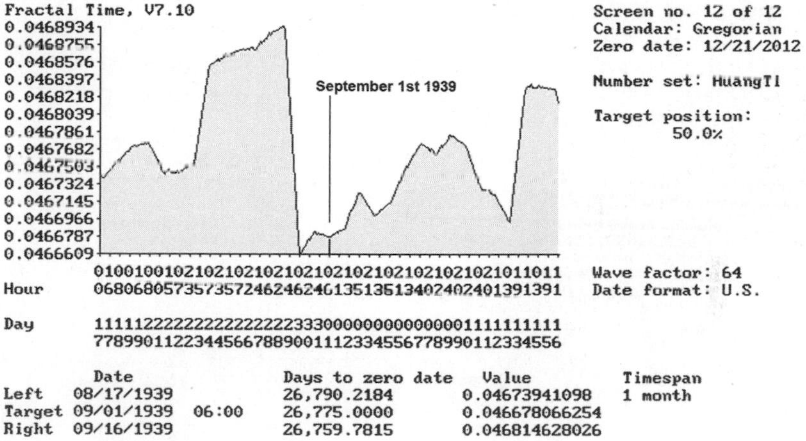

Fig. 4.11. The Huang Ti wave showing a target date of September 1, 1939, exhibiting resonance to the 9/11 event in figure 4.10. As in figure 4.10, if the end-date is moved to December 23, the novelty spike moves on two days—in this case to September 1, 1939, when Hitler invaded Poland. Screen shot from Peter Meyer's Fractal Time software.

suggests that perhaps the original November 17, 2012, date of maximum novelty really is the end of the timewave after all.

McKenna admitted in 1975 that the "best-fit" method was only one of the ways to determine the "possible dates of a future concrescence."[2]

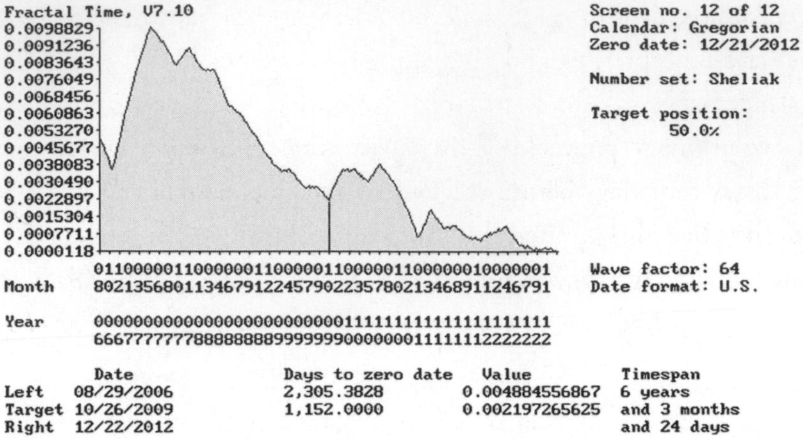

Fig. 4.12. The Sheliak wave: the resonance period of 6.31 solar years (6 x 384 days) leading up to the date of concrescence in 2012. Screen shot from Peter Meyer's Fractal Time software.

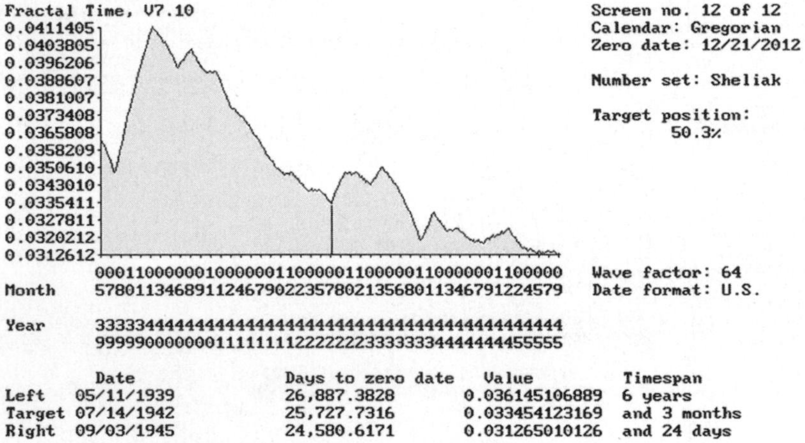

Fig. 4.13. The 6.31-year period leading up to September 1945, encapsulating World War II, which shows an almost exact resonance (on the Sheliak wave) to the same period leading up to 2012 (see fig. 4.12). Screen shot from Peter Meyer's Fractal Time software.

The second method, he said, "is more subtle and takes account of the precession of the equinoxes." He went on to suggest that the winter solstice node is approaching Galactic Center, and that when that node eclipses Galactic Center, it "might be an event unusual enough to signal an onset of concrescence."[3] He also pointed out that although the solstice node is currently "only about 3° from Galactic Center, a degree covers a large area in space, and the galaxy may be presumed to have a gravitational center, a radio center, and a spatial center."[4]

In other words, not only did McKenna's timewave predict that the evolutionary pinnacle of this space-time dimension would occur just thirty-four days before the end of the thirteen-baktun cycle, but McKenna had also intuited the importance of the 2012 alignment of the winter solstice Sun with a point near Galactic Center, not knowing that the Maya had foreseen this very event well over two thousand years ago.

9/11 RESONANCE

Following the terrorist attacks on the World Trade Center and other targets in the United States on September 11, 2001, Zyzygyz, the founder of the Time Wave Zero 2012[5] Internet discussion group, checked the Timewave software to see if there was a novelty spike on that day, but there was nothing obvious. When he looked for resonances on the timewave, he found a partial resonance to the Nazi invasion of Poland on September 1, 1939.

When Zyzygyz mentioned this to Peter Meyer, he took a look himself and pointed out that there was an exact resonance using the Huang Ti numbers (see below) and the December 23, 2012, end-date.* The resonance is also exact using the December 21, 2012, end-date, but both dates are on the bottom of a large novelty spike when the 23rd is used as an end-date.

Since Meyer had only calculated the Huang Ti number set to test timewave theory, and the Huang Ti sequence has a dubious history,† plus the fact that this only works with the Lounsbury end-date (December 23, 2012), it started looking as if none of the existing number sets were quite right, since they didn't show a novelty spike on the 9/11 event using the original end-date in November 2012 or the winter solstice end-date of

*December 23, 2012, is the second most popular end point for the thirteen-baktun cycle, and is known as the Lounsbury correlation, the 584285 correlation, or the GMT3.

†Emporer Huang Ti was a book burner and there is no record of him having developed an I Ching sequence. It seems likely that Peter Meyer produced this sequence as a test for timewave theory, and just arbitrarily named it after any old emperor.

the Maya. However, this brought to light the fact that there is a very close resonance on Timewave One (the Sheliak wave) between the six years and one day of World War II and the six years and one day leading up to December 21, 2012 (see color plate 6). (This is less so for the Kelley and Watkins waves, and even less than that in the case of the Huang Ti wave). This is partially explainable in terms of the resonances mentioned above.

At this point, the Timewave software allowed the user to choose from several number sets, each of which generated a different version of the timewave. The original number set had been generated from the King Wen sequence of the I Ching around 1974 by mathematicians Leon Taylor and Royce Kelley (who had been recruited for the job by Terence McKenna), leading to the first wave, which is known as the Kelley Wave. In 1996, Mathew Watkins, a British mathematician who had just completed his doctoral dissertation, highlighted a problem (the "half-twist") with the mathematics of the original number set. Watkins's paper was known as the Watkins Objection, and although it almost derailed the whole timewave concept, it did lead to a new number set—the Watkins set. As already mentioned, this is when John Sheliak saved the day by correcting the problem and developing another number set—the Sheliak set. Finally, Peter Meyer developed a fourth number set called the Huang Ti set, in order to "test" the timewave theory, but it was named after an emperor who certainly was not the originator of an I Ching sequence, since he banned the I Ching! Yet it was this one that provided the 9/11 novelty spike.

FRANKLIN TIMEWAVE—ZYZYGYZ'S BRAINWAVE

Magic squares are grids of numbers arranged in rows and columns in a square in such a way that all columns, rows, and diagonals add up to the same number. Though often found in today's puzzle books, they were highly regarded by ancient mathematicians, since they were considered to illustrate cosmic laws. Traditionally there is a magic square associated with each planet: a 3 × 3 square is associated with Saturn; a 4 × 4 square with Jupiter; 5 × 5 with Mars; 6 × 6 with the Sun; 7 × 7 with Venus; 8 × 8 with Mercury; and 9 × 9 with the Moon.[6] Benjamin Franklin (who was a Freemason and so may have been privy to Maya calendrical information—see "The Eye in the Pyramid," in chapter 8), one of the Founding Fathers of the United States, devised a different arrangement of the magic square of Mercury (an 8 × 8 square, consisting of the numbers from 1 to 64). However, the numbers in the Franklin square add to the same number in so many ways (at least seventy-five)[7] that the square has become a mathematical legend. When you take into account the fact that the numbers in the 8 × 8 magic square of Mercury add up to 260—the number of days in the Maya Sacred Calendar, the tzolkin, and also the number of katuns in the "macroscopic tzolkin" of the thirteen-baktun cycle—then we have the suggestion of links connecting the Franklin 8 × 8 square, the I Ching (an 8 × 8 square of sixty-four hexagrams), and the Maya calendars.

								260
8	58	59	5	4	62	63	1	260
49	15	14	52	53	11	10	56	260
41	23	22	44	45	19	18	48	260
32	34	35	29	28	38	39	25	260
40	26	27	37	36	30	31	33	260
17	47	46	20	21	43	42	24	260
9	55	54	12	13	51	50	16	260
64	2	3	61	60	6	7	57	260
260	260	260	260	260	260	260	260	260

Fig. 4.14. The magic square of Mercury, showing how rows, columns, and diagonals add up to 260. Any quadrilaterally symmetrical pattern of eight squares will also add up to 260, e.g., two squares mirror-imaged in each corner.

This connection was discovered and published by José Argüelles in 1984 in *Earth Ascending,* before he knew that Terence McKenna had used the I Ching to arrive at 2012 as the year in which the concrescence (evolutionary omega point) of the universe would be achieved.

Zyzygyz (without knowing that Argüelles had already published the Franklin–I Ching connection) figured that if the numbers in the Franklin 8 × 8 square were each replaced with the hexagram of that number (in the King Wen sequence), then the Franklin 8 × 8 square could be converted into a timewave. Peter Meyer[8] helped out by generating the 384 number set from Zyzygyz's sixty-four Franklin hexagram numbers, and

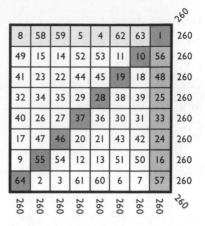

52	61	4	13	20	29	36	45	
14	3	62	51	46	35	30	19	
53	60	5	12	21	28	37	44	
11	6	59	54	43	38	27	22	260
55	58	7	10	23	26	39	42	260
9	8	57	56	41	40	25	24	260
50	63	2	15	18	31	34	47	260
16	1	64	49	48	33	32	17	260

Fig. 4.15. The Franklin square does everything that the Mercury square does, plus many more bilaterally symmetrical patterns, such as these.

together they found that a true timewave was possible (this is not guaranteed—an earlier attempt to make a timewave from the earliest I Ching arrangement, the Fu Hsi sequence, wouldn't work).* The largest level of the wave could then be mapped against time, and Zyzygyz noted various correlations to events in the solar system, with peaks and troughs on the Franklin timewave.

With help from Brendan Boerner, John Sheliak, and myself, Zyzygyz persisted until he was able to generate a Franklin timewave that would map the last seven billion years, but it

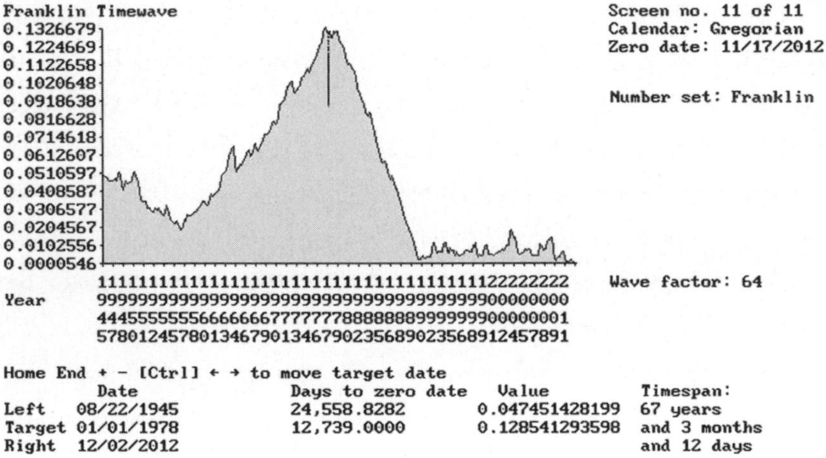

Fig. 4.16. The Franklin wave showing the 67.29-year resonance period leading up to 2012. Saddam Hussein was made president on July 17, 1979, which is just to the right of the vertical line, at the top of the steep descent into novelty. Screen shot from Peter Meyer's Fractal Time software, adapted for the Franklin number set.

*The 384 quantification values did not begin or end with a zero, so the resulting graph did not descend to the point of maximum novelty. Therefore the wave has no beginning or end point. For the timewave calculations, see the Time Wave Zero 2012 discussion group message thread from message 1722: http://tech.groups.yahoo.com/group/TIMEWAVEZERO2012/message/1722.

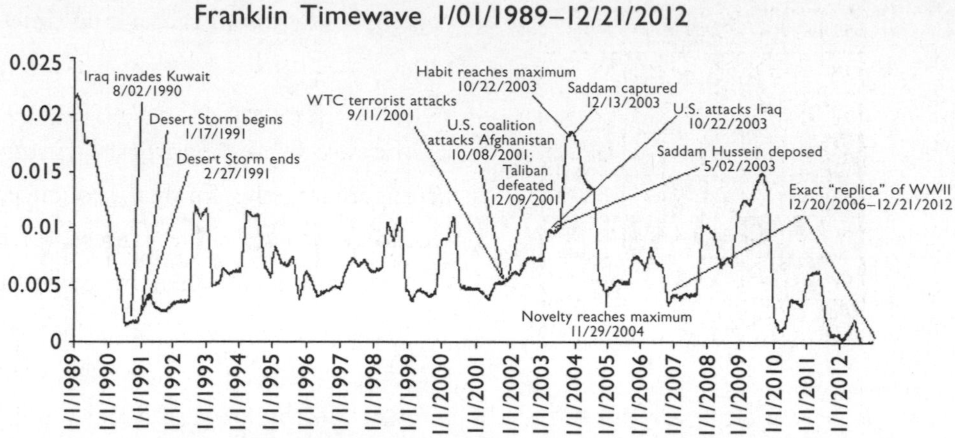

Fig. 4.17. The Franklin wave showing how recent events in Iraq coincided with peaks and troughs on the wave. Diagram by Zyzygyz.

doesn't answer all the questions after all. It doesn't show the rush into novelty that all the other waves (except Huang Ti) show between the late 1960s and early 1990s. It shows more resonance between World War II and the 2006-to-2012 period than the Huang Ti, Watkins, and Kelley waves, but not as much as the Sheliak wave, though it doesn't show a novelty peak for the Hiroshima date or any of the first nuclear tests. It does, however, seem to show a resonance between Saddam Hussein's assumption of power in Iraq and Sargon of Akkad's takeover of Mesopotamia (Iraq) in 2334–2279 BC.

5

ABRAHAMIC RELIGIONS

Disclaimer: Before going any further, I need to point out that I do not subscribe to any of the judgmental or right-wing attitudes expressed on any of the literalist religious sites that I quote from; nor do I sympathize with any fanatical or tyrannical attitudes expressed on any websites. If there is any judging to be done, then I would suggest that our higher selves judge our lower selves as individuals. I also suggest that religion and race are irrelevant.

CHRISTIANITY: THE RAPTURE AND 2012

There have been several Christian groups and individuals claiming on the Internet and in books that in 2012 the "Day of the Lord" will arrive. For instance, in a recent novel by Ivan and Dora Cain, *The Year 2012,* a character named Elbib Ehtdrow is a reporter in Jerusalem covering the descent of a hooded humanoid on a cloud in 2012. The person is taken to be the returned Messiah and calls himself Natas Reficul, and starts putting the world in order. This is a predictable replay of the prophecies of the book of Revelation—just reverse the spelling of names to get the plot.

There is also *Magic Music Myth,*[1] a rock opera, CD, and live concert by Alchemy about the judgment of the Devil in 2012, and a proposed film, originally known as *Phantasm 2012 AD: The Mormon Mausoleum—Apocalypse and Hell on Earth,* now retitled *Phantasm's*

End. The more positive side emerges in the form of a music CD called *2012* by Soul Rising, who call themselves "humble seekers of Christ." Yet another example of the Christian–2012 connection is a book called *God's New Millennium,* by Richard Henry Whiteside, and his website, GodsWeb,[2] which announces that New Jerusalem is to be situated in central California and fully functioning by 2012. So are these authors and groups just jumping on the bandwagon, or are there scriptural reasons for assuming that Christianity's Day of Judgment will happen in 2012?

Without getting into complex theological doctrines of premillennialism, amillennialism, and postmillennialism, or pre-tribulational, mid-tribulational, or post-tribulational positions, let's just see if we can find any reason why concepts such as the Rapture, the Second Coming, and Judgment Day are being connected with 2012. The various approaches mentioned above can be summarized by saying that there are biblical reasons for expecting the seven-year period of Tribulation (troubled times), and that either before, after, or halfway through this period the Rapture will occur, when 144,000 "chosen ones" will be taken up to heaven. Judgment Day (usually taken to be the Second Coming, though some refer to this as the Rapture) will then occur either at the end of the seven-year period or after an additional thousand-year period.

Fundamentalist preacher Jack Van Impe says that the Second Coming of Christ will occur between 2001 and 2012,[3] and gives some interesting clues as to the reasoning in the summary of his video *Left Behind*.[4] The clues point to a three-part answer. The first part is in the first three gospels (Matthew, Mark, and Luke); the second part is in the book of Daniel; and the third part is in the rabbinical lore of Judaism. The first two are covered in the next two sections; the third is covered later in this chapter, in the discussion on Judaism.

MATTHEW, MARK, AND LUKE

Many people assume that all the end-time prophecies originate in the book of Revelation—the last book in the bible, accredited to the apostle John—but there are many other references, including a "*mini-apocalypse*" in three of the four gospels: Matthew 24–25, Mark 13, and Luke 21. The first thing explained by the online authors[5] is that the first sections of the mini-apocalypses describing the Tribulation are actually about the events leading up to the destruction of the Temple of Jerusalem in 68–70 AD, not about Jesus's Second Coming. This initial reference to a mini-apocalypse was in response to the first part of a question put to Jesus by the disciples, who asked when the Temple would be destroyed. The second part of the question—"And what will be the sign of your coming and of the close of the age?"—was answered next (the questions are fully listed in the Matthew version, and the answers fully listed in the Luke version). In Luke's gospel, this question is answered in 21:24: ". . . and Jerusalem will be trampled by the Gentiles until the times of the Gentiles are fulfilled." In 68 AD (often put at 70 AD), when the Romans destroyed the Temple, the Jewish people had to wait another nineteen hundred years until they regained control of Jerusalem in 1967 during the Six-Day War, at which point "the times of the Gentiles" (non-Jewish people) were fulfilled. The verses of Luke chapter 21, from verse 24 onward, may describe the physical, religious, and moral signs that would take place between the first sign, or pre-sign, in 1967, and Jesus's return. Most theologists believe that these signs have now all been fulfilled.

THE BOOK OF DANIEL

Nebuchadnezzar, king of Babylon, had a dream of a statue with a head of gold, arms and chest of silver, a belly and thighs of bronze, and legs of iron. The feet were part iron and part clay. Then a boulder smashed into the feet of the statue and it crumbled to dust, leaving the stone, which "became a mountain and filled the whole Earth." The prophet Daniel decoded the dream as follows: There would be four empires or world kingdoms, symbolized by gold, silver, bronze, and iron, and representing the Babylonian, Persian, Greek, and Roman empires. The last of these is to be followed by the heavenly Kingdom of God . . . except that there is actually another empire in between—iron mixed with clay—possibly referring to the "empire" of the Roman (Catholic) Church. So we have a similar version to the Maya and Aztec vision of a set of four or five eras, ending in a dramatic transition to a new state of being.

Fig. 5.1. The statue of Nebuchadnezzar's dream, depicting four or five empires preceding the coming of the Kingdom of God.

Before looking at the encoded numbers of years, we must first take note of the biblical code in which one day represents one year. Earlier in his book, Daniel speaks of "seventy weeks of years,"[6] which represents a 490-year period (70 × 7 days = 490 days) and gives us a clue about a day-year code.[7] There is, however, a direct connection to the words of Jesus when Daniel overheard a "holy one," or angel, ask how long the sanctuary would be "trampled under foot" (Daniel 8:13–14), and the reply was, "For two thousand and three hundred evenings and mornings; then the sanctuary shall be restored to its rightful state." Using the day-equals-year rule, we find that 2,300 years after the invasion of Israel by the Greeks around 333 BC leads to 1967 AD. However, closer investigation reveals a discrepancy of up to two years, since the calculation fails to take into account the missing year zero between BC and AD dates, and although Alexander the Great invaded Israel in 333 BC, he captured Jerusalem the following year, 332 BC.

In the last three verses of the book of Daniel, it says the period preceding the end of days can be figured as 1,290 "days" or 1,335 "days": "From the time that the continual burnt offering is taken away and the abomination that makes desolate is set up, there shall be a thousand two-hundred and ninety days. Blessed is he who waits and comes to the thousand three-hundred and thirty-five days . . . the end of days." The difference between these periods is forty-five days, and using the day-equals-year rule, this gives the forty-five years between 1967 and 2012.

However, this begs the question, When was the burnt offering taken away, and what is the "abomination of desolation"? Jerusalem fell to Nebuchadnezzar around 597–602 BC, when the burnt offering would have been taken away. About 1,290 years later, in 691 AD, the Dome of the Rock was built on Temple Mount—an obvious contender for "abomination" as far as the Hebrews were concerned. But the passage could be interpreted to mean that the burnt offering ended at the same time that the abomination was set up.

I have heard a rumor that there was an earlier mosque on the site of the Dome of the Rock, with a suggested construction date of 677 AD; 1,290 years after this would bring us to 1967, and then the extra forty-five years would bring us to 2012.[8] However, this could be an attempt to manufacture evidence.

So to summarize, the recapturing of Jerusalem by the Israelis in the Six-Day War of 1967 is a key date encoded in the words of both Jesus and Daniel, and starts a forty-five-year period ending in 2012. The Second Coming will occur during this forty-five-year period, according to the beliefs of some Christians. There are many possible interpretations as to exactly what the Second Coming might be.

THE FIG TREE PARABLE

Some sources that don't connect the abomination with Islamic mosques have said the parable of "the leaves of the fig tree"[9] has a 2012 connection. This version is from Luke:

> Look at the fig tree, and all the trees; as soon as they come out in leaf, you see for yourselves and know that the summer is already near. So, also, when you see these things taking place, you will know that the Kingdom of God is near. Truly, I say to you, this generation will not pass away till all has taken place.[10]

A biblical generation is threescore years and ten, or seventy years.[11] The modern nation of Israel was established by a United Nations resolution in 1947 (though not proclaimed until the following year), so the Second Coming would happen by the sixty-ninth year after 1947, in 2016. If an "abomination of desolation" was to be set up at the Temple Mount on December 21, 2012, then 1,335 days later would be August 17, 2016, which is forty-nine years, or seven weeks of years, after 1967. That would then be the Day of Judgment.

REVELATION

The book of Revelation repeatedly mentions (five times) a certain period that comes straight from the book of Daniel. In Revelation, the "nations" will "trample over the holy city for 42 months."[12] This period is exactly 1,260 days (42 × 30 = 1,260), and it also appears as "a time, two times and half a time,"[13] where it is based on a 360-day year (3 × 360 + 180 = 1,260). This form of referring to the period of 1,260 days (a time, two times and half a time) appears in the book of Daniel twice.[14] The third form of referring to 1,260 days is "one thousand two-hundred and sixty days," which appears in Revelation twice.[15] Some researchers have connected this 1,260 figure with the Feast of Trumpets (Rosh Hashanah, the Jewish New Year) 2005, which started at sundown on October 3 that year (it always occurs on the day 1 Tishri in the Hebrew calendar). There was an annular solar eclipse on October 3, 2005, and there is an interval of exactly 2,550 days before the day of Yom Kippur in 2012 (September 26, 2012), also known as the Great Trump (there will be more on this later in this chapter). Note that 2,550 is the sum of 1,260 and 1,290.

However, the seven-year Tribulation, if taken as 360-day years, is two times 1,260 days, and the books of Daniel and Revelation emphasize the halfway point as significant ("a time, two times and half a time"). Seven 360-day years after October 3–4, 2005, leads to August 27–28, 2012.

Fig. 5.2. Tarot trump card number three, the Empress, shows a woman with a crescent moon under her feet and a circle of stars above her head—some packs actually show twelve stars. The word *trump* is short for "trumpets of Armageddon," the last of which will wake up the dead on Judgment Day, and the twenty-two tarot trumps correspond to the twenty-two chapters of the book of Revelation. From a deck designed by Oswald Wirth in 1896.

Following twenty-five years of study, Bible prophecy researcher Andy Alcon calculated independently that October 4, 2005, was to be the date of the "sign in the heavens" indicated in Revelation 12:1: "And a great portent appeared in heaven, a woman clothed with the Sun, with the Moon under her feet, and on her head a crown of twelve stars; she was with child and she cried out in her pangs of birth, in anguish for delivery."

The Moon would be at the feet of Virgo, one of the twelve constellations, with the Sun at her shoulders and Jupiter and Mercury in her belly, says Alcon.[16] He expected a red comet to appear at this time, corresponding to the red dragon that Revelation mentions in the next verse. Alcon says this would be the start of the seven-years of the Tribulation that will end on October 3, 2012, which would be "the start of the millennium and the end of this current age," when there will be an alignment of Venus with the star Regulus, the "foot" of Leo—an alignment that corresponds to several biblical passages. However, this seven-year Tribulation period is seven 365-day years plus two leap days, so does not relate directly to the 360-day years described in Daniel and Revelation. Another problem is that nowhere in Revelation, or in any of the biblical passages describing the Tribulation, is there a period of seven years mentioned.[17] It is an assumption based mainly on Daniel 9:24–27, which is actually about the timing of the "first coming" of the Messiah. The

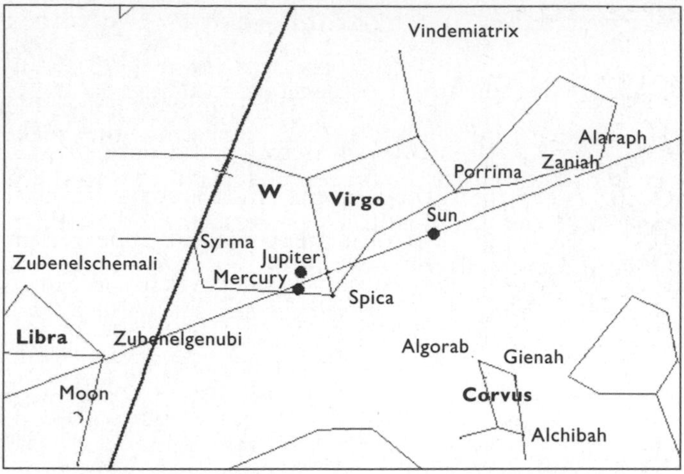

Fig. 5.3. On October 6, 2005, Virgo had the crescent moon beneath her feet, the Sun at her shoulders, and Jupiter and Mercury "in her belly." Image created using CyberSky software—http://cybersky.com.

idea is usually supported by the notion that the seven seals, trumpets, and plagues encode a seven-year period, and that the 1,260-day period describes the halfway point. Another problem is that when I checked October 4, 2005, on astronomical software (SkyGlobe and Cybersky), the Moon was with Jupiter and Mercury, and not beneath the feet of Virgo until October 6.

The interesting thing about this is that visionary novelist Joel J. Keene independently pinpointed October 3, 2012, as being the day of the arrival of an electromagnetic pulse, which he calculated from the Nostradamus quatrain 1:16. He included this in his coincidence-inspired novel *Cosmic Locusts* (about which we shall say more in chapter 11), and was amazed when Alcon contacted him to tell him of his own conclusions. As it happens, October 3, 2012, will be in the middle of Sukkot, the Feast of the Tabernacles in the Jewish calendar, which celebrates the end of the harvest. On October 4, 2005, no red comets were reported, but Hurricane Stan hit Mexico and Central America, killing more than 1,620 people. On October 8, 2005, the Kashmir earthquake killed about 80,000 people.

EZEKIEL

The book of Ezekiel contains a precessional code, according to author Stephen E. Crockett in *The Prophet Code,* but although Crockett interprets the code to signify that the precessional cycle ends in the year 2144 AD, when he says the vernal equinox precesses into Aquarius, the year 2012 still figures strongly in the code, as the years between 1944 and 2012 are "the time of the end." The period between 1948 and 2012 is significant, and the 2005-to-2012 period signifies a seven-year cleanup period from "the destruction of Gog's military to 2012." (Gog is here interpreted as one of the nations involved in warfare at the end of the age (see Revelation, chapter 20). The years 2000 to 2012 signify "an extinction level event," but Crockett also claims that the third of the four "horsemen of the apocalypse" mentioned in the book of Revelation is due between 2010 and 2012. This is the "Black Horse" that brings famine, and is followed by the "Pale Horse" of death, sometime between 2013 and 2016.

JUDAISM AND 2012

At the beginning of the book of Genesis, the first book of the five comprising the Torah (the first five books in our Old Testament), we are told that God created the world in six days and on the seventh he rested. The Jewish rabbis had a specific interpretation of this, which shows a macrocosmic Creation event extending through history. This echoes the Maya belief, as uncovered by Jenkins, of a forthcoming Creation day:

> The Rabbis taught that there were 7,000 years of time. They divided time into a 6,000-year segment followed by the 1,000-year Messianic Kingdom. They based this on Psalms 90:4, which says that to God a thousand years is as a day. What the Rabbis taught was that as the week of creation was seven days, that is seven 24-hour days, so each day stood for a thousand years, with the seventh day as a day of rest—the day of the Messianic kingdom.[18]

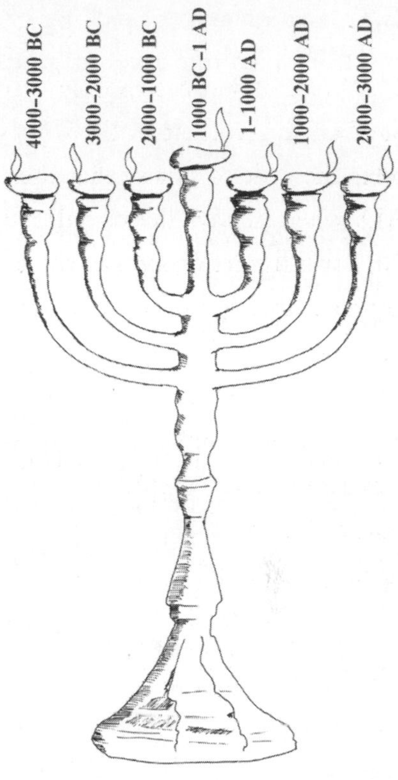

Fig. 5.4. The rabbinical interpretation of the menorah, or seven-branched lamp holder, is that it represents the seven days of Creation as described in the book of Genesis, but on a macroscopic scale, in which each "day" lasts 1,000 years.

The Jewish calendar starts with the birth of Adam, and by sixteenth-century Anglican Archbishop James Ussher's famous analysis, in which he consulted all the male genealogies in the Old Testament (adding given ages of fathers at the birth of their first sons, working back from the zero point of Christ's birth, on which our Gregorian calendar is based), he gives 4004 BC as Adam's birth date. Calculated thus, we arrive at autumn 1997 as being six thousand years after the birth of Adam (the Jewish New Year, Rosh Hashanah, usually falls between autumn and winter). It seems that Ussher's calculation has a three-year discrepancy according to the Byzantium[19] website, which says that the birth of Adam occurred four thousand years before the birth of Christ, so Gregorian year 2000 should be the Jewish year 6000. However, Jewish year 5760 started on September 30, 1999, so this means there is a 240-year discrepancy between the Jewish and Gregorian calendars.

In a four-part online article, "Rosh Hashanah—Is It the Rapture?"[20] the author explains how the year 5760 is kabbalistically encoded in the phrase "six days" (of Creation), thereby implying that Gregorian year 2000 is the end of the Jewish six-thousand-year period. The article concludes that on "some Rosh Hashanah between 1999 and 2012, or shortly thereafter, might be the completion of the 6,000 years of man and the start of the millennium." The seven years between Adam and

Eve's creation and the picking of the apple from the Tree of Knowledge must be taken into account, putting the start of the Day of the Lord at 2007 AD, but the author goes on to acknowledge that it could be "as late as 2012" due to "the calendar reforms and modifications that have taken place over the years." This is really an admission that the case for 2012 sounds convincing, but no further details can be found to explain the discrepancy between the termination points of the seven days of Creation and the thirteen-baktun cycle.

Rosh Hashanah starts at sunset and continues to the following sunset, but it is also the start of a ten-day festival called Teshuvah, "which means 'returning.' In the spiritual sense, Teshuvah means returning to the Creator of the universe, who we call HaShem. It means turning away from the secular, from the worldly way of life, and turning back to His way of life."[21]

The definition of this return sounds very gnostic: "The concept of Teshuvah as 'return' emphasizes the fundamental spiritual potential of every person. Chassidic thought teaches that within each of us resides a Divine soul, a spark of God. This infinite Godly potential represents the core of our souls, our genuine 'I'."[22]

In Judaism, there are three days that are heralded by a trumpet. These are Shavuot (First Trump), proclaiming God's betrothal to Israel; Rosh Hashanah (Last Trump), which is the New Year, and the start of the ten-day Teshuvah, or "return"; and Yom Kippur (Great Trump), heralding the return of the Messiah.

Rosh Hashanah takes on even more significance when we see a full list of its titles:

1. Teshuvah (repentance)
2. Rosh Hashanah (head of the year, the birthday of the world)
3. Yom Teruah (the day of the awakening blast/the Feast of Trumpets)
4. Yom HaDin (the day of Judgment)
5. HaMelech (the coronation of the Messiah)

6. Yom HaZikkaron (the day of remembrance, or memorial)
7. The time of Jacob's trouble (the birth pangs of the Messiah)
8. The opening of the gates of Shomayim (heaven)
9. Kiddushin/Nesu'in (the wedding ceremony)
10. The resurrection of the dead (Rapture, Natzal)
11. The last trump (shofar)
12. Yom Hakeseh (the hidden day)

In the year 2012, Shavuot falls on May 27, Rosh Hashanah falls on September 17–18, and Yom Kippur falls on September 26. The conclusion is that the last trump is synonymous with Rosh Hashanah, when the gates of heaven will open and the "righteous will enter in rapture." However, Yom Kippur also has another title—the closing of gates; the gates of heaven will therefore close on September 26, 2012.

As a chilling postscript, if we look at "Phobos Flyby of Earth" in chapter 7 we find that amateur astronomer Glen Deen's low-precision orbit model gave Earth an impact with Mars's moon Phobos on September 18, 2012. This is reminiscent of the Bible code prediction (later in this chapter), although some Bible-code scholars interpret it as Mars being crumbled by the Earth. However, there are new developments in Glen Deen's theory, and the latest one does not involve Phobos or September 2012 (discussed in chapter 7).

ISLAM AND 2012

Like Judaism and Christianity, Islam also expects a Last Day, or Judgment Day. Islam also has *mala'ika* (angels), including Mikail (Michael) and Jibril (Gabriel). There is an evil jinn (Earth spirit) called Shaytan. At death, the *ruh,* or spirit, leaves the body and waits in *barzakh,* purgatory, and returns to the body on Yawm al-Ba'th, the day of "Rising from the Grave." The destination is then Janna—the Garden—for Muslims only—or *nar* (the fire). They also expect an Antichrist figure called Dajjal, which means "deceiver," and before the

Last Day, Jesus (who is considered a prophet) will reappear and kill Dajjal. Most Muslim sources regard prediction and prophecy as a bad idea (like the Bible, the Qur'an gives signs, but says only God, Allah, knows *sa'*, "the hour.")

However, some have stuck their necks out. According to coauthor of *The 80 Greatest Conspiracies of All Time* John Whalen:

> Several Sufi sects have already declared 2076 as the Year of the Haj (end day), because it will coincide with the year 1500 in the Islamic Calendar. An article by Abdal Hakim Murad, "Islam and the New Millennium," confirmed this, quoting Imam al-Suyuti, the greatest scholar of medieval Egypt. At first, al-Suyuti believed the end of the world would occur in the year 1000 of the Islamic Calendar. He later changed his prediction to 1500 of the Islamic Calendar.[23]

However, Safar Ibn 'Abd Al-Rahman Al-Hawali, author of *The Day of Wrath,* says that he agrees with the Christian date interpretation I have already mentioned. Speaking of Jerusalem, he says:

> When Daniel specified the period between its distress and relief, between the era of anguish and the era of blessing, he put it as 45 years! We have already seen that he specified the time of the establishment of the "abomination of desolation" as the year 1967, which is what in fact occurred. Therefore, the end—or the beginning of the end—will be 1967 + 45 = 2012, or in lunar years 1387 + 45 = 1433.[24]

From an Islamic perspective, the United States is Babylon, and the "abomination of desolation" was the gaining of Jerusalem by Israel in 1967. All three religions, Christianity, Judaism, and Islam, see themselves as the ones who will be chosen and saved on Judgment Day.

THE MASONIC CALENDAR

I recently bought some Masonic documents and discovered that they were dated in the Masonic calendar of Anno Lucis, or Year of Light. Anno Lucis 5950, the year in which one of the documents was produced, is also given as the Gregorian date AD 1950. Thus the Gregorian year 2000 AD would equate to AL 6000 in the Masonic calendar. So the Masonic calendar corresponds to the "corrected" version of the Hebrew calendar. The capping of the Great Pyramid that was planned for New Year's Eve 1999 but was canceled by the Egyptian authorities can now be seen as a symbol of the completion of man-as-microcosm; or as the descent of the New Jerusalem; or as the expected end of the sixth day of Creation and the start of the seventh day, or golden age (see chapter 8, "The Eye in the Pyramid").

Further investigation showed that not all Masonic brotherhoods use the Anno Lucis system. The Ancient and Accepted Scottish Rite, for example, uses Anno Mundi, or the Year of the World, which corresponds to the current Hebrew calendar and is 240 years behind ours. *Encyclopédia Judaica*[25] agrees that this discrepancy is due to an error, and that the Jewish year 6000 should coincide with our year 2000 AD (2000 AD + 4000 BC = 6000). This still leaves unexplained the twelve-year discrepancy between the transition of the ages marked on the one hand by the Judeo-Christian calendar and the age transition that is marked on the other hand by the Long Count calendar of the Maya, apart from the seven years spent by Adam and Eve in the Garden, which brings the date forward to 2007. For this, we have corroboration coming from Ethiopia.

THE ETHIOPIAN CALENDAR

The Ethiopian calendar, which originated in the Coptic Egyptian calendar, puts the date of Creation at 5500 BC. However, it puts the birth of Christ in the Gregorian year 7 AD. Since the Ethiopian calendar (like the Gregorian) uses the birth of Christ as a baseline, this means that the Ethiopian year 2000 will start on the Gregorian date September 12, 2007 AD (which is the start of Coptic year 1724). Thus, although the Ethiopian date of Creation is linked to Egyptian ideas, the transition of the ages, signified by the two-thousandth year after the birth of Christ, is over seven years closer to the end point of the Maya than is the same date in the Gregorian calendar.

The Ethiopian year starts on September 11, except during leap years (which occurs in the years preceding Gregorian leap years), when it starts on September 12. So September 11, 2012, will be the start of Ethiopian year 2005. The first month in the Coptic year, which also starts on September 11, is called Tout—from the Egyptian god Thoth, god of calendrical cycles, whose sacred number was 52, the same number that obsessed the Aztecs.[26] Although the Dogon people of Mali also currently celebrate the New Year on September 11, this is not the heliacal rise of Sirius, which currently happens on August 1.

THE BIBLE CODE

Michael Drosnin's *The Bible Code* tells the story of a code discovered in the original Hebrew version of the Old Testament, which has now been verified by various mathematicians and code breakers, and apparently predicts all major news items "from the holocaust to Hiroshima." The discovery would have been impossible before the computer age, since special software is able to apply a skip code to the whole text, skipping a certain number of letters, then extracting a letter and repeating the process over and over in a matter of minutes—a task that previously would have taken many lifetimes.

Using such software, some recent events have been decoded before they happened, including the assassination of Israel's prime minister Yitzhak Rabin on November 4, 1995. Drosnin actually warned Rabin about the prophecy fourteen months before it happened, but the warning did not avert the killing. The Hebrew year 5756 that would start in September 1995 was found with "Rabin assassination" and the name "Yitzhak Rabin" was also found with "name of assassin who will assassinate." It was only after the event that they found the name of the assassin, Amir, and the place, Tel Aviv, where it took place. The collision of the comet Shoemaker-Levy 9 with Jupiter in 1994 was decoded two months before the event, along with the names of the astronomers who discovered it, and the date, 8 Av (one of the Hebrew month names), which equates to July 16. Although the astronomers had predicted this event a few months beforehand, the Bible code apparently predicted it three thousand years before that, Drosnin points out.

The code also mentions the asteroid that wiped out the dinosaurs sixty-five million years ago, but Drosnin mentions no more comets or asteroids until the Shoemaker-Levy impact. Then he found that the Hebrew year 5766 (October 4, 2005, to September 22, 2006) was also encoded with the word *comet,* plus the words *"Its path struck their dwelling"* and *"year predicted for the world."* Again in 5770, or 2009–2010, the word *comet* combines with the words *days of horror,* plus *darkness* and *gloom.* If this isn't bad enough, the year 5772 (or September 29, 2011, to September 16, 2012) combines *comet* with the words *Earth annihilated.*

Only three years are mentioned by Drosnin after this. They are: the year 5774, or 2014 (an encoded year that combines with *the great terror* and coincides with the predicted close passage of known asteroid 2003 QQ47; the year 5873, or 2113 (which combines with *for everyone the great terror, fire, earthquake,* and *desolated, empty, depopulated;* and the year 5886, or 2126, the last date he mentions, which is encoded along with the word *swift* and the phrase *in the seventh month it came* (although this would refer to the seventh month of the Gregorian calendar, rather than the Hebrew one). Astronomers have, in fact, predicted the return of Comet Swift-Tuttle to the solar system in July 2126.

This all sounds depressing, to say the least (and I have tried to say

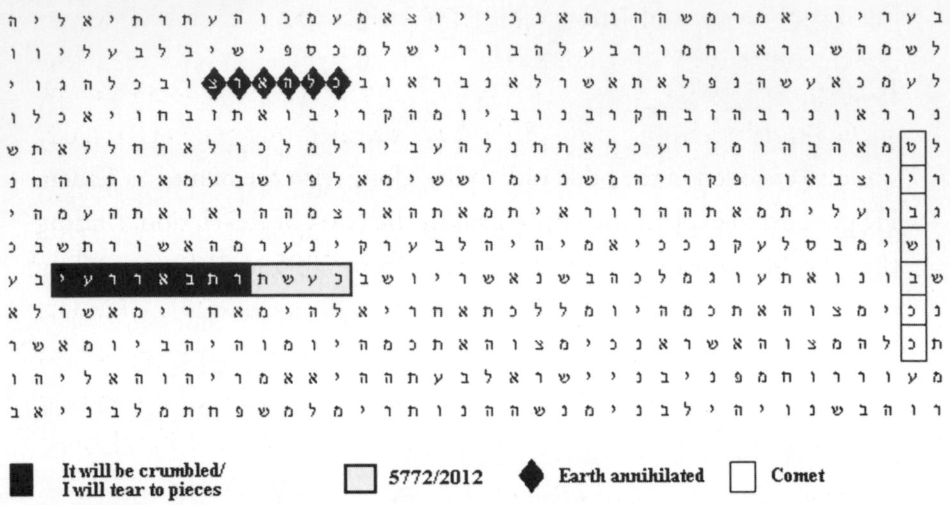

Fig. 5.5. A skip code applied to the text from Exodus 34 to Deuteronomy 1 reveals the combination given here. Adapted from Torah4U software—http://exodus2006.com/torah4u.htm.

the least), and it all seems to back up the even more depressing book of Revelation. However, Drosnin says that the finding of the code enabled humankind to alter fate, and the code implies that there are now at least five possible futures at any one time. Drosnin himself has been heavily involved in trying to alter the future by talking to Israeli and Palestinian politicians.

So what to make of these comet impacts? Well, Earth can't be annihilated in 2012, because there are still people left to be terrified by the remaining events, but also because the number *2012* and the word *comet* are encoded with *It will be crumbled, driven out, I will tear to pieces,* which means the comet will break up like the comet Shoemaker-Levy. The pieces would thus, hopefully, be diverted from their collision course with Earth.

The Bible code continues to be challenged. Some think the codes are imaginary, some think they are real but cannot predict the future. Some just object for religious reasons. Many others object because it is just too depressing to contemplate seriously.

NOAH PREDICTS FLOOD

Further exploration of the Bible code by Joseph Noah in *Future Prospects of the World According to the Bible Code* revealed that the 9/11 tragedy was also encoded in the book of Exodus, along with the name bin Laden. He goes on to explain the symbolism in the book of Revelation, finding that the "seven seals" are the seven years from 2005 to 2012 (according to the Hebrew calendar, i.e., October 2005 to September 2012). He also finds that Mars's moon Phobos* will be hit by an American missile in 2010 and break into seven pieces. The seven pieces break up further into ten pieces—these are the seven heads and ten horns of Revelation.† Three of these ten pieces will collide with the United States in 2010, another three will hit Russia in 2011, and three more will hit China in 2012. The final asteroid (Wormwood, in Revelation) will hit the Gulf of Arabia in 2012, resulting in a world axis shift of twenty-five degrees, and the submergence of Japan, the Philippines, and the U.S. West Coast. (Note that Cassandra Musgrave also saw Japan slip into the sea in 2012, in her near-death experience; see "Near-Death Experiences" in chapter 14.)

Drosnin's sequel, *Bible Code II: The Countdown,* only goes up to 2006, since that was the predicted date for World War III (which has not occurred). Andy McCracken, webmaster of the Exodus 2006 website,[27] using Bible code software, has found that the same piece of code that Drosnin found, where the word *comet* is encoded with *It will be*

---

*Gordon Michael Scallion said in *Notes from the Cosmos* (1997) that he had seen Phobos loosened from its orbit (see chapter 19), and Glen Deen actually predicted its arrival on September 18, 2012, which is Rosh Hashanah, or Judgment Day, in the Hebrew calendar. However, Deen predicted this by working from a presumed "loosening" date of June 5, 2000, not 2010. He has since said it was loosened in 2001, rather than 2000. Glen Deen now says Planet X could impact Mercury around March 2, 2012, and Mercury could impact Venus around June 6, 2012, resulting in a rendezvous of Venus/Mercury with Earth on December 11, 2012.

†Elliott Rudisill, in his book *E.Din: Land of Righteousness,* sees the seven heads as one celestial body—Leviathan—and the ten heads as another—Behemoth. See chapter 7.

crumbled . . . I will tear to pieces, is also accompanied by words that imply the comet is a "60-mile blunderbuss" that splits up, part of it hitting Canada in 2012 and causing a huge earthquake.

FINAL WORD ON PART ONE

For those of us indoctrinated by Christian dogma from an early age and who later broke free from the guilt and fear that have been used as a control mechanism for two thousand years, it comes as a bit of a shock to realize that the ancient Vedic system is also based on eras represented by metals in the same order as the metallic eras of Daniel—the oldest being gold, then silver, bronze, then the current age of iron—except that in Daniel's version there is a fifth era, iron and clay. It is also a surprise to find that in the Hopi myths and in some Maya myths, we also seem to be in the Fourth Era, or World, while in the Aztec and variant Maya myths we are in the Fifth World, or Sun. The precession of the equinoxes (or solstices) is also a recurrent theme, cropping up in a gnostic text called The Sophia of Jesus Christ, as well as in some Maya, Egyptian, and Vedic myths. But what is most amazing is that all these calendars and prophecies seem to culminate in 2012.

PART 2

2012 Theories

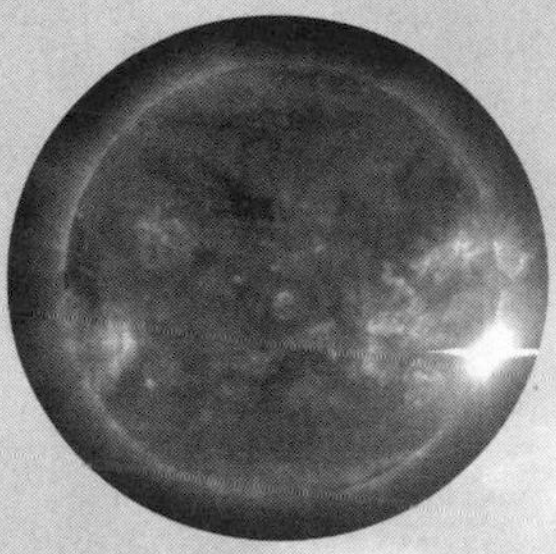

The Creator, when he looked upon the things that happened, established his design, which is good, against the disorder. He took away error, and cut off evil. Sometimes, he submerged it in a great flood; at other times, he burned it in a searing fire; and at still other times, he crushed it in wars and plagues, until he brought . . . [4 lines missing] . . . of the work. And this is the birth of the world.

Asclepius 21–29, from the
Nag Hammadi Library

6

SUNSPOT CYCLES

COTTERELL'S MAGNETIC FLIP

Adrian Gilbert and Maurice Cotterell, in their book *The Mayan Prophecies,* say that the end of the thirteen-baktun cycle is the culmination of a series of long-term sunspot cycles, which will flip the Sun's magnetic field and thus that of Earth, too, causing earthquakes and worldwide flooding. Moreover, the changing magnetic field will alter the hormone production of the pineal and other endocrine glands. The book, which was a bestseller, makes some interesting points, and has raised serious fears for a lot of people. However, when I studied the text more closely, I could not see how the many pages of graphs and calculations related to the end of the thirteen-baktun cycle. I was beginning to think that some vital parts of the manuscript must have been omitted when I discovered that I was not alone in my misgivings.

John Major Jenkins had also found fault with *The Mayan Prophecies,* and in an online review he detailed many errors in the book.

> Cotterell's theories as they relate to Mayan astronomy and calendrics require major adjusting, correcting or even abandonment. The sunspot cycle information in itself was interesting to learn about, but the subsequent argument for "field reversals," even when carefully read, is unconvincing. Even less convincing is the implication of conscious intent on the part of the Maya in calculating these things.[1]

The detailed graphs of the cycles do not actually show significant termination points at the end of the thirteen-baktun cycle, so even though Cotterell's original theory of a sunspot-cycle-driven astrology postulated in part 1 of his first book, *Astrogenetics,* was quite feasible, the mistakes he made in trying to apply it to the Maya calendrical systems in *The Mayan Prophecies* means it is convincing only if you don't dig too deeply into the reasoning behind it. The original theory is repeated in his follow-up books: *The Supergods, The Tutankhamun Prophecies, The Lost Tomb of Viracocha, The Terracotta Warriors,* and *The Celtic Chronicles.* We can see a pattern emerging here: every two or three years Cotterell publishes another book that applies the same theory to another lost culture, examining archaeological relics and finding connections to the theory, often by examining detailed jewelry and counting the spots, lines, and markings until he finds a number that fits into his sunspot theory. We can thus expect more of the same in 2009 and 2011.

To summarize: Cotterell's theory examines the different rotation periods of the Sun's polar magnetic fields and its equatorial magnetic fields—thirty-seven and twenty-six days, respectively*—that drive the eleven-year sunspot cycle. When a third variable of Earth's position is added, a 187-year sunspot cycle emerges. Further extrapolations seem to imply a set of five larger cycles of either 3,553 years' or 3,740 years' duration. He correlates these to the five eras, or Suns, of the Aztecs. Since the 3,740-year period amounts to 1,366,040 days, which Cotterell says is "close to" a number that appears in the Dresden Codex, this shows that the Maya were tracking sunspot cycles. However, the period in the Dresden Codex is 1,366,560 days, which is a difference of 520 days. Since

*I have found varying timings given for the differential rotation of the Sun, and none of these agrees with Cotterell's figures of twenty-six days at the solar equator and thirty-seven days at the solar poles; most quote twenty-five days and thirty-five days, respectively. These figures would alter all the findings of Cotterell. For example, if we substitute the rotational periods of 25.38 and 35 for the 26 and 37 of Cotterell, then his 87.454545-day bits come out at 92.3385 days. This means that all his larger cycles that multiply upward from this are wrong.

the Dresden Codex contains Venus tables, Cotterell suggests that since 2,340 Venus cycles of 584 days equal 1,366,560, the Maya must have been using Venus as a calibrator to adjust the 1,366,560 figure back to 1,366,040.

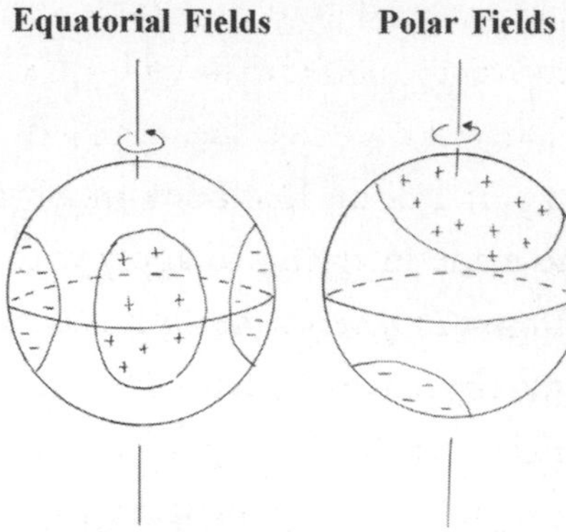

Fig. 6.1. The four equatorial magnetic fields of the Sun rotate faster than the two polar fields. This causes the intertwining field lines that erupt as sunspots every 11.3 years on average. After Gilbert and Cotterell's book *The Mayan Prophecies: Unlocking the Secrets of a Lost Civilization.*

However much of this number crunching is done (and *The Mayan Prophecies* has a one-hundred-page appendix full of it), it doesn't explain what will happen in 2012. The reader's mind is numbed by the complexity of the theory, so that she just has to take the authors' word for it that the world will be destroyed in 2012 by a full solar magnetic reversal that causes Earth's magnetic field to reverse polarity.

Even though this theory is seriously flawed, it is a fact that the scientifically accepted records of sunspot activity do seem to be heading for a climax in the near future, and there has been some research that indicates the existence of megacycles (longer overall cycles). Geomagnetic reversals do occur, and we are overdue for one, but *The Mayan Prophecies* has led many authors and researchers into an irrelevant maze. Those who found the book convincing would be well advised to read John Major Jenkins's review.[2]

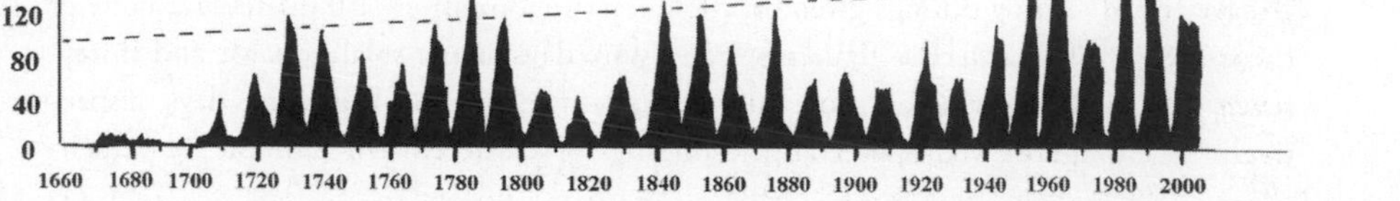

Fig. 6.2. Increasing solar activity gives an upward trajectory when graphed. Adapted from several Internet sources.

CHANGE IN THE SOLAR MAXIMUM

The sunspot cycle reaches its maximum approximately every eleven years, and then sunspot activity switches to the opposite hemisphere of the Sun. The recent maximum was due in March–April 2000, and the following one was therefore expected in 2011. However, the maximum didn't actually occur until February 2001, so the following one is now expected in 2012. Some people have suggested that this is evidence of a sunspot megacycle, possibly connected with the end of the Maya thirteen-baktun cycle, but the recorded cycle length has varied between eight and fourteen years in the past, so the lengthened sunspot cycle may not be so significant in itself. Having said that, we have also seen a continual escalation of solar activity over the last fifteen years, with ever-larger X-ray flares, record-breaking coronal mass ejections, and solar storms that have caused the northern lights to descend farther and farther south. When the last cycle finally peaked in 2001, George Withbroe, science director of NASA's Sun-Earth Connection Program, said, "This is a unique solar maximum in history. The images and the data are beyond the wildest expectations of the astronomers of a generation ago."[3]

After the solar maximum, the activity continued, with a record-breaking X-20 solar flare on April 2, 2001 (some reports said it was an X-22, since the dial only goes up to X-20 and the needle was stuck at the upper limit). The northern lights, or aurora borealis, are an aerial display of shifting colored-light patterns usually only visible from the Arctic Circle, and caused by the interaction of solar wind, Earth's magnetic field, and the upper atmosphere. In April 2000, a solar storm caused the aurora to be seen as far south as North Carolina, but in June 2001, a solar storm caused a geomagnetic effect that pulled the aurora as far south as Mexico. In 2002, there was the most complex coronal mass ejection, or CME, ever recorded. A CME happens when the gas envelope of the Sun explodes outward. In 2003, we had a record-breaking nine major solar eruptions in the space of twelve days, and then, on November 4, 2003, there was the largest solar flare ever recorded, estimated at the time to be between X-28 and X-35 but now known to have been even bigger. Following these events some scientists said the Sun was more active than it had been for a thousand years. However, in 2008–2009 the Sun should have become active again after solar minimum but instead became quieter than it has been for almost a century. We shall return to this discussion later, in chapter 12.

AUSTRALIAN HAM THEORY

Kev Peacock is an Australian amateur radio operator, or ham, who set out to study how Earth's atmospheric and weather conditions affected his radio signal. He became immersed in a detailed investigation of the sunspot cycle, solar flares, and solar wind. He noticed that the graphs of solar activity show a clear incline over the last few hundred years since records began in 1611 AD, and that geomagnetic and seismic activity is also increasing. In fact, there is a direct correlation between the peaks and troughs of all three graphs. Peacock concluded that in June–August 2011 there is a possible coincidence of the following four factors that could, if they occur at the same time, endanger Earth:

1. The height of summer in the northern hemisphere is June 2011 (summer solstice is June 21), when Earth's magnetic North Pole is closest to the Sun.
2. There will be three eclipses that will occur between June 2, 2011, and July 1, 2011 (two solar and one lunar), causing geomagnetic effects (in the same way that a major earthquake in Turkey followed the eclipse in August 1999).
3. The solar maximum is due at this time (see the preceding material), when the Sun's magnetic field reverses (every eleven years, on average).
4. The peak of a solar megacycle, or long-term sunspot cycle, indicated by the incline in graphs of solar activity may coincide with the solar maximum.

The combination of aligned Sun, Earth, and Moon with the eleven-year sunspot cycle maximum, with Earth's north magnetic pole pointing to the Sun, hasn't happened for sixty-five million years, according to Peacock. He says the coincidence of the above four factors will cause Earth's magnetic field to collapse and reverse, the "resulting earthquake and volcano activity bringing about the end of our civilization by the end of 2012."[4] A new ice age would probably follow due to a global temperature drop, as volcanic ash would cut out sunlight.

As for Peacock's solar maximum calculations, he devised this theory when the solar maximum was still due in 2011, but since it has now shifted to 2012, these factors will coincide in 2011 only if the current sunspot cycle is around ten years and four months in length. Even so, his discovery of the coincidence of the peaks and troughs of the seismic, geomagnetic, and sunspot graphs could be very significant (see color plates 7, 8, 9, and 10).

MASS TELEPATHY IN 2012

Jazz Rasool, a biophysicist and corporate coach, gave a talk in 2001 at the Mysteries of the World conference, in Marlborough, UK, in which he pointed out that we are currently experiencing the most active solar cycle on record, and that the eleven-year sunspot cycle has just jumped forward a year, so that the current phase will terminate in 2012 instead of 2011. That the Schumann resonance (electromagnetic waves in the cavity between Earth's crust and the ionosphere) is said by some researchers to be rising in frequency (see chapter 9), combined with recent discoveries by the Japanese scientist Hiroshi Motoyama that the frequency around psychics increases to a level of seventeen hertz, leads Rasool to the conclusion that these environmental factors will lead to a "switching on" of some of the massive amount of genes with officially unknown function (so-called junk DNA), which he calls "species DNA," leading to mass telepathy and telekinesis. This would be the birth of a new species: *Homo spiritus*.[5]

PI IN THE SKY

Michael Poynder, a jeweler and dowser specializing in geopathic stress, says in his book *Pi in the Sky* that the Stone Age priesthood used to work with and manipulate Earth's magnetic energy grid. They knew about the twenty-two-year binary sunspot cycle, and they used the value of pi (the ratio of the circumference of a circle to its diameter), approximated to twenty-two over seven, in the layout of many sacred sites. The sunspot part of the theory was inspired by Maurice Cotterell's first book, *Astrogenetics,* as Poynder confirmed to me.[6] However, Poynder seems to have been one of the first to look for, and find, astronomical meaning in ancient jewelry and artifacts. Having himself found a sixteenth-century German silver chalice with a built-in gnomon (a rod, the shadow of which moves with the Sun) that threw a shadow onto the inside surface, thereby acting as a sundial, Poynder suggested that a ninth-century chalice and paten (Communion plate) discovered in 1980 on the island of Derrynaflan, in County Tipperary, Ireland, were designed to also serve as a sundial, allowing "differential dialling of the Sun's magnetic fields and of the equinoxes, solstices and quarter days, i.e., the 'pattern' of the Druidic year."[7]

This book is something like a missing link between Cotterell's *Astrogenetics* and Cotterell and Gilbert's *The Mayan Prophecies.* Cotterell suggested in part 2 of *Astrogenetics* that the 260-day tzolkin of the Maya is based on the coincidence of the solar polar and equatorial magnetic fields (according to Cotterell's figures, they coincide at 262-plus days, but after 260 days the equatorial field has made exactly ten revolutions—however, these figures are incorrect.

He has also said that the "Mayan collapse" (i.e., when they abandoned their cities) was due to a decline in fertility caused by the proposed magnetic-field reversal of 672 AD (Jenkins argues that this is far too early).[8] In 1988, Cotterell didn't mention the Long Count calendar, or 2012. Meanwhile, in 1992, Poynder must have asked Cotterell's permission to use the sunspot graphics from *Astrogenetics,* bringing Cotterell's attention to *Pi in the Sky,* in which Poynder finds sunspot knowledge encoded in ancient artifacts and concludes that the ancients were tracking the sunspot cycle in preparation for 2012. Then, in 1997, Cotterell launched *The Mayan Prophecies* and its sequels, attempting to show how sunspot knowledge was encoded in ancient artifacts as a message to future humans about 2012. However, in the Cotterell version, it was as a warning of catastrophe.

Having established in his book that Stone Age humans dowsed Earth energies, and showing many examples of pendulums, Poynder traces a series of Earth energy lines and "Earth stars" covering the globe. He takes a close look at Newgrange, one of the Irish passage graves, the major structures of which "were completed sometime between 3200 and 3700 BC," according to Martin Brennan, author of *The Stones of Time.*[9] The inner chamber is illuminated by the Sun every winter solstice. Poynder demonstrates how Newgrange was designed around the golden ratio, denoted by the Greek letter ϕ (phi), and quotes Cotterell as saying, "The Golden Mean [i.e., golden ratio] is the solar radiation frequency."[10]

Poynder's research can be thus summarized: Around or just prior to 3113 BC (3114 BC by current protocol), at the end of the last thirteen-baktun cycle, the "Shining Ones" (priests of light) of Irish mythology entered the stone chambers, never to reemerge. The winter solstice Sun entered the cairns through polarizing crystal skylights, accelerating the auric body of a cairn through the speed of light, when it massified into a "vehicle of electromagnetic resonance," or an extradimensional UFO. The priests, who were out of body, and whose subtle bodies melded with that of the cairn, have time- traveled in their UFOs to the present, to communicate the message that we should raise our consciousness for 2012, which will be the start of a new age of enlightenment. In 2004, Poynder published a follow-up book titled *Lost Science of the Stone Age: Sacred Energy and the I Ching,* in which he finds evidence that the ancient priests also understood the energy concepts that underlie the I Ching and govern the Chinese arts of feng shui and acupuncture.

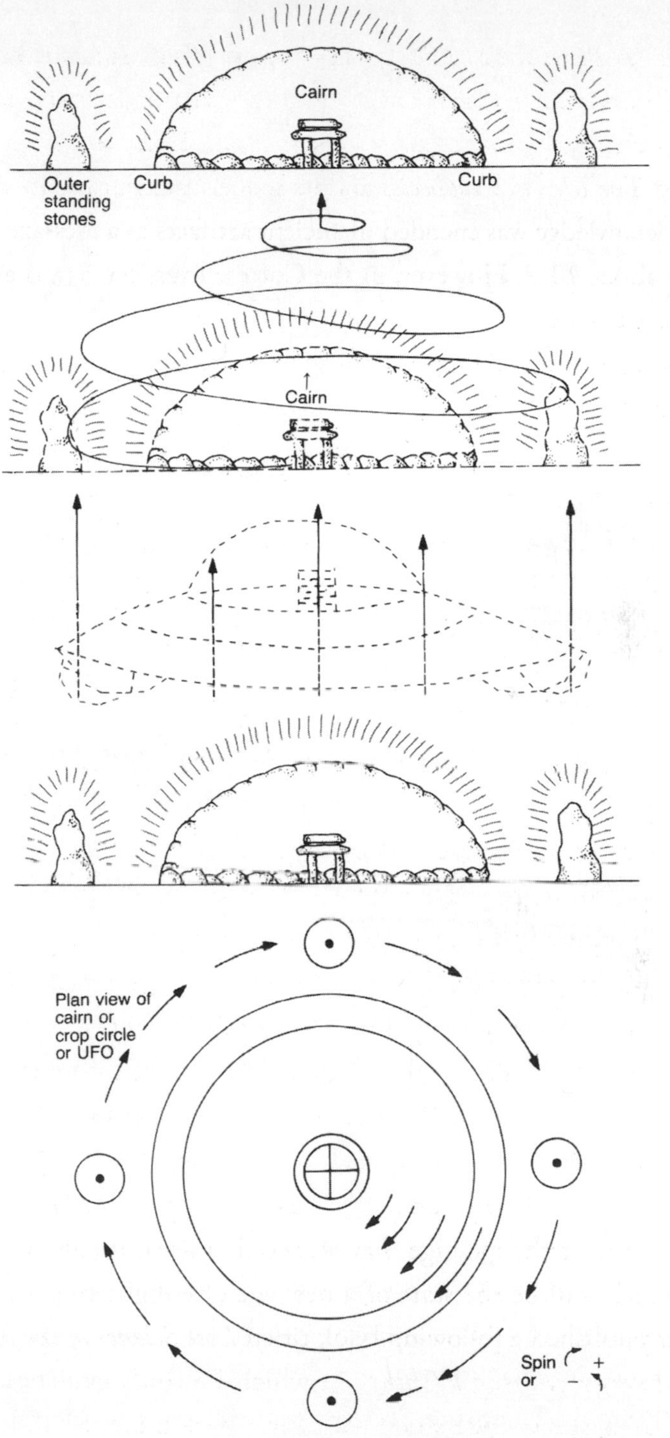

Fig. 6.3. The auric body of a cairn becomes a UFO. From *Pi in the Sky: A Revelation of the Ancient Wisdom Tradition,* by Michael Poynder.

7

ASTRONOMICAL CLAIMS

ASTRONOMICAL PHENOMENA

Astrologer Raymond Mardyks has pointed out five unusual astronomical phenomena that are due to occur in 2012.

1. A transit of Venus, in which Venus will pass directly in front of the Sun (from our perspective on Earth)—something he says happens only every 120 years.
2. A solar eclipse, in which the Sun and Moon conjunct the Pleiades, on May 20, 2012.
3. A second solar eclipse, in which the Sun and Moon align with the head of the constellation Serpens, the serpent (held by Ophiuchus, the Snake Handler, who stands with one foot on the body of Scorpio and the other on its stinger, right next to Galactic Center).
4. The alignment of the solstice Sun with the galactic equator, which Mardyks says started in 1986[1] or "around the time of Harmonic Convergence in 1987"[2] and will finish in 2012 (he also puts the mid-alignment point, when the center of the Sun is on the galactic equator, as 1998–2001).
5. From an Earth perspective, Venus returns to the same area of the sky every eight years and each time is getting nearer and

nearer to the Pleiades. In 1972, Venus was one degree from Alcyone, the central star of the Pleiades, but in 2012 Venus will be so close to Alcyone that it will be closer to it than some of the other stars in the constellation. Mardyks then connects this to the 2012 termination point of the thirteen-baktun cycle, since he thinks that the Pleiades were very important to ancient Maya daykeepers. Mardyks sees the Pleiades as a kind of transformer that steps down energy and information from Galactic Center and transmits it to Earth in a less potent form. After 2012, he says the Pleiades will no longer act as a galactic transmission center for Earth—we will receive energy and frequencies directly from Galactic Center.

Comments:

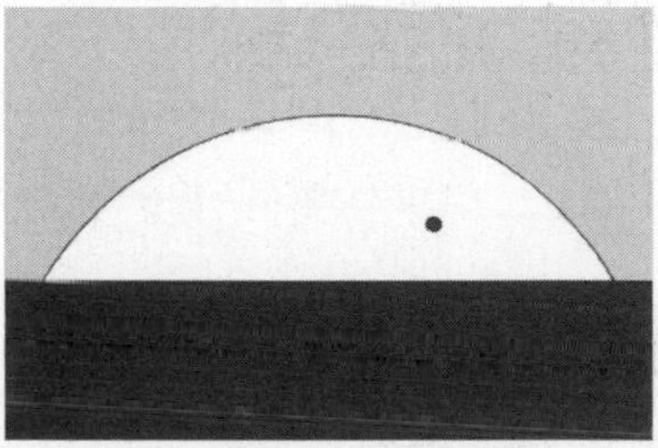

Fig. 7.1. This is how the 2012 Venus transit will appear through a filtered telescope.

1. Some sources have said that Venus transits (or alternatively, Venus "passages") occur once per century, but this is a little misleading. They do occur once per century, but in a pair, eight years apart. In other words, they occur twice per century (but not necessarily every century, since there were none in the twentieth century). They occurred in 1761, 1769, 1874, 1882, and 2004, and one will take place in 2012. Here are the approximate 120-year intervals (113 and 130 years in this case—the total is 243 years) mentioned by Mardyks, but between pairs of transits: The first of a pair occurred on June 8, 2004. The second will occur on June 5–6, 2012, shortly after 10 PM Universal Time on June 5, continuing until about 4:45 AM Universal Time on June 6.

2. The May eclipse that conjuncts the Pleiades is also a conjunction of the Pleiades with the zenith passage of the Sun (May 20, 2012), the event that Jenkins says is being pointed out by the Pyramid of Kukulcan at Chichén Itzá (see "A Precessional Alarm Clock" in chapter 8). The conjunction will happen at a local time (at Chichén Itzá) of noon.
3. The second of these solar eclipses is on November 13, one day before the day of the Aztec New Fire ceremony, which takes place every fifty-two years—a Calendar Round—and (thanks to John Major Jenkins's book *Tzolkin*) we know there is a Maya Calendar Round that ends in 2012; in the Tikal Calendar Round series, it falls on April 2, 2012. The Serpens constellation that Mardyks says will be aligned with this eclipse is not actually on the ecliptic, where the Sun and planets travel, but its head is about the width of a zodiac constellation away from the ecliptic.
4. The alignment of the solstice Sun with the galactic equator was astronomically exact in 1998–1999, and since the Sun is a half-degree wide, and precession moves at about one degree every seventy-two years, that means it takes thirty-six years for the Sun to cross the galactic equator. In other words, the process started in 1980–81 and will finish in 2016–17. Jenkins has shown that the Maya were targeting the end of the thirteen-baktun cycle in this galactic alignment process. Mardyks's attempt to connect the 1987 Harmonic Convergence with galactic alignment is incorrect.[3]
5. The synodic revolution of Venus occurs five times in eight haabs (a haab being a 365-day cycle). These five points where Earth is overtaken by Venus (which orbits the Sun in only 224.7 days), or Earth–Venus inferior conjunctions, are called retrograde loops, since Venus appears, from our perspective, to reverse its direction for about a month. After eight haabs it returns to the same part of the sky as seen from Earth (though its posi-

> tion actually precesses round the zodiac by about 2.4 degrees every eight haabs).* The period from one loop to the next averages at close to 584 days (583.92 days), the Venus period closely watched by the Maya: 5 × 584 = 2,920; 8 × 365 = 2,920. We will have more to say about the Venus–Pleiades conjunction later in this chapter.

Swedish Maya calendar researcher Carl Calleman says there is a Maya prophecy that "the new world of consciousness will be born" during the June 5–6, 2012, Venus transit.[4] He doesn't give a reference for the prophecy (but as we shall see later, in chapter 19, this was his own spin on Hebrew prophecy as a "solution" to the Maya calendar) and has sometimes changed his focus to 2011, since his own theory seems to require a 2011 end point. Yet Calleman has recently restated the importance of the 2012 transit.

There will also be two lunar eclipses in 2012.[5]

VENUS TRANSIT THEORY

Researcher Will Hart, author of *The Genesis Race,* remarks that since some sources say the current thirteen-baktun cycle began on the "Birth of Venus," and since it will conclude in 2012, which is the

*It takes 243 years for Venus to return to the exact point where it started in relation to the stars, which is 152 of the 584-day cycles (the true average is around 583.92). This is equal to 30.4 pentagrams, or 30 pentagrams plus two extra 584-day cycles (or two points of a pentagram), bringing it into alignment with the original five points. Note, this is not a complete revolution of the pentagram, but a 72-degree revolution. It takes more than a multiple of the eight-year pentagram, because the constellations have also precessed over 240 years, by about 3.3 degrees. One complete revolution of the pentagram takes 1,199 years. I have checked all these movements on astronomical software; thus the pattern that starts with a standstill on May 14, 2012 AD, is a repeat of the pattern that started 243 years previously, on May 12, 1769 AD, and these are both repeats of the pattern that occurred 1,199 years prior to 2012, on April 26, 813 AD. For more information on the retrograde loops of Venus, see the review of the Orion prophecy at www.diagnosis2012.co.uk/orp.htm.

same year as one of the rare aforementioned Venus transits, then it must have something to do with the meaning of the forthcoming end point. Reviewing past Venus transits, Hart notes that the 1518–1526 pair corresponded to the landing of Cortés in Mexico; the following transit pair, in 1631–1639, was followed by the Maunder Minimum, a period when there were no sunspots seen for seventy years; and the 1761–1769 pair is connected to the American Revolution. He points out that Earth has been slowly warming up since the last ice age ended 13,000 years ago, and we are now nearing the end of an interglacial period. He also notes how solar activity has been increasing for the past three hundred years. Although sunspot numbers reached their highest so far in 1960, solar activity is still much higher than predicted, with record-breaking temperatures in August 2003, and solar storms causing an electrical failure across the United States and Canada, affecting thirty million people. Following 1960, there has been a steady increase in earthquakes and volcanic activity. Since the next predicted solar maximum has now shifted forward to 2012, Will Hart concludes that "the Maya knew that the Venus transit acted like a circuit breaker switching off the sunspot cycle and impacting the Sun-Moon-Earth-Venus system."[6] He predicts that the increased volcanic ash in the atmosphere will start a cooling of the planet that will be the end of the Fifth Sun of the Aztecs and the start of the Sixth Sun, between June and December 2012.

Carl Calleman has pointed out that previous Venus transits seem to have coincided with jumps in global communications. The 1518 transit coincided with Magellan's planning for the first circumnavigation of the globe; the 1631 and 1639 transits coincided with the first postal services; the 1761 transit inspired a cooperation of astronomers all over the world. (The Venus transit of 1761 had long been prepared for by astronomers, who were planning to use their observations as a means of measuring the distance to the sun. This goal could, however, only be accomplished if observations were made all around the world, and for this to happen astronomers in different countries needed to collaborate.)

The 1874 transit coincided with the laying of the first transatlantic telegraph cable and Alexander Graham Bell's first plans for a telephone. Calleman thinks the Venus transits of 2004 and 2012 will somehow trigger the first major outbreak of human telepathy, resulting in a kind of global brain, as Gaia, the Earth goddess, wakes up. Peter Russell put this concept forward in his 1982 book, *The Awakening Earth: The Global Brain,* though not in connection with Venus transits, but with human population (see chapter 12).

VENUS-PLEIADES CONJUNCTION

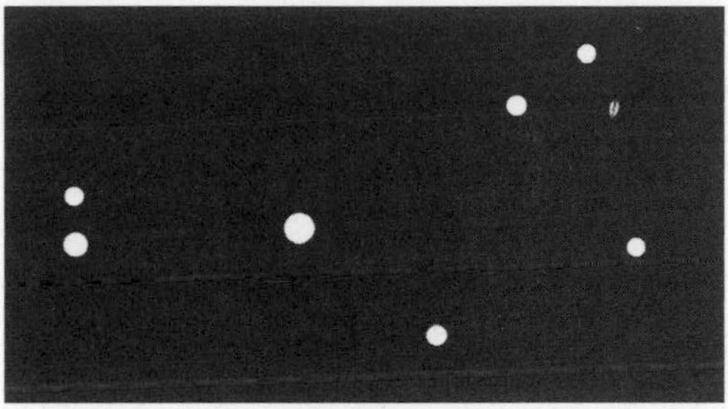

Fig. 7.2. The Pleiades star cluster, which is located in the shoulder of Taurus. The large central star is Alcyone.

As we have noted, Mardyks has said that "Venus returns to the same area of the sky every eight years, and each time is getting nearer and nearer to the Pleiades." He points out that in 1972, Venus was one degree away from Alcyone, the central star of the Pleiades, but in 2012 Venus will be right next to Alcyone. Venus and the Pleiades were both very prominent in the Maya time-keeping systems, so I decided to do some investigation myself, using CyberSky software.[7]

The first thing I discovered was that Venus is very close to Alcyone on April 3, 2012. This is connected to the retrograde loops of Venus, as mentioned above, when Venus appears to curl back on itself every 584 days and completes a pentagram of five loops around the zodiac every eight years; but this return is actually 2.4 degrees away from its original position eight years previously. It takes 243 years for Venus to return to the exact point where it started, which equals 152 of the 584-day cycles (the true average is around 583.92). This is equal to 30.4 pentagrams, or 30 pentagrams plus two extra 584-day cycles (or two points of

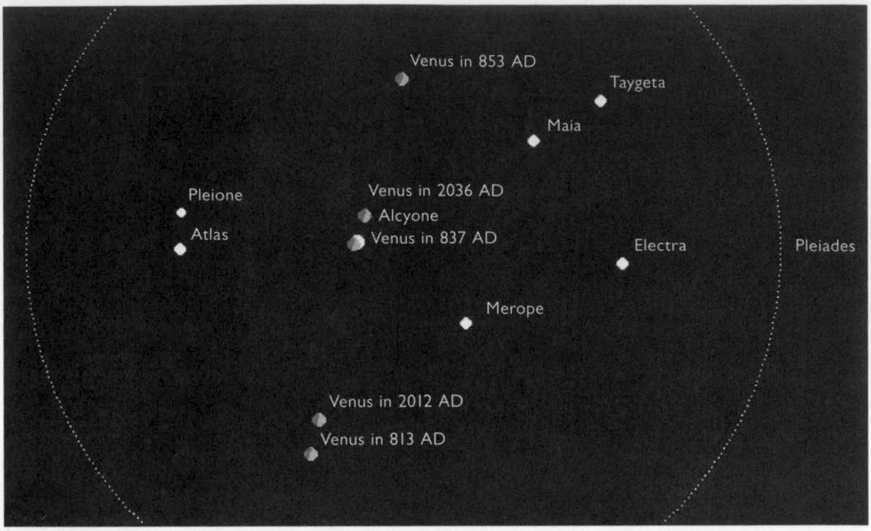

Fig. 7.3. Diagram showing the proximity of Venus to Alcyone, the central star of the Pleiades star cluster, on March 13, 813 AD; on March 15, 837 AD; on March 16, 853 AD; on April 3, 2012; and on April 4, 2036. An amalgamation of five screen shots from CyberSky software—http://cybersky.com.

a pentagram), bringing it into alignment with the original five points. This is not a complete revolution of the pentagram, but a seventy-two-degree revolution. It takes more than a multiple of the eight-year pentagram, because the constellations have also precessed over 243 years, by about 3.3 degrees. One complete revolution of the pentagram takes 1,199 years.

The astronomy software enabled me to reverse the movement of the planets and stars by 1,199 years, the time it takes for the entire Venus pentagram to get back to its starting point. This took me to the year 813 AD. However, further research showed that the 1,199-year cycle of Venus's proximity to the Pleiades runs from 837 AD to 2036 AD. Venus moved farther away from the Pleiades at each eight-year interval, remaining closer to Alcyone than its 2012 position, until the year 853 AD. Then it continued moving farther from Alcyone until about 1436 AD, when it started to move back toward Alcyone again. After the 1,199-year cycle completes, in 2036 AD, Venus is at its closest to Alcyone again.

The conclusion is that in 2012, although Venus will be closer to Alcyone than it has been since about 853 AD (an interval of 1,159 years), it continues to get closer to Alcyone until 2036 AD, so the conjunction does not seem to be overly significant to 2012 after all.

BIRTH AND DEATH OF VENUS

In his 1950 book *Worlds in Collision,* Immanuel Velikovsky proposed that the birth of Venus occurred around the fifteenth century BC, when a comet or comet-like object (now called the planet Venus) was ejected from Jupiter, leaving the famous red spot where it emerged—just as some cosmological theories say that the planets were thrown out of the Sun and later solidified. This is clearly illustrated in the Greek myth of Athena, who sprang from the head of Zeus (see chapter 14, fig. 14.4), making Mt. Olympus tremble and stirring up the sea (Athena, reasons Velikovsky, is the planet Venus, and Zeus is the planet Jupiter). Velikovsky dated this event as contemporary with the biblical Exodus, in which the Israelites fled Egypt, thus explaining the plagues and parting of the Red Sea. Several Maya and Aztec myths were also cited as evidence, including the fact that Quetzalcoatl, the Feathered Serpent, is the well-known name for the planet Venus, also called "the star that smoked." Velikovsky quotes from Charles-Étienne Brasseur de Bourbourg's *Histoire des Nations Civilisées du Mexique:*

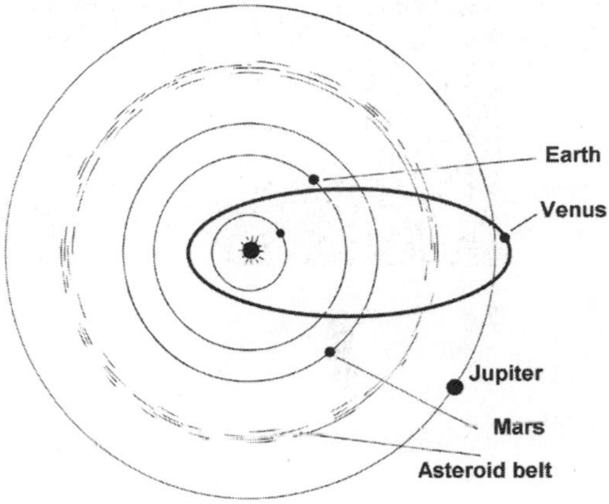

Fig. 7.4. The early orbit of Venus, according to Velikovsky. After *The Venus Legacy: When the Moon Turns to Blood* by Shane O'Brien.

> The Sun refused to show itself and during four days the world was deprived of light. Then a great star . . . appeared; it was given the name

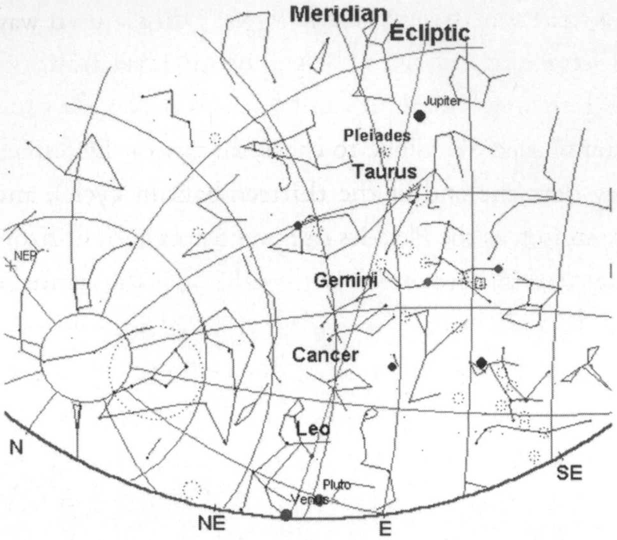

Fig. 7.5. The sky on August 12, 3114 BC, with the Pleiades on the meridian and Venus appearing just on the horizon. Adrian Gilbert believes this represented the "birth of Venus." Screen shot from CyberSky software—http://cybersky.com.

> Quetzal-cohuatl . . . The sky, to show its anger . . . caused to perish a great number of people who died of famine and pestilence . . . It was then . . . that the people [of ancient Mexico] regulated anew the reckoning of days, nights, and hours, according to the difference in time.[8]

In *The Mayan Prophecies,* Gilbert and Cotterell say that they used astronomical software called SkyGlobe to look at the sky as it appeared at dawn on August 12, 3114 BC (where they date the beginning of the thirteen-baktun cycle). They found that as the Pleiades were positioned on the meridian, the rising Sun was preceded by Venus as the morning star. This they took to be the mythical birth of Venus, associated by Velikovsky with the start of the new calendar implied by Brasseur. However, whereas Velikovsky dated the event at around 1500 BC (coincidentally, when Izapa, the origin point of the Long Count calendar, was first inhabited), Gilbert and Cotterell associated it with the start of the thirteen-baktun cycle in 3114 BC. According to Gilbert's previous book, *The Orion Mystery* (with Robert Bauval), in ancient Egypt the dawn rising of Sirius heralded the flooding of the Nile and the beginning of a new year, and the meridian transits of Orion correspond to the "First Time" and "Last Time" of Osiris/Orion (see chapter 8, fig. 8.7). In a similar way, Gilbert and Cotterell reasoned that to the Maya, the meridian transit of the Pleiades announced the birth of Venus, as it arose from the horizon.

Gilbert and Cotterell used SkyGlobe to check the sky on December 22, 2012 (where they date the end of the thirteen-baktun cycle), and found that just before sunset, as the Pleiades rise over the eastern horizon, Venus sinks below the western horizon. This, they think, is the symbolic "death of Venus," indicating the start of a new precessional age.

Chris Morton and Ceri Louise Thomas have repeated this conclusion in *Mystery of the Crystal Skulls* (see chapter 19), and have also supplied a death-of-Venus sky chart, which Gilbert and Cotterell did not include in *The Mayan Prophecies*. Morton and Thomas have also adjusted the start- and end-dates of the thirteen-baktun cycle by one day, to the True Count correlation—dates that Jenkins has confirmed.

However, Jenkins has checked the sky chart for August 12, 3114 BC, on SkyGlobe and two other astronomy packages, and found that Gilbert and Cotterell omitted double-checking their findings, since Venus actually "made its last appearance as the morning star almost two weeks before the Long-Count zero-date."[9] I have looked at the SkyGlobe[10] and the Burden of Time Maya calendar software,[11] which gives Venus phases, and found that on August 12, 3114 BC, Venus was approaching the end of its 263-day period of visibility and was five days away from its heliacal set when it disappeared into the glare of the Sun.*

*On the SkyGlobe software, the astronomical configurations are correct for the given dates, but it is not that simple. The 584-day Venus cycle is split up into an eight-day period of invisibility (when it nears and passes in front of the Sun at inferior conjunction), then the heliacal rise happens (the first visible rising, close to the Sun). This is followed by a 263-day period of visibility, followed by heliacal set, then a 50-day period of invisibility (when it nears and goes behind the Sun at superior conjunction), and then another 263-day period of visibility. If we check on the Burden of Time Maya calendar software, which gives Venus phases, we find that, as Jenkins pointed out, Venus's last appearance as morning star was on August 1, 3114 BC. Venus was actually in the phase when it was approaching the end of a 263-day period of visibility, and was nearing the Sun again, for heliacal set, which was due on August 17–18, 3114 BC.

So Adrian Gilbert (who was the coauthor who investigated the birth of Venus) saw on the SkyGlobe program that Venus was close to the Sun at the start of the thirteen-baktun cycle, and concluded that it was at heliacal rise, when in fact it was close to heliacal set, which happened about six days later. This means that the sinking of Venus below the horizon while the Pleiades are on the meridian (as found by Morton and Thomas) on winter solstice 2012 loses any significance as a "death" of Venus.

NIBIRU

As we have noted, Velikovsky dated the astronomical event corresponding to the birth of Venus at around 1500 BC, with the Jewish Exodus, but historical detective and author Graham Phillips, in his book *Act of God,* has also convincingly explained the events of the Exodus as side effects of the eruption of the volcano Thera. The founder of the disbanded Egypt excavation program, Operation Hermes, and an author accused of plagiarism, Nigel Appleby has, in his book *Hall of the Gods,* unified both these ideas with that of Nibiru, the tenth planet.

The existence of Nibiru was first postulated by Zecharia Sitchin in his book *The Twelfth Planet* (twelve planets, counting the Sun and Moon), since his interpretation of ancient Sumerian mythology implies that there is a large planet—Nibiru—on an elliptical orbit, which enters the solar system every 3,600 years, and it is populated by the Annunaki, who are the biblical "Sons of God" (Genesis 6) and are called "the Watchers" in the book of Enoch. Velikovsky and Sitchin actually used the same Sumerian myth (the Marduk and Tiamat myth) as partial evidence for their respective theories, so they can't both be completely right. According to Sitchin's chronology, the next return is due in 3400–3600 AD.

Appleby suggests that it was Nibiru that caused the disasters of the Exodus, as it "shunted both Mars and the Earth dangerously close to each other's orbit,"[12] and it was this that triggered the eruption of

Fig. 7.6. An Akkadian cylinder seal, apparently showing a star with eleven planets around it.

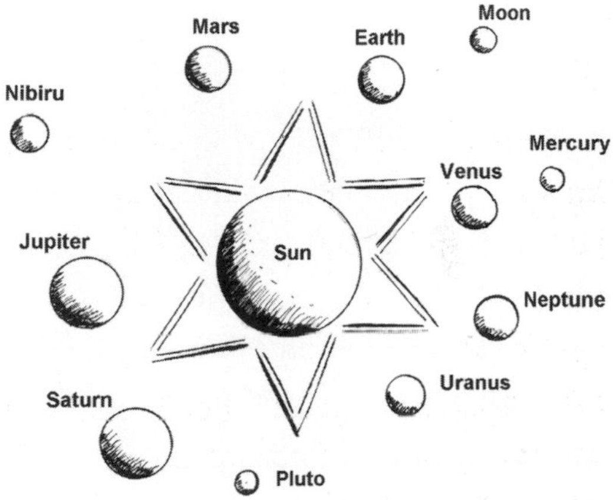

Fig. 7.7. According to Zecharia Sitchin's interpretation, the orbs on the cylinder seal correspond to the Sun and nine planets, plus the Moon and Nibiru. Pluto appears out of sequence, since it is shown in its previous position as a moon of Saturn.

Thera. However, Velikovsky didn't say Mars was involved in the original birth of Venus incident, but rather two later events that occurred around 700 BC. Nonetheless, the theory does provide the mechanism of a large passing body interfering with Jupiter and causing the ejection of Venus. In addition, it would mean that Nibiru is due to return soon (1500 BC to 2000 AD = 3,500 years); Appleby suggests that the next return of Nibiru is due between 2012 and 2036, as we shall see.

If there is anyone who is still in favor of correlating the birth of Venus incident and the date of 3114 BC, despite the errors of the Gilbert and Cotterell interpretation, then it would follow that the thirteen-baktun cycle is measuring the period of return of some other huge planetoid (as suggested by Willaru Huayta, Credo Mutwa, and the Bible code) in 2012.

RETURN OF NIBIRU

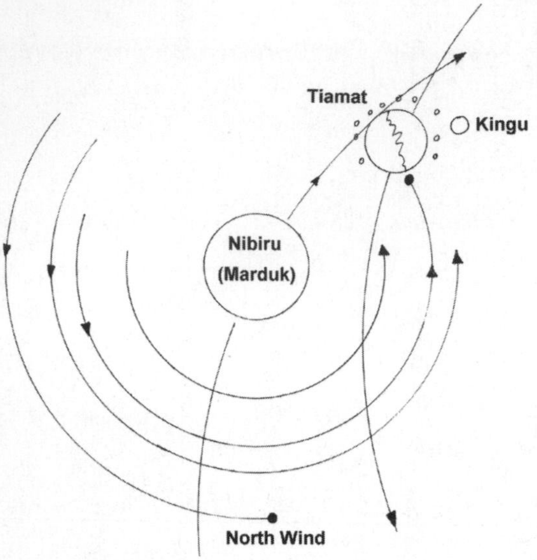

Fig. 7.8. Sitchin's interpretation of the Marduk and Tiamat myth is a series of collisions between the moons of Nibiru and the planet that once orbited between Mars and Jupiter—Tiamat. The moons of Tiamat became comets, he says; the shattered half of Tiamat became the asteroid belt, the other half became Earth, and Tiamat's main moon, Kingu, became our Moon. Appleby thinks Nibiru also lost some mass in these collisions, becoming similar in size to Earth. After *The Twelfth Planet,* by Zecharia Sitchin.

Appleby quotes the *Washington Post* (December 30, 1983), stating that "a heavenly body possibly as large as the giant planet Jupiter"[13] has been spotted by the Infrared Astronomical Satellite (IRAS) in the direction of Orion, and, says Appleby, this is Nibiru, and it is due to "become visible to Earth and appear as a new star sometime between 2012 and 2036."[14] However, Appleby says it is now near the size of Earth, since it broke in two, as described in Babylonian myths (this somewhat contradicts the statement of IRAS chief scientist Gerry Neugebauer about it being the size of Jupiter). Thus any cataclysms as it comes close to Earth again will be less severe than previous ones.

There is a possible connection here to Egyptian and Greek mythology. According to alternative historian and author Murry Hope, the "Eye of Ra, which was the exclusive weapon of Sekhmet, is described as a bright and burning object (a planetessimal or comet perhaps?) . . . we are seeing Ra in terms of the Sirian System, and Sekhmet as his daughter."[15]

Typhon, the father of Orthrus (Sirius), is shown in *The Twelfth Planet* as a winged serpent, like Quetzalcoatl (who promised to return), and was "a kind of comet" that was invisible.[16]

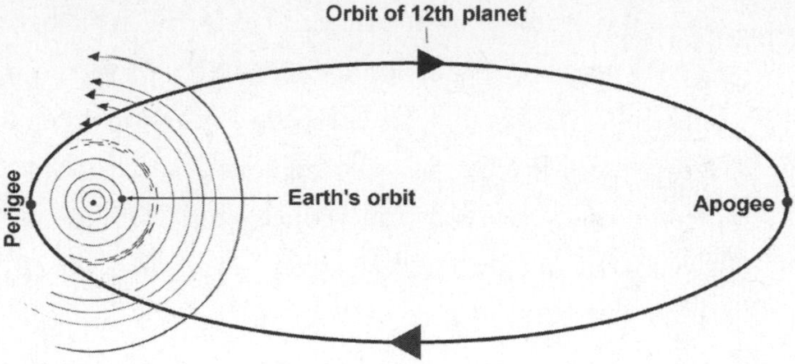

Fig. 7. 9. The eventual orbit of Nibiru. After *The Twelfth Planet,* by Zecharia Sitchin.

Fig. 7.10. Zeus and Typhon. After *The Twelfth Planet,* by Zecharia Sitchin.

If we take a detailed look at the zodiac of Dendera, it seems as if the Eye of Ra has just come out of the Square of Pegasus, between the two fish of Pisces (see chapter 8, fig. 8.1). In the epic book on astromythology, *Hamlet's Mill,* there is a collection of myths that together imply that Nibiru/Marduk (identified variously as Jupiter, a comet, or "the ark") entered the Square of Pegasus, which is said to be the "entrance to the abyss."[17] Channeler Jelaila Starr claims to have received messages from "the Nibiruan Council" saying that an anomalous NASA photo of an object "near the vicinity of EQ/Pegasi in the Pegasus star system" is actually Nibiru.[18] Finally, author and astrologer Barbara Hand Clow connects these threads by saying that Nibiru's 3,600-year orbit takes it near Sirius.[19]

PHOBOS FLYBY OF EARTH

In an article by James van der Worp[20] about the flyby of Mars by Comet 76P on June 5, 2000, evidence was presented that the orbit of Phobos (see color plate 11) was affected, resulting in it being shunted toward Earth's orbit, on a collision course, with a projected impact time around September 12, 2000. An Internet discussion ensued between Van der Worp and Glen Deen involving complex astrophysics calculations, and Deen came to the following conclusion: "My low-precision orbit model gave Earth impact of Phobos on September 18, 2012, with an escape velocity of 3.99731 Km/s."[21]

However, Glen Deen later modified his theory. He has since said it was not loosened in 2000, but in 2001 (by Comet 75P, Kohoutek), and has also stated that "Planet X" could impact Mercury around March 2, 2012, and Mercury could impact Venus around June 6, 2012, resulting in a rendezvous of Venus/Mercury with Earth on December 11, 2012,[22] like a game of cosmic billiard balls—which sounds very unlikely.

BEHEMOTH AND LEVIATHAN

E.Din: Land of Righteousness is a massive work by El (a.k.a. Elliott Rudisill), with over 350 pages of very small print. The author has used the Bible, as well as the Dead Sea Scrolls, the book of Enoch, the book of Jubilees, and other apocryphal works, combined with the works of Zecharia Sitchin, to form his interpretation of human history. He also lists the Popol vuh, Bhagavad Gita, Ramayana, Talmud, Upanishads, Kabbalah, and Qur'an in his list, as well as some of the most influential modern books on 2012.

The planet Nibiru is said by the author to have passed through the solar system around 5100 BC (no reference is given), and is predicted to return around 2012 AD, possibly as the "New Jerusalem."[23] The book of Job is analyzed in detail, since Rudisill is convinced that the family of Job was killed by a meteor in the form of "fire from heaven" and an accompanying "great wind," and that the two biblical "monsters of the deep," Behemoth and Leviathan, are actually huge asteroids swimming in the sea of space. Daniel, Psalms, and Enoch all mention these monsters, too, and, when cross-referenced with the book of Revelation, their consistency and route can be surmised.

By combining the Great Pyramid timeline, the "Enochian Jubilee Calendar," and the thirteen-baktun cycle of the Maya, the author came up with a "Galactic Calendar" at the root of them all,[24] and was able to plot an unfolding, if now outdated, scenario:

Between autumn 2004 and spring 2005, an asteroid or comet would impact Earth and penetrate its crust.[25] Then a star would appear in Pisces. This is Leviathan, and its apparent route through the sky toward Orion[26] is mapped. Around the same time that Leviathan appears in Pisces, Behemoth would travel another route from the "southern constellations" starting at Hydras, and its apparent route would also take it toward Orion.[27] Since Behemoth is "the one that hides," it might remain hidden until it got to Orion. Behemoth has "ten horns" and Leviathan has "seven heads," but when meeting in Orion, "somewhere in the middle of 2008," they would combine to become the "seven-headed ten-horned 'Red Dragon'" of the book of Revelation. The Red Dragon would be "cast into the pit" around summer solstice 2008[28]—or the Leviathan part, at least—while the Behemoth component would bypass Earth but return in 2012 due to being thrown off orbit by the galactic superwave that Paul LaViolette (see the following) has predicted.

Thus as 2012 approaches, Mercury (Archangel Michael) collides with the Red Dragon (Leviathan), and Venus and Jupiter may also be hit, but the Moon will take a severe hammering and may blow apart, with the pieces hitting Earth along with Leviathan.[29] Three and a half years later (forty-two months, or 1,260 days), Behemoth will return, becoming Osiris reborn, or the reborn First Father of the Maya.[30] The galactic superwave will come into view in Sagittarius, where Galactic Center is located, appearing as a blue star, or the Eye of God, "shortly before the winter of 2012,"[31] and will arrive at the same time as Behemoth and Nibiru, which will be the "descent of the New Jerusalem."[32]

VASCILLATION FILTRATION

A detailed review of *E.Din* was posted on the Diagnosis 2012 website and Rudisill responded, posting a reply and corrections at his own website.[33] Among several problems with the book, I noticed that the diagrams show the timing of the winter solstice 2012 event to within an accuracy of one second, while a four-year error due to the Jubilee calendar being neglected during the time that the Israelites were in Egypt is called "a calculated error," that is, "close to exact."[34] (The Jubilee period was originally a span of seven times seven or forty-nine lunar years, followed by a fiftieth, or Jubilee year; when later reckoned in solar years, it equated to a total of forty-nine solar years; the Jubilee year was in this case the forty-ninth year.) A forty-year error is said to be more accurate than the Great Pyramid, and the 1,260-day period from the book of Revelation is miscalculated by 384 days, which displaces the whole impact scenario by the same period, from mid-2008 to July 2009.[35]

Rudisill agreed on his reply-and-corrections page that most of the points I raised were valid, but he didn't mention this 384-day error. Unfortunately, *E.Din* provides no references or index, so checking out the information is very tedious, but having performed that drudgery, we conclude that the complex theory is error riddled, reducing it to an interesting but unsubstantiated idea. The full review is posted online, at www.diagnosis2012.co.uk/new4.htm#leviathan, where there is a link to the author's reply and his corrections.

OBLIQUITY HARMONICS

Keith Hunter is the author of several articles posted on the Internet at www.2near.com/edge. In the first of these, "Earth Pole Shift 2012 13.0.0.0.0.,"[36] Hunter theorized that a pole shift is coming that will reinstate Earth's 360-day rotation period and eventually restore it to its original spherical state, raising the consciousness of humankind in the process. In the second essay, "Sacred Geometry, the Earth, and the Mayan Calendar 'End-Date' of 2012,"[37] he continued the theme, explaining the harmonic relationships he had discovered between Earth's oblateness (ratio of its equatorial diameter to its polar diameter) and the angle of Earth's obliquity of the ecliptic (the tilt of Earth's axis).

Following a request from Hunter to comment on his theory, I informed him of some problems I had found with it. After this communication, he removed his original essay, "Earth Pole Shift 2012," and made several alterations to his second essay, "Sacred Geometry," so I will comment only on the second version of the second essay.

Hunter says that Earth's axis tilt angle, or obliquity, is diminishing every year, and he calculates that in 2012 it will be in a harmonic relationship to its oblateness. However, it is plain that the numbers do not have a harmonic relationship to each other, in the same way that (for example) the numbers 85 and 1,085 do not have a harmonic relationship.* When I pointed this out to Hunter, he said, "It is not important that they are not harmonics. They are obviously related. Just to look at them, this is obvious."[38] Another problem emerged when I checked more than ten Internet sources on the polar and equatorial measurements of Earth. None of the resulting oblateness figures was anywhere near Hunter's figure, but his explanation was even more perplexing than the previous one. He said that his figures were "not based upon the actual physical dimensions of the Earth that we measure."[39]

Details of several other errors were e-mailed to Hunter so that he might post corrected versions, or remove the essays . . . or not.[40]

GALACTIC CORE EXPLOSION

Astrophysicist Dr. Paul LaViolette, in *Earth Under Fire,* combines recent scientific data with myth and legend to rediscover the reason

*The numbers are not harmonics of each other. Even if we ignore the error between the calculated obliquity figure for 2012, 1406.263785 minutes and the ideal number it should be (1406.25 minutes), this is still not a harmonic of the 1.0040625 figure for oblateness.

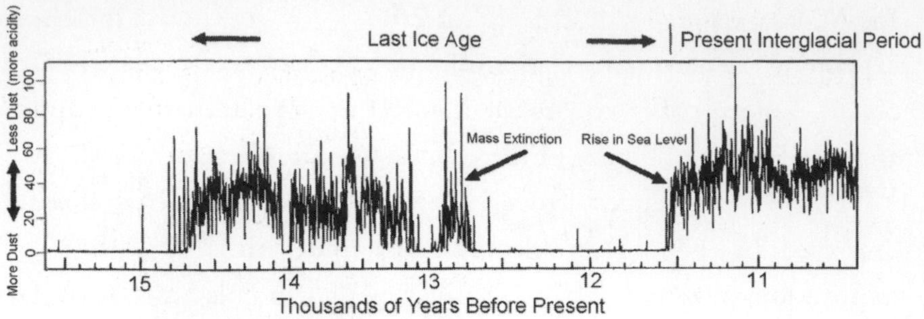

Fig. 7.11. A graph showing changes in acidity around the close of the last ice age, deduced from the Greenland Summit ice cores. Adapted from Paul LaViolette's book *Earth Under Fire: Humanity's Survival of the Apocalypse.*

why ancient civilizations tried to warn us of impending disaster. LaViolette has an impressive record of achievement: he lists twelve groundbreaking discoveries to his credit, which were later vindicated by other researchers.

LaViolette's new model predicts that the core of our Milky Way galaxy is not a black hole, but a massive object that explodes periodically. These "galactic core explosions," he says, happen approximately every 26,000 years, "with the possibility of a 13,000 year recurrence interval,"[41] sending out a "galactic superwave" of cosmic rays. He theorizes that the superwaves could have entrained Earth's cycle of precession into its approximate 26,000-year pattern.[42] This is especially likely if a wave arrived when Earth's polar axis was facing Galactic Center, as it was around 12,850 years ago. This is very close to the 12,700-year BP date (Before Present; conventionally, before 1950 AD) indicated for the superwave through ice core examination (see diagram above, as well as chapter 10), and also close to the date at which the Greeks set the beginning of the Platonic year (the cycle of the precession of the equinoxes), which was "when the vernal equinox was just about to enter the constellation of Leo, i.e., around 10,800 BC"[43] or 12,750 years BP.

This is also the approximate date that Robert Bauval, Adrian Gilbert, and alternative Egyptologist Graham Hancock found encoded at the Giza Plateau in Egypt (see the next chapter). LaViolette follows up on

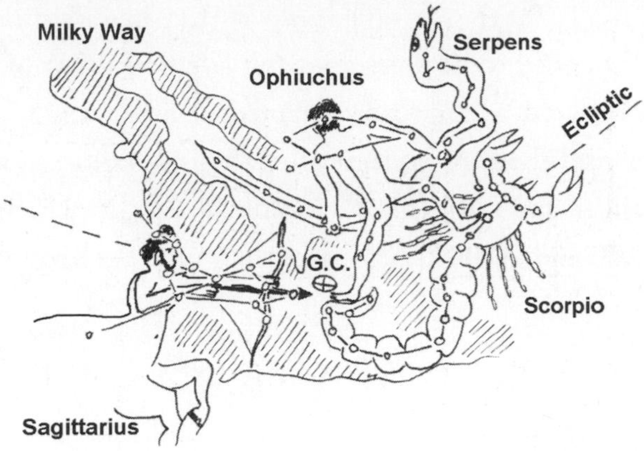

Fig. 7.12. The arrow of Sagittarius and the sting of Scorpio, as well as the foot of Ophiuchus, the Snake Handler (see chapter 8), all define the position of Galactic Center. After Paul LaViolette's book *Earth Under Fire: Humanity's Survival of the Apocalypse.*

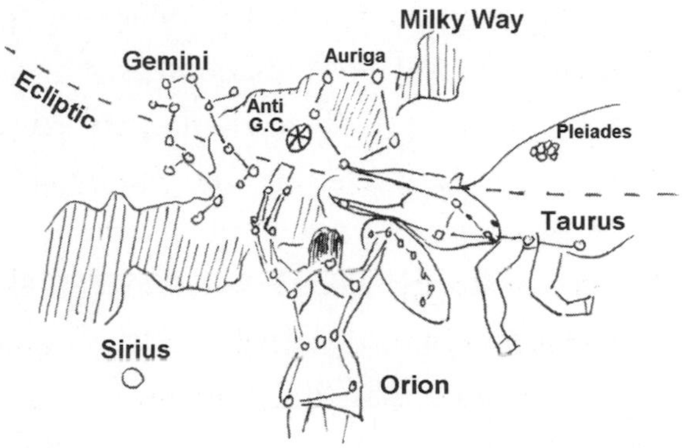

Fig. 7.13. The club of Orion and the horns of Taurus define the galactic anticenter, which is the direction in which the galactic superwave departed. After Paul LaViolette's book *Earth Under Fire: Humanity's Survival of the Apocalypse.*

this, showing that Giza, as well as the zodiac and the tarot, were all time-capsule systems encoded with information about the last galactic superwave that swept through the solar system, ending the ice age about 13,000 years ago. This knowledge may still be preserved in the myths of the Eye of Ra and the Eye of Horus, which were sent to punish mankind.

Geologist Gregg Braden (see chapter 9) predicts a geomagnetic reversal linked to a 13,000-year cycle. LaViolette confirms that around 13,000 years ago "the intensity and declination of the Earth's magnetic field underwent major variations in step with the eleven-year sunspot cycle. The amplitude of these cycles was hundreds of times larger than modern geomagnetic solar cycles, suggesting that solar-flare activity at that time was also hundreds of times more intense, approaching levels normally observed in T Tauri stars."[44]

T Tauri stars are those with a mass similar to the Sun (though expanded to between two and five times that of the Sun), but that have collected a large amount of dust along their equatorial planes. This causes them to emit most of their radiation in the infrared part of the spectrum and exhibit constant flaring, in which the flares are between 100 and 1,000 times larger than typical flares of the Sun.

ELECTROMAGNETIC PULSE WAVE

I recently discovered online a fictional work called *Galactic Tsunami,*[45] in which it becomes apparent to the characters in the story that the event that governed the timing of the 2012 termination point of the Maya thirteen-baktun cycle was the arrival of an electromagnetic pulse, or EMP wave. In the story, this wave is said to be due every five hundred years, and currently overdue by two hundred years. Although this is fiction, it is based on the discoveries of astrophysicist Paul LaViolette. The emphasis, in LaViolette's *Earth Under Fire,* was on a galactic core explosion that happens at intervals of 13,000 and 26,000 years and sends out a galactic superwave of cosmic waves that bring dust and cause solar activity to increase, while blocking out the sunlight and causing major geomagnetic variations.

However, when I reread pages 334–35 of LaViolette's book, I found that he did say smaller EMP events happen on average every five hundred years. The following is a quote from an online essay by LaViolette.

> Galactic superwaves may also produce an intense electromagnetic pulse (EMP) whenever a cosmic ray front happens to strike the Earth's atmosphere. Galactic superwaves such as those that arrived during the last ice age could have generated pulses delivering tens of thousands of volts per meter in times as short as a billionth of a second, comparable to the early-time EMP signal from a high-altitude nuclear explosion. . . .
>
> Galactic Center activity occurs frequently between major super-

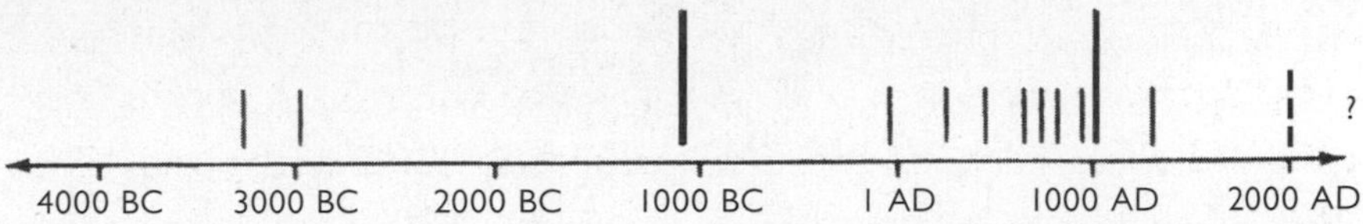

Fig. 7.14. The timing of minor galactic core explosions over the last 6,000 years and consequent cosmic ray pulses would have caused an electromagnetic pulse through Earth's atmosphere. Adapted from Paul LaViolette's book *Earth Under Fire: Humanity's Survival of the Apocalypse.*

> wave events. Astronomical observation indicates that during the last 6,000 years, the Galactic Center has expelled 14 clouds of ionized gas . . . These outbursts may have produced minor superwave emissions with EMP effects comparable to those of major superwaves. About 80% of these bursts took place within 500 hundred years of one another . . . With the most recent outburst occurring 700 years ago, there is a high probability of another one occurring in the near future.[46]

Nevertheless, the diagram in *Earth Under Fire* reveals that although the average may be five hundred years, there was one period of almost two thousand years without a wave, and the pattern certainly doesn't look predictable.

8
ALTERNATIVE ARCHAEOLOGY

THE DENDERA ZODIAC

Astrologer John Lash gave a lecture at the Alternative Egypt Questing Conference, held at Logan Hall of the Institute of Education, London, on October 23, 1999, at which the theme was "Alternative Egypt." Lash claimed that there were originally thirteen signs in the zodiac, the thirteenth being the Snake Handler. Instead of splitting the ecliptic into twelve equally sized sections, there were thirteen sections of different sizes, depending on the sizes of the constellations. Once this is taken into account, Lash says that the Sun's vernal equinox will not transit from Pisces to Aquarius for another seven hundred to eight hundred years.

The circular zodiac on the ceiling of the Temple of Dendera in Egypt shows the twelve signs, the Snake Handler without the snake, several other constellations, the five visible planets, and the thirty-six decans (ten-degree sections of the sky). The Egyptologist R. A. Schwaller de Lubicz, in his book *Sacred Science,* has shown that the Dendera zodiac demonstrates that the Egyptians knew about the precession of the equinoxes. The axes marked on the zodiac show the movement of solstices and equinoxes through the constellations between the foundation of Egypt and the time of the building of the Temple of Dendera. Sirius

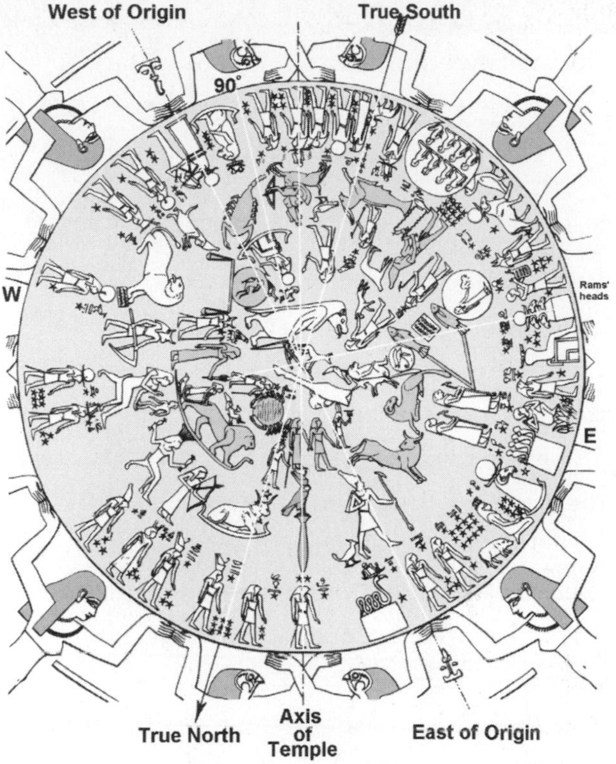

Fig. 8.1. The circular zodiac on the ceiling of the Temple of Dendera, with the zodiac constellations shaded. It seems that the current era is encoded in the carving. After Lucie Lamy.

appears twice—once on the true north–south axis, above the horns of the "cow of Isis," and again on the axis of the temple, as Horus on a papyrus stem. The light of the star would illumine the inner temple on New Year's Day, at the heliacal rising, when the temple was built. In that era, this would have coincided with the summer solstice in Cancer, which is why Sirius also appears aligned with Cancer on the north–south axis.

John Lash refers to another axis on the Dendera zodiac that goes from the four rams' heads, through Pisces, across the pole, to the star Spica, which is being held by Virgo. This, he says, is pointing out the position of the vernal equinox today. The line goes through the Square of Pegasus, which seems to have writing on it—the "Programs of Destiny" (though cross-cultural comparisons made by Santillana and Von Dechend in *Hamlet's Mill* suggest it represents a checkered game board). At exactly ninety degrees to this axis, another line can be drawn through the tip of the arrow of Sagittarius, which points to Galactic Center, close to where the solstice alignment will occur in 2012.

CRUSTAL DISPLACEMENT

In his book *Fingerprints of the Gods,* Graham Hancock suggests that an unusual planetary alignment at the end of the thirteen-baktun cycle will cause a gravitational effect on the crust of Earth, which is top-heavy with polar ice. The crust will slip, wiping out most of life on Earth as a result of colossal tidal waves. However, this planetary alignment is actually due on December 24, 2011, which is the date given by Frank Waters in *Mexico Mystique* as the end of the thirteen-baktun cycle. Since the end point has now been confirmed as December 21, 2012, then if we are due for crustal displacement, this planetary alignment will probably not be the trigger. However, Hancock finds convincing evidence that the event that triggered the flooding at the end of the ice age, around 13,000 years ago (he puts it at 10,500 BC), is connected to the precession of the equinoxes, and is due again right about now.

As we have previously mentioned, it seems that Frank Waters got the December 24, 2011, end point from Michael Coe's 1966 book, *The Maya,* but in later editions and other works Coe used the Lounsbury correlation, which fixes the thirteen-baktun end point (13.0.0.0.0) at December 23, 2012. This correlation is second only to the final Goodman (GMT-2) correlation in popularity—the one that supports the True Count, and the one that ends on winter solstice, December 21, 2012.

THE HALL OF RECORDS

In his 1998 book, *Gods of Eden,* Andrew Collins studied the legends of an undiscovered underground complex at the Giza Plateau in Egypt, and through an examination of the much overlooked Edfu texts he has concluded that the complex takes the form of twelve chambers surrounding a thirteenth.

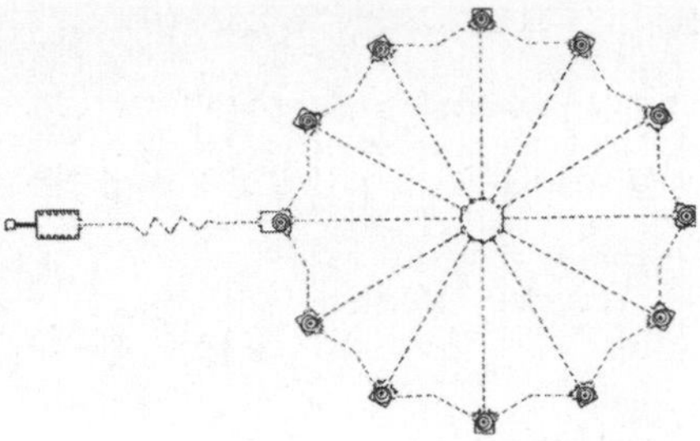

Fig. 8.2. Andrew Collins's plan of the probable arrangement of the Chambers of First Creation, based on the Edfu texts and books of the Duat (underworld). Drawing by Bernard G. From *Gods of Eden* by Andrew Collins.

He calls the complex the Chambers of First Creation, and says that in each of the twelve chambers were kept "Iht-relics," and in the central, thirteenth chamber was the "Bnnt-embryo," also known as a "seed" or "egg of creation."[1] Collins imagines the embryo to be "a large conical-shaped omphalos or benben stone, plausibly a crystal-like structure" and the Iht-relics, "smaller, hand-held sacred stones, or crystals."[2] This is such a similar idea to that of the Ark of Native Americans (see chapter 19, "Crystal Skulls") that it implies a 2012 connection—which is made by Nigel Appleby.

Appleby came to the same conclusion as Collins after studying the Edfu texts, though he differs in his prediction of the location of this underground complex. There has been some controversy about Appleby's book *Hall of the Gods,* as several authors have accused him of plagiarism, but since it was published at the same time as *Gods of Eden,* this probably doesn't apply to Appleby's following conclusions: Since the three Giza pyramids in Egypt correspond exactly to the layout of Orion's Belt,[3] if the image of the Orion constellation were superimposed onto the pyramids, the position of Sirius, argues Appleby, would overlay the "Hall of Records" that American psychic Edgar Cayce predicted would be discovered at Giza around the turn of the millennium. From his study of the Edfu texts, Appleby says that the Hall of Records is a central chamber surrounded by twelve others. He even conjectured that the central chamber contains a small pyramid, as opposed to just a pyramidion, or capstone. Inside should be

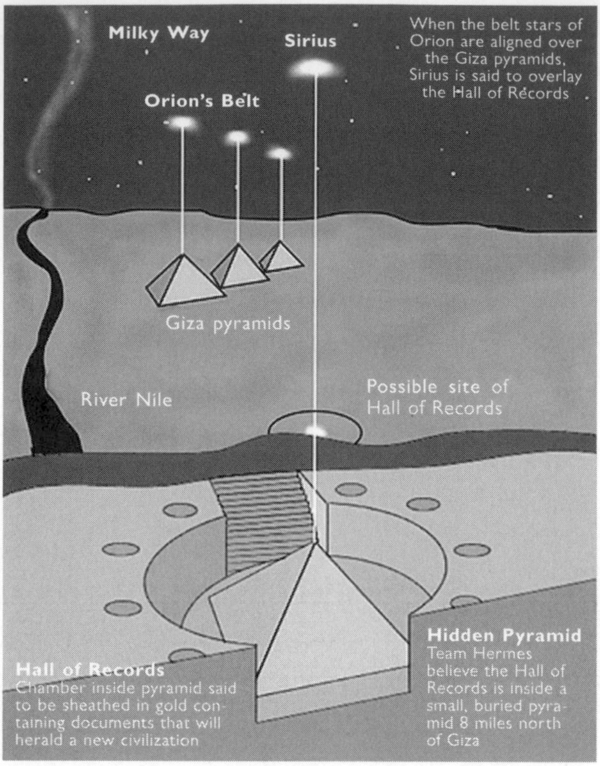

Fig. 8.3. The location of the lost Hall of Records according to Nigel Appleby and the Operation Hermes team.

information concerning the history of humankind, and, says Appleby, crucial information to prepare us for the coming of the "Fifth Age of Man," which will begin in 2012.

THE RETURN OF THE PHOENIX

The Phoenix myth relates to the periodic reversal of Earth's magnetic field due to the long-term sunspot cycle that culminates in a solar magnetic reversal, says Appleby. This magnetic reversal is due in 2012. Here Appleby has obviously been influenced by Cotterell and Gilbert's *The Mayan Prophecies*. Appleby refers to Michael Mandeville's self-published book *The Return of the Phoenix*,[4] which predicts that the "fifth flight of the Phoenix" is a movement of Earth's lithosphere, in which it "takes flight," moving as much as thirty degrees latitude. This is otherwise known as crustal displacement, or crustal slip (as predicted by Hancock, mentioned earlier). Appleby says, "If Mandeville is correct, then it will fly again on December 22, 2012."[5]

Fig. 8.4. The phoenix or Bennu bird. After Robert Bauval and Adrian Gilbert's book *The Orion Mystery: Unlocking the Secrets of the Pyramids.*

Thus it was a surprise when I found that Mandeville asserted[6] that the fifth flight will most likely occur in the year 2000 or 2001. More recently I looked at his website, and now that we have passed 2001 and the fifth flight failed to happen, he says that it cannot be predicted ("except by those with 'otherworldly' connections") and denies that the Maya calendar could be connected with it. However, Mandeville has seemingly misunderstood the whole concept of galactic alignment, associating the alignment with the spring equinox of 2012, rather than the winter solstice.

Swedish scientists have determined that the date of the most recent magnetic shift (not a full reversal, say the Swedes—others vary—see

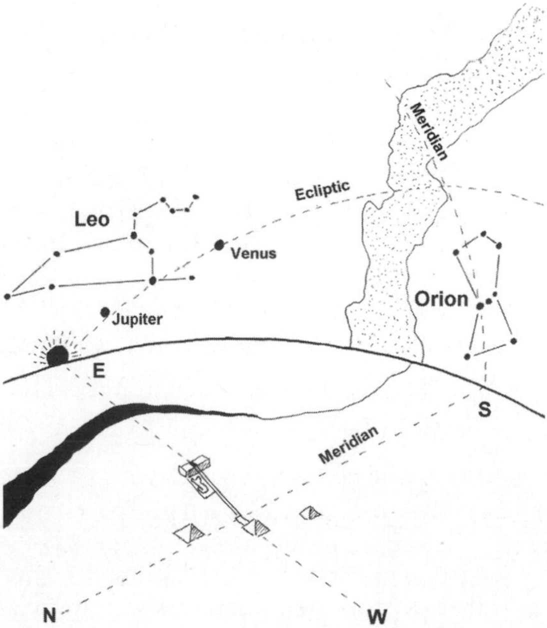

Fig. 8.5. At vernal equinox 10,500 BC, the Sun in Leo lines up with the Sphinx, while Orion's Belt culminates at the meridian—a pattern mimicked on the ground by the Pyramids of Giza. After Robert Bauval and R. J. Cook, in Bauval and Hancock's book *Keeper of Genesis: A Quest for the Hidden Legacy of Mankind.*

note 10) occurred in 10,500 BC, says Appleby.* This agrees, "with extraordinary exactness," says Appleby, with the date revealed by Edgar Cayce's visions as the time the Great Pyramid was built (10,490 BC). It is also the date that alternative Egyptologist John Anthony West[7] and geologist Robert Schoch have determined as the construction date of the Sphinx. Thus the Sphinx and the Great Pyramid may have been designed as clues to last 12,500 years, to help warn us of the coming changes in 2012, and to lead us to the Hall of Records.[8]

According to Appleby,[9] the Sphinx was designed as a phoenix (man + bull + lion + eagle), and is aligned to a point on the horizon that gives us a clue as to the meaning of the Phoenix legend. The Sphinx, as pointed out by Robert Bauval and Graham Hancock in *Keeper of Genesis,* faces the point on the horizon where on the vernal equinox the sunrise was in the constellation of Leo and Orion was at its lowest point in the sky. This occurred in 10,500 BC, and is known as Zep Tepi—the First Time of Osiris (Orion), a golden age when the gods ruled Egypt. The Last Time of Osiris will occur when Orion is at its highest point in the sky, and Bauval and Hancock give the date variously as 2450 AD[10] and 2500 AD.[11] Appleby gives it as 2012 AD. Having checked, I have found reason to agree with him.† To arrive at this conclusion,

*Appleby's source for the date of the last magnetic reversal (10,500 BC) was probably Alan Alford (*Gods of the New Millennium,* pp. 225, 625), who says the Swedes reported in *New Scientist* (June 1, 1972, p. 7) that "a reversal of the Earth's magnetic field had occurred 12,400 years ago," a date that equates to 10,428 BC. This event is known as the Gothenburg Magnetic Flip. Paul LaViolette says it occurred 12,700 years ago (10,700 BC); other sources say it happened 12,500 years ago and 12,600 years ago. There seems to be some disagreement as to whether this event was a magnetic reversal or just a local fluctuation. LaViolette calls it a "substantial geomagnetic disturbance." Gregg Braden quotes a later Swedish paper that drew on new evidence from the southern hemisphere to say that it was a 180-degree shift, and it occurred 13,200 years ago. However, Alford has since written a book, *The Phoenix Solution,* theorizing that the Phoenix myth is about exploded planets.

†Bauval and Hancock calculated these dates with SkyGlobe software, but when I checked them with the same software, I found that from the 1960s on to 2800 AD, the altitude of the Orion star Al Nitak remains at fifty-eight degrees (from Cairo, looking due south). This means that we are in the Last Time of Osiris/Orion now.

Appleby needed to place Zep Tepi around five hundred years before Bauval and Hancock's Zep Tepi date, thus bringing their projected date for the Last Time back another five hundred years. He thus placed Zep Tepi at 11,500 BC,[12] but does not seem to realize that this means the Last Time would have happened around 1400–1500 AD, which is half a precessional cycle later (around 12,900 years).

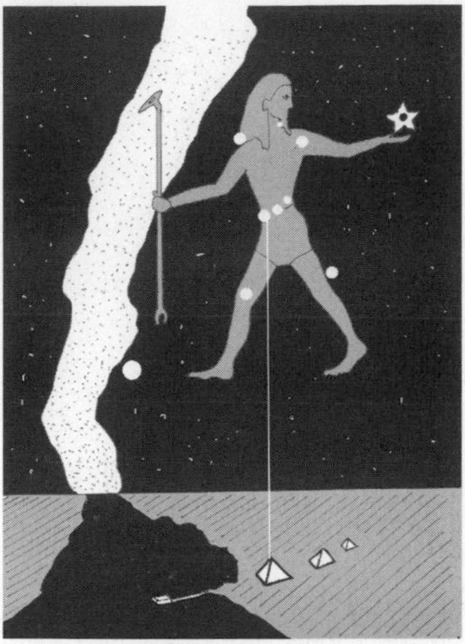

Fig. 8.6. Orion's Belt and the Giza pyramids. The star Al Nitak aligns with the southern shaft of the King's Chamber of the Great Pyramid. After Robin Cook, in Robert Bauval and Adrian Gilbert's book *The Orion Mystery: Unlocking the Secrets of the Pyramids.*

Appleby also says that at the time of Zep Tepi, the planet Venus performed a retrograde loop over the top of the Orion constellation: "This phenomenon only occurs once every 12,500 years, and the next time it will occur happens to be in the year 2012."[13] However, as we shall see, this information on the retrograde loops of Venus was supplied to him by astronomer Gino Ratinckx and his coauthor, Patrick Geryl, and on closer examination turns out to be totally misleading. But more than that, the date that Geryl and Ratinckx gave for the retrograde loop was actually 9792 BC, not 10,500 BC, or his revised Zep Tepi date of 11,500 BC. This is an error of 1,708 years. As we shall soon see, when I consider Geryl and Ratinckx's claims, they also made an error, and the best match for the 2012 loop happened in 9389 BC, according to the results I got with SkyGlobe software. This would bring Appleby's error up to 2,111 years. It is a pity that he tried to incorporate the Venus loop into his theory (which is mostly irrelevant, as we shall see), since it has obscured the point that we are now in the Last Time of Osiris.

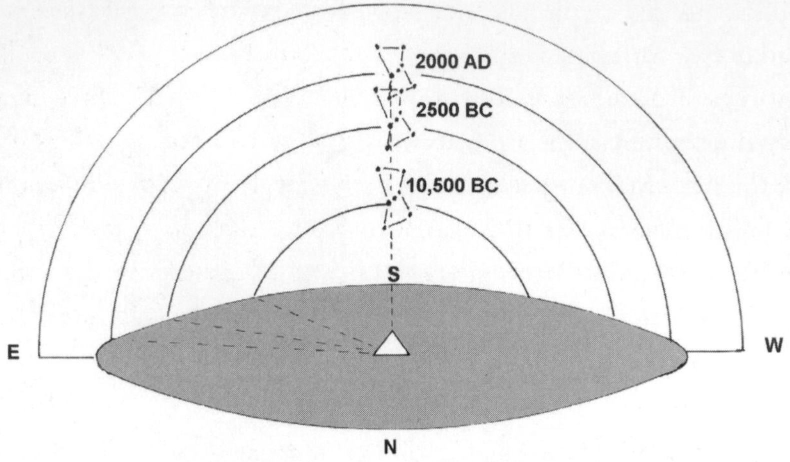

Fig. 8.7. The Pyramids of Giza represent Orion's Belt around 10,500 BC, the First Time of Osiris according to Bauval and Gilbert, and Hancock. Orion is currently approaching its highest point—the Last Time of Osiris. After Robin Cook, in Robert Bauval and Adrian Gilbert's book *The Orion Mystery: Unlocking the Secrets of the Pyramids.*

THE PYRAMID OF KUKULCAN: A PRECESSIONAL ALARM CLOCK

Fig. 8.8. Quetzalcoatl, the Feathered Serpent

When the Toltec people from northern Mexico moved south to Chichén Itzá, on the Yucatán Peninsula, they merged their own zenith cosmology with that of the Maya system, and the result was the Pyramid of Kukulcan. This has been designed so that every year, on spring equinox, the afternoon sun causes a shadow play, so that it appears that a huge serpent is descending from the sky, down the pyramid. However, John Major Jenkins shows that the pyramid is much more than an equinox indicator. It is a "precessional clock with its alarm set for the twenty-first century."[14]

Jenkins says that Kukulcan (the Maya name for Quetzalcoatl, the Feathered Serpent) was the symbol of a Sun-Pleiades-zenith conjunction. The zenith passage of the Sun takes place over Chichén Itzá on May 20, around sixty days after the spring equinox. *Crotalus durissus,* the rattlesnake, whose pattern is constantly used in Mesoamerican art, has a marking on it that is identical to the solar Ahau glyph of the Maya, and its rattle was called *tzab,* which is the same word in the Yucatec-Mayan language used to designate the Pleiades, a small cluster of seven stars in the Taurus constellation.

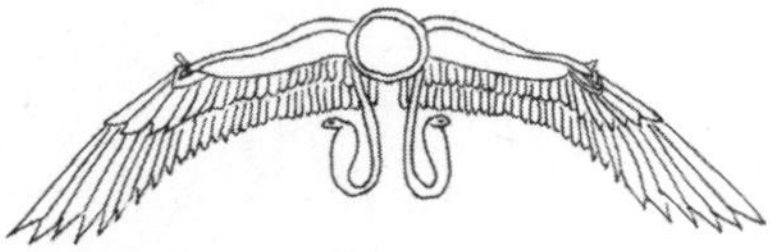

Fig. 8.9. In Egypt, the Sun was depicted as a feathered snake.

The moving snake on the Pyramid of Kukulcan is an annual reminder of the conjunction of the zenith Sun with the Pleiades that occurs directly over the pyramid. This conjunction will occur in a 360-year window of time that is just about to open. Right at the start of the 360-year window is 2012, and on May 20, 2012, the zenith passage of the Sun will combine with a solar eclipse, on the tzolkin day 10 Chicchan, which means "serpent."

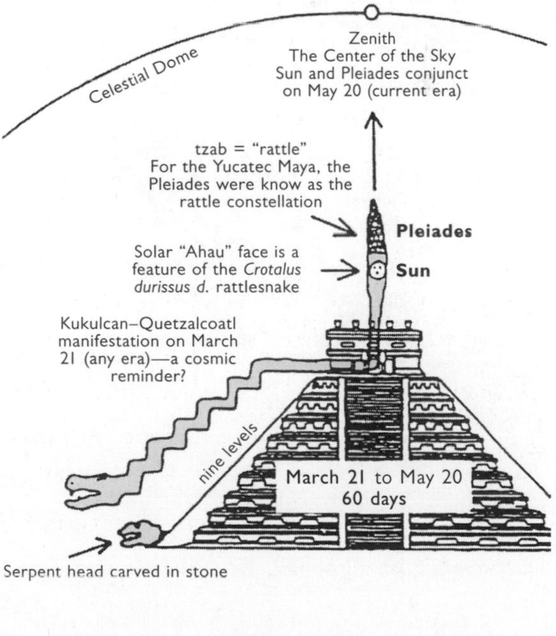

Fig. 8.10. The shadow play occurs sixty days before the Pleiades-Sun-zenith conjunction in 2012. From *Galactic Alignment: The Transformation of Consciousness According to Mayan, Egyptian and Vedic Traditions,* by John Major Jenkins.

Thus Jenkins has discovered the meaning of the Toltec New Fire ceremony[15] that was bequeathed to the Aztecs. Every fifty-two years, when the combination of haab and tzolkin calendars—the Calendar Round—approached its end point, the priests would wait to see if the Pleiades reached the zenith at midnight. They were actually measuring the precessional movement of the Sun–Pleiades conjunction toward the zenith on the zenith-passage day that occurs exactly six months later. The conjunction was calculated at night, six months previously, since the Pleiades are not visible in daylight.

THE RATTLING OF THE PLEIADES

In his book *Gateway to Atlantis,* Andrew Collins has amassed evidence that the civilization of Atlantis was destroyed by a tidal wave following the collision of a comet (known to geologists as the Carolina Bays comet), and that Cuba is a remnant of the former landmass. The proximity of Cuba to Mesoamerica would explain the Aztec and Maya fascination with cataclysm. The oracular text known as the Chilam Balam of Chumayel, Collins tells us, "spoke of an almighty cataclysm during which the "Great Serpent" was "ravished from the heavens together with the rattles of its tail" so that its "skin and pieces" of its bones "fell here upon the Earth."[16] The Calina Carib tribe of Suriname, South America, retains myths of a fiery serpent that came from the Pleiades and "brought the world to an end" with "a great fire and a deluge."[17]

Fig. 8.11. The serpents on the rim of the Aztec Sunstone have the seven stars of the Pleiades on their snouts, according to Jenkins. Adapted from Roberto Sieck Flandes.

There is also a Hebrew legend, says Collins, that holds that "the Great Flood was caused after 'the upper waters rushed through the space left when God removed two stars out of the constellation Pleiades.'"[18]

Collins, along with Jenkins, has pointed out that the Pleiades were seen as the rattle of a celestial snake. However, Collins adds that the Pleiades, as the snake rattle, "warns its potential victim of an imminent strike."[19] Collins thinks this may have come from the direction of the Pleiades.

Astronomers call one class of Earth-crossing asteroids "Apollo objects." This brings to mind Philip K. Dick's enigmatic statement in *Valis* that "the Head Apollo is about to return."[20]

THE EYE IN THE PYRAMID

Fig. 8.12. The Great Seal of the United States, found on every dollar bill.

Raymond Mardyks has decoded[21] the Great Seal of the United States, which appears on every dollar bill. The seal shows a thirteen-step pyramid with the date *1776* in Roman numerals on it. Just as the Pyramid of Kukulcan has ninety-one steps on each of the four sides, making 364 in all, plus the top level, making the number 365, the Great Seal pyramid also has an encoded calendrical meaning. Like some Maya pyramids, it has a date on it, but it is given in the Gregorian calendar, and four sides of thirteen levels adds up to fifty-two, which is the number of weeks in our year. However, 13 and 52 are also the key numbers in the Maya calendar systems.

The thirteen-baktun cycle consists of twenty katuns each; each katun consists of twenty tuns, so there are 5,200 tuns in the thirteen-baktun cycle. There are also fifty-two haabs in a Calendar Round. Some Maya groups named cycles after end-dates rather than start-dates. They also used a thirteen-katun cycle known as the Short Count. The year 1776 was not only the year that the American Declaration of Independence was signed (on July 4), but was also a special year in the Maya Long Count calendar.

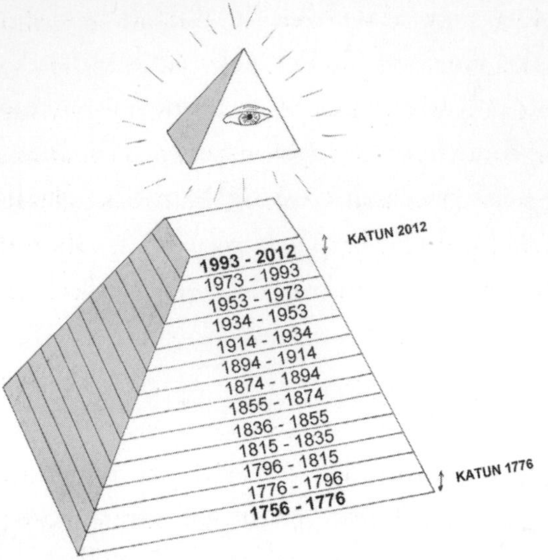

Fig. 8.13. The Great Seal as an encoded single cycle of the Short Count calendar. The thirteen-katun Short Count continued to be used after the Long Count and its thirteen-baktun cycle fell into disuse. Adapted from *Maya Calendar: Voice of the Galaxy*, by Raymond Mardyks and Stacia Alana-Leah.

Just as the last katun in the thirteen-baktun cycle could be called "katun 2012," the first katun in the cycle of thirteen katuns would be called "katun 1776." In fact, the katun ended thirty-three days before the signing of the Declaration. So 1776 is the bottom level of the pyramid, where the date is actually inscribed—making the top of the pyramid 2012.

The top would also be 2012 if each level represented one of the thirteen baktuns in the thirteen-baktun cycle, with 3114 BC at the bottom. The top of the Great Seal pyramid shows an eye in a triangle, which has been variously associated with Sirius, God, the pineal gland, and the Illuminati. Mardyks goes on to point out that not only was the Egyptian calendar based on the rising of Sirius, but that "the Sun is astrologically conjunct Sirius every year on July 4th, for the birthday of the United States of America" (this means they are plotted on the same degree of the same zodiac sign, even though Sirius is some distance off the ecliptic). Also, some Maya groups in the Yucatán froze (with the 365-day haab, the new year had previously wandered back through the seasons) their New Year to July 26, "when Sirius rises in that part of the world" (after its period of invisibility). On January 1, at midnight, Sirius culminates, reaching its highest point in the sky at the only time of year when it is visible all night long, according to Mardyks (about 12:30 AM, according to SkyGlobe). It is thus also a key marker in the Gregorian calendar.

Sirius was of great calendrical importance to the ancient Egyptians, since its heliacal (first dawn) rising marked the New Year and the annual Nile flood. In 2004, Robert Bauval and Graham Hancock released their book *Talisman: Sacred Cities, Secret Faith,* concerning Masonic secrets about Sirius, which were encoded into the design of Paris and Washington, D.C. (Masons were behind the gaining of America's independence and the French Revolution; Benjamin Franklin, George Washington, and Napoleon Bonaparte were all known Masons.)

Paul LaViolette's 13,000-year cycle would also fit the Great Seal pyramid on the dollar bill, as would Adrian Gilbert's concept of the Age of Adam (later in this chapter). With each level representing a thousand years, 2012 would still be on the top.

THE GREAT PYRAMID: CALENDAR IN STONE?

An essay by Paul White, "The Secrets of Thoth and the Keys of Enoch" (which can be found online), contains the following: "According to the 'calendar in stone' of the Great Pyramid, which describes the so-called Phoenix Cycle of our galactic orbit, the present time period ends (converted to our present calendar) in the year 2012 AD. The Greek word *phoenix,* derived from the Egyptian word *pa-hanok,* actually means 'The House of Enoch.'"[22]

J. J. Hurtak, author of *The Keys of Enoch,* and mentioned in White's article, does not refer to 2012 as far as I know (I have searched his book and the website www.keysofenoch.org). In fact, Hurtak's book says that the keys "were given on January 2–3, 1973, to prepare mankind for the activation of events that are to come to pass in the next 30 years of 'Earth time.'"[23] Those thirty years expired in January 2003. We have already established that Michael Mandeville could not be the source of the Phoenix–2012 connection, at least not based on the current version of his website. But what about Appleby's claim that Mandeville's 1997 self-published book *The Return of the Phoenix* stated that 2012 was the end of the Phoenix cycle? Appleby's wording could have meant that "if Mandeville is correct," then according to Appleby's interpretation "the Phoenix will fly again on 22 December 2012."[24]

Since the earliest posting of White's essay "The Secrets of Thoth and the Keys of Enoch" seems to be October 1999, it is possible that it could have been inspired by Appleby's *Hall of the Gods,* which came out the previous year (albeit only for six days, before it was withdrawn due to accusations of plagiarism; I managed to get myself a copy during those six days). I wondered if the information could have come from archaeologist Howard Middleton-Jones or researcher James Wilkie, who worked together on the Ambilac website (now Ambilac-UK). But the 2012 prediction from Middleton-Jones and Wilkie comes from an analysis of the whole Giza Plateau, not just the Great Pyramid.

STAR MAP-TO-GROUND CORRELATION AT GIZA IN 2012

Howard Middleton-Jones showed a sky map for December 21, 2012, at his lecture at the Glastonbury Symposium in 1999, saying that the satellite pyramids (the miniature pyramids that accompany each of the three Giza pyramids—some of these are sometimes called Queen's pyramids) at Giza reflect the positions of the planets at the end of the Maya thirteen-baktun cycle. Here is an extract of an article from his website, Archaeology—Archaeoastronomy—Ancient Civilizations:

> Are we going to have a new neighbour (Sun giving birth?) on 21st December 2012 . . . A satellite pyramid of Khufu will be the centre point for the complete Giza alignment on 21st December 2012 at a precise time (22.00 hrs, 18 mins, 13 secs, local Cairo time) [*sic*] and Venus will be completely aligned—connections will be made to our previous civilization on Mars, and our Sun, which was born of Alcyone in the Pleiades, our nearest relatives.[25]

I spoke to Middleton-Jones on the phone shortly after the 1999 symposium, and he said that the alignment in 2012 would involve a planet or body currently beyond Pluto, which had just been discovered.

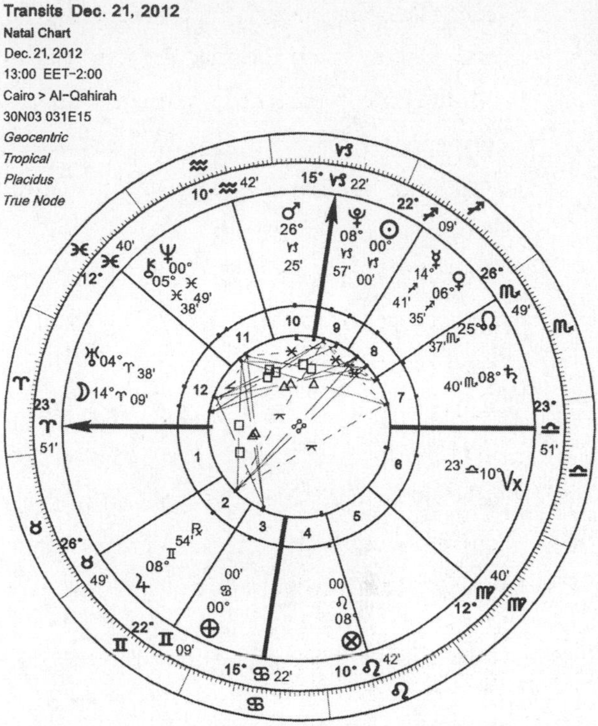

Fig. 8.14. An astrology chart for Cairo on December 21, 2012, according to the tropical zodiac. Chart prepared by Paul Wright, Orionstar Charts.

Struggling to find an astronomical justification for this, I asked astrologer Paul Wright to calculate an astrological chart for Cairo on December 21, 2012. He prepared a chart for early afternoon, which coincides with dawn in Mexico, and although there are no obvious conjunctions, Mercury and Venus are both in Sagittarius, with the Sun rising just after them. However, this is the position in the tropical zodiac, as used by astrologers, based on the zodiac as it was 2,300 years ago, since the constellations were chosen to represent twelve partitions in the seasonal year. The zodiac has precessed through many degrees since then, so that in the sidereal zodiac—the positions of the constellations as they are now—Mercury and Venus will be in Scorpio, with the Sun rising just after them.

Since the event that occurs on this date is the conjunction of the solstice Sun with the galactic equator, close to Galactic Center, also in Sagittarius, I suddenly remembered that Zecharia Sitchin predicted that when Nibiru returns it would reappear "in the regions of the constellation of Sagittarius."[26] Could this be the new planet that Middleton-Jones

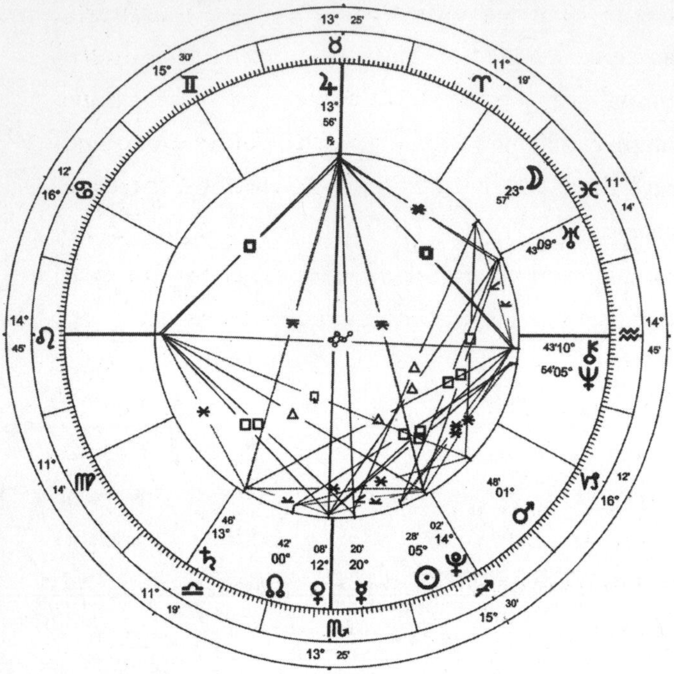

Fig. 8.15. The sidereal zodiac for Cairo on December 21, 2012. Chart prepared by Helen Sewell.

predicts will be "born of the Sun"? Comet Hale-Bopp was discovered in Sagittarius in 1995, having an orbital period of three thousand to four thousand years. Some people suggested that the comet was Nibiru, but it was too small. However, Appleby believes that Nibiru has fragmented. Could Hale-Bopp be a forerunner of Nibiru, on the same orbit?

GIZA-GENESIS

When I had a look at the various websites of James Wilkie[27] and Howard Middleton-Jones (now unified at Ambilac-UK[28]), I found the information so complex and convoluted that it was hard to know where to start. When Middleton-Jones and Wilkie finally produced a book, *Giza-Genesis: Best Kept Secrets,* I was relieved that the confusion would finally be over and I could follow the whole story logically from start to finish. I followed the story, but there was nothing logical about it.

In the book, the characters of the Old Testament have been translated into a star chart using their ages at fatherhood, as well as at death, where their ages at death are divided by 360; the result is then multiplied by the age at fatherhood to arrive at an angle. The same is done for the son, and so on, for six generations, and the points are joined together by equal-length lines. After the first six generations, there are four remaining generations until Noah, but the formula results in an angle over 360 degrees for the first of these—Enoch. Instead of continuing around the circle to get an acute angle, the authors decide to go back to the "Adam point" and draw the line from there. The logic breaks down further at this point, and eventually the reader has to take their word for it that this diagram represents the Giza Plateau and helps locate the hidden Hall of Records. Noah's Ark is then mixed up with the Ark of the Covenant to find that the biblical recipe for these is really the instructions for building the Great Pyramid, and there is a secret room accessed from the Grand Gallery, leading to a giant crystal underneath one of the Queen's Pyramids at the Khufu complex.[29] Further contortions lead to the conclusion that the Giza Plateau is a map of the sky on December 21, 2012, when everything goes interdimensional.

Somehow, Wilkie and Middleton-Jones have arrived at the idea that the Great Pyramid is misaligned from true north by almost four degrees, and even give the figure to seven decimal places. But in *The Complete Pyramids,* Egyptologist Mark Lehner reveals that the misalignment is close to a mere three minutes (or a twentieth of a degree) from true north.[30] Wilkie and Middleton-Jones's Giza alignment on December 21, 2012, is calculated using their supposedly super-accurate, but in fact totally inaccurate, number, which means that the whole theory is invalidated. Even their diagrams of the Giza pyramids show the Khufu, or Great Pyramid, as being smaller than the Khafre Pyramid. The Giza-Genesis theory is thus a prime example of "pyramidiocy" and unworthy of any further discussion. A better title would have been *Giza-Genesis: Best Kept a Secret!* For a more detailed review, see www.diagnosis2012.co.uk/new2.htm#giza-gen.

THE REBIRTH OF OSIRIS

In an e-mail debate between astromythologist Rush Allen and James Wilkie, posted on Allen's SiLoaM.net website,[31] in which Allen challenges Wilkie to explain his theory and Wilkie fails to, Allen claims that although he sees the Ambilac articles as disinformation, they are correct in their conclusion that the Hall of Records will be opened on December 21, 2012, at 10:18:13 PM local Cairo time (22:18:13 local Cairo time by twenty-four-hour clock). However, he says the opening of the Hall of Records is a kind of mass illumination of humankind. He concludes that Wilkie and Middleton-Jones must have discovered the actual time by intuitional methods, since their reasoning is impossible to follow, while Allen himself found the time by astronomy and astrology.

Rush Allen also found that "the zenith meridian at Giza runs through Jupiter in the Hyades star cluster in Taurus. The metaphor for this alignment is that the kingdom (Jupiter) will be restored to Egypt (Taurus) at that precise moment. Thus Osiris, the bull of Egypt, will return from the dead at 22:18:13 (local Cairo time) on December 21, 2012."[32]

SECRETS OF THE ALCHEMISTS

In 1926, an enigmatic alchemist known by the pseudonym of Fulcanelli published a book called *Le Mystère des Cathédrales* (The Mystery of the Cathedrals), which explains how Gothic cathedrals have hermetic and alchemical secrets encoded into their architecture and sculptures. In 1957, a second edition appeared that included an extra chapter on the significance of a stone monument in the town of Hendaye, in the Pyrénées. The monument—an engraved pedestal with a pillar and engraved cross—was built around 1680 (see color plate 16). Fulcanelli said that on the cross is encoded a warning of an imminent trial by fire for the northern hemisphere.

More recently, in Kenneth Rayner Johnson's 1980 book, *The*

Fulcanelli Phenomenon, Paul Mevryl contributed an afterword titled "The Cyclic Cross of Hendaye." Mevryl suggests several possible interpretations of the cross. These scenarios include solar flares triggered by the Grand Cross planetary conjunction of August 1999—which didn't happen—or the approach of an elliptically orbiting planet between Sirius and our Sun. The second scenario will crop up later in this book. Suffice it to say that it is somewhat incredible that Mevryl should have come up with this interpretation, considering that he makes no mention of Robert Temple's *The Sirius Mystery,* which had been in print for four years when this book came out. When I saw the drawings of the cross, I noticed that on one of the panels on the pedestal there is a quincunx with a central large, angry Sun, reminiscent of the Aztec Sunstone and its five eras. However, Fulcanelli said that only four ages of the world were represented on the cross, but these can be correlated to the four Worlds of the Hopi and the Maya.

Fig. 8.16. The four sides of the base of the Cross of Hendaye. The shieldlike design faces south, the star faces east, the Sun faces west, and the Moon faces north. From *The Mysteries of the Great Cross of Hendaye: Alchemy and the End of Time,* by Jay Weidner and Vincent Bridges, used by permission. Photos © Darlene, 2004.

In 1999, Jay Weidner and Vincent Bridges published a thorough exposition of the meaning encoded in the Great Cross of Hendaye, *A Monument to the End of Time: Alchemy, Fulcanelli, and the Great Cross.* The authors traced the alchemical thread back to Egypt, to the first gnostic groups, and reexposed the fact that behind alchemy there lies a triple transmutation: of the inner, of the outer, and of time. The inner transmutation is the refining of the psychosexual energies and fluids; the outer is using the inner change to transmute physical states; the third is the same transmutation applied to the whole Earth, changing the Age of Iron to the Age of Gold. The four ages—Gold, Silver, Bronze, and Iron—relate to the 26,000-year cycle of precession, say the authors, with the Fall occurring when the "Tree" (the celestial axis, or Earth's rotational axis) points away from the center of the galaxy. "Resurrection" or "redemption" occurs when the Tree points toward the center of the galaxy. This is the point when the "sparks of light"—our souls as fragments of God—return to the source: "the Gnostic concept of the Great Return of the lesser lights to the Light."[33]

The thread is traced from ancient Egypt, through the Coptic Church, the Gnostics, the Hebrews, Islam (the Sufis), the Order of Sion, the Knights Templar, the grail romances, the tarot, and, with increasing persecution of heretics, through to the engraving of cathedrals, and, finally, to the Rosicrucians.

The four engravings on the pedestal are found to refer to the four tarot trumps: the Star, the Moon, the Sun, and Judgment. The pillar represents the Djed, or Earth's rotational axis (Moira Timms has already pointed this out in her essay "Raising the Djed").[34] The two "X" engravings refer to the winding of the ecliptic and equinoctial points, plus trump number twenty—Judgment. The authors decode the meaning of the engraved phrase *"ocruxaves pesunica,"* which means "the secret concerns a cross and a snake which somehow measures the twelfth part."[35]

When these clues are combined (along with clues from the kabbalistic Tree of Life, the twenty-two paths that correspond to the twenty-two tarot trumps), the authors conclude that half a precessional cycle ago, the spring equinox fell on the cusp of Leo–Virgo; 12,960 years later, in 2002, the fall equinox would be in the same position. The planetary attributions of the tarot trumps, combined with the orientation of the faces of the monument, confirm that September 23, 2002, was the date in question. Trump 14, Temperance (Alchemy), corresponds to the direction of visual Galactic Center, while trump 21, the World, corresponds to the opposing galactic edge (Galactic Anticenter).

Fulcanelli suggests that the Cross of Hendaye signifies a "fatal

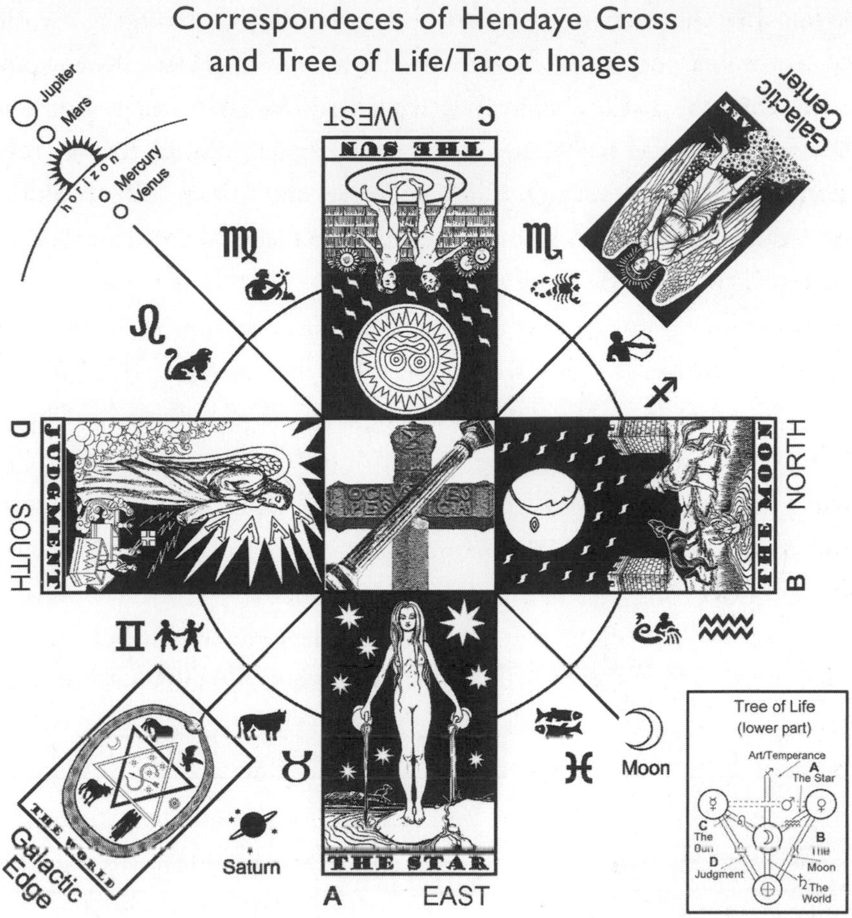

Fig. 8.17. The Star, the Moon, the Sun, and the Last Judgment represent the four faces of the Cross of Hendaye. From *The Mysteries of the Great Cross of Hendaye,* by Jay Weidner and Vincent Bridges, used by permission. Illustration © Darlene, 2004.

period" of a "double catastrophe." If the fall equinox of 2002 is the midpoint of a twenty-year period (the *XX* engraving would suggest this), this may be taken to mean the last katun of the thirteen-baktun cycle, since this is the nearest fall equinox to the katun midpoint (February 12, 2003; Long Count date: 12.19.10.0.0.).* Weidner and Bridges were

*José Argüelles gives 1992 as the start of the final katun in the last baktun of the thirteen-baktun cycle, and this date is mistakenly quoted by Weidner and Bridges.

stunned to discover that Paul LaViolette has been warning the world of just such a double catastrophe, in the form of a galactic-core explosion, which he already tentatively linked to the 2012 end point. The first effects would be "electromagnetic shifts . . . crustal torque, pole shifts, tidal waves and high winds." The second catastrophe would be "an explosion of the Sun's corona caused by the influx of cosmic dust pushed by the galactic superwave."[36]

Fulcanelli also said that the inscription *ocruxaves pesunica* revealed a place of refuge. By clues from Fulcanelli, the authors arrived at two anagrams: Inca cave, Cuzco, Peru; and Hail to the Cross, at Urcos. The authors were again amazed to find not only caves at Cuzco but also a nearby town called Urcos, with a cross. Sadly, the original cross had been destroyed, but they concluded that the place of refuge may be either the caves at Cuzco or the legendary tunnels under the Andes, if someone finds an entrance.

Fig. 8.18. The Latin inscription means "Hail, O Cross, the Only Hope," but also conceals anagrams. From *The Mysteries of the Great Cross of Hendaye*, by Jay Weidner and Vincent Bridges, used by permission. Photo © Darlene, 2004.

Weidner and Bridges suggest that we should learn to "weave [our] Bardo or transitional body into nicely fractal flows of self-awareness"[37] simply by meditating on compassion, as suggested by the work of scientist, mystic, and author Itzhak Bentov.[38] We shall thus be ready to be "harvested," to become a soul inhabiting a star, like the pharaohs of Egypt.[39] The Great Paris Magical Papyrus (Papyrus 574 in the Bibliothèque Nationale of France, a Greek magic text probably modified by an Egyptian priest and discovered after Napoleon's conquest of Egypt) gives some clues that the transformational process may be triggered by the light from the glow of the exploding Galactic Center, triggering a cerebro-chemical outpouring, which fuels an internal light that externalizes as "the shining light or star body of imperishable quality."[40]

DESCENT OF THE NEW JERUSALEM CUBE

Monument to the End of Time concludes that the Cross of Hendaye encodes information on the formation of a "cube of space" over the twenty-year period between 1992–93 and 2012. This cube is formed by the alignment of three axes: the ecliptic axis, the galactic axis, and the "planetary equinox axis." On each equinox, an Earth cube is formed: "As the equinoxes precess backward against the fixed stars, the Earth cube slowly rotates. The fixed locations of the Cube of Space form a space/time constant with which the Earth cube comes into alignment once every 13,000 or so years."[41]

The midpoint in the alignment of these cubes, when they are perfectly coordinated, was on the autumnal equinox of 2002. The twenty-year period ending on winter solstice 2012 is the "eschatological moment when the New Jerusalem descends to Earth,"[42] as described in the book of Revelation (which does describe it as a cube, but with sides of 12,000 stadia, or 1,500 miles,[43] though 12,000 is probably symbolic). The process must be aided by a ceremony every equinox, in which the kabbalistic Tree of Life is "projected onto the Celestial Sphere,"[44] a technique developed by the Hermetic Order of the Golden Dawn during the late 1800s. The method is combined with the Enochian magic system devised by Sir John Dee in Elizabethan times. Both of these have geometrical connections to these nesting cubes, described as the "Gnosis Gnomon," like the throne of Osiris. The results of a failure (of the ceremonial aiding of the cube alignment process) would be "dire": "The New Jerusalem Cube is the switch box, and, to follow the analogy, if it is wired wrong or not at all, then it explodes when the juice is turned

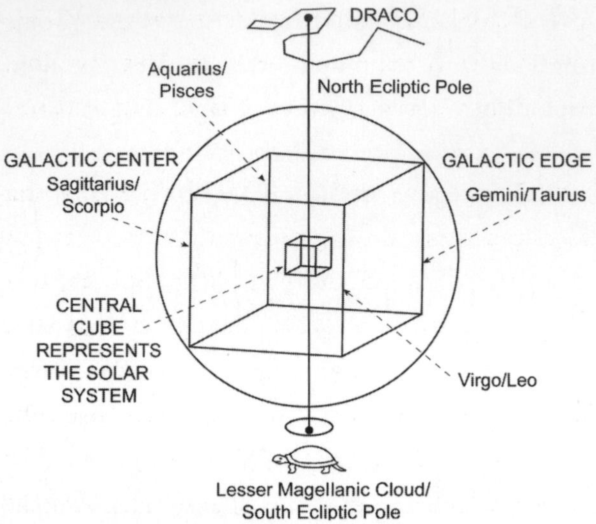

Fig. 8.19. The alignment of the Cubes of Space—the solar cube is inside the large galactic cube. From *The Mysteries of the Great Cross of Hendaye,* by Jay Weidner and Vincent Bridges, used by permission. Illustration © Darlene, 2004.

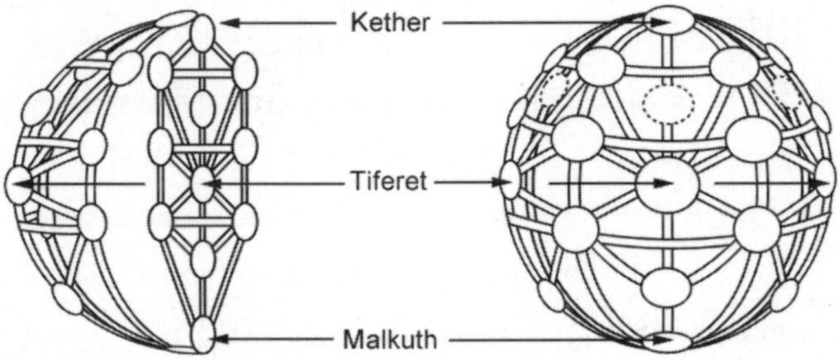

Fig. 8.20. Projecting the Tree of Life onto a sphere, with Kether corresponding to the North Pole. From *The Mysteries of the Great Cross of Hendaye,* by Jay Weidner and Vincent Bridges, used by permission. Illustration © Darlene, 2004.

on. The current is already running, and the sparks are popping all over the planet."[45]

Success would mean that "New Jerusalem then becomes the unitary model which allows the Earth cube and the solar hyper-cube to nest within the higher dimensional flow coming from galactic central. It would mediate the centropic to entropic collapse of light into matter by stabilizing the oscillation between 3 and 4-D, in other words, a virtually eternal reality constructed from mind mirroring matter with its own light. It will connect celestial & terrestrial grids allowing 4-D existence."[46] The "Enochian transformer cube" had to be resonant by 2002, when it would start crystalizing. If not, it would start to fragment over the ten years to 2012.

On September 23, 2002, in Cairo, at 5:30 AM, Nicki Scully, shamanic teacher and healer, and those on her shamanic tour of Egypt, performed the Raising the Djed ceremony in front of the Sphinx, thus effectively starting the cube crystalization process. The Diagnosis 2012 website had been an important catalyst in bringing about this event, by posting an essay written by astromythologist Rush Allen that commented on the website coverage of *A Monument to the End of Time,* which in turn inspired its authors, Weidner and Bridges, to write an article[47] that culminated in the event. My friend Ian Crane was independently on the Giza Plateau at that time and said, "The alignment of Sirius, Orion and the full Moon was absolutely spectacular, with Orion's belt aligned directly over the pyramids at dawn. I have never before experienced quite such a vivid enactment of 'As above, so below.'"[48]

SIGNS IN THE SKY

Fig. 8.21. Osiris as the Orion constellation, carved on the pyramidion (capstone) of the pyramid of Amenemhet III in the Cairo Museum.

Adrian Gilbert's book *Signs in the Sky* explores biblical prophecy and traces Christian origins to Egypt, in trying to find an answer to the question of what exactly the prophesied "signs in the sky" preceding the end of the age will be. Gilbert has been influenced by Santillana and von Dechend's epic work on mythology, *Hamlet's Mill,* which finds most mythology to be cosmologically or astronomically based. In support of Paul

White's previously quoted statement, Gilbert concludes that Giza is a "map of time," and that "this position of Orion's Belt could therefore symbolize the idea of the returning phoenix at the end of one age and the beginning of another."[49]

Building on his discoveries in *The Orion Mystery,* Gilbert has found that the Orion cult has spread into Turkestan, with the Sabian "star-worshippers" and the Sarmoung Brotherhood, of which the spiritual teacher Gurdjieff was a member. Many biblical stories contain coded astronomical and astrological messages revealing connections to Egyptian beliefs, which came out of Egypt with Moses, says Gilbert.

We know about the alignment of the winter solstice Sun with the galactic equator, between Sagittarius and Scorpio, which John Major Jenkins has identified as the 2012 end point of the thirteen-baktun cycle of the Maya. This is caused by the approximate 26,000-year precession of the equinoxes, except that the Maya were measuring "precession of the solstices," which is the same cycle but measured from a different point—the precession of the winter solstice, as opposed to the vernal (spring) equinox. What Gilbert seems to have discovered is that the Egyptians knew about the precession of the summer solstice, and the subsequent encoding of this knowledge in the Bible. The vernal equinox, as we know, precesses from Pisces to Aquarius at the same time that the winter solstice precesses from Sagittarius to Scorpio, but at the same time the summer solstice precesses from Gemini to Taurus.

To recap part of chapter 3: The ancient Greeks, who received much of their knowledge from Egypt, believed that souls reside in the Milky Way between incarnations, and that there are two "gates" in the Milky Way. These are the Silver Gate of Gemini, through which souls descend to Earth; and the Golden Gate of Sagittarius, through which souls ascend. Other versions say the souls of humans can ascend by either gate, but that the Silver Gate leads to reincarnation and the ancestors while the Golden Gate leads beyond reincarnation. The Golden Gate is also that through which the gods descend.

The Silver Gate is just above the hand of Orion, whom the Egyptians associated with Osiris, and they depicted him holding a star in his hand. In *The Orion Mystery,* Bauval and Gilbert assumed this was the star Aldebaran. Now Gilbert has found that the Egyptian word for star, *s'ba,* also means "door"; so Osiris is holding a star gate. In some depictions, Osiris is holding an ankh toward the gate, so this must be the key that unlocks the star gate.

Fig. 8.22. Osiris holding the ankh, or key to the star, or door. After Robin Cook, in *The Orion Mystery: Unlocking the Secrets of the Pyramids,* by Robert Bauval and Adrian Gilbert.

After studying Bible prophecies, visiting Israel, and computing star and planet positions on SkyGlobe software, Gilbert came to the conclusion that the astronomy of summer solstice 2000 (June 21–22, 2000, and also on June 29, 2000) represents the beginning of the apocalypse, when the Silver Star Gate opens. This is the "Omega position" of Orion, rather than 2380 AD, when "the Belt of Orion will reach its most northerly position."[50] It is the end of a 13,000-year cycle that Gilbert calls the "Age of Adam," or the "Age of Orion."*

In August 2012, says Gilbert, Venus (symbolizing Isis and Mary as the "Bride of Heaven") will be "stationary in the hand of Orion." He is talking here about the infamous retrograde loop of Venus mentioned by Appleby. Orion, as he has explained convincingly in his book, symbolizes Christ, and this is his mystic marriage (there is a hint that we should be prepared with our "wedding garment," which to the Egyptians was a "stellar, Sahu body" that needed to be "crystallized").[51] This could also be the return of Quetzalcoatl, since he was connected with Venus, and thereby the sign preceding a "second coming."

*In *The Orion Mystery,* by Gilbert and Bauval, the First Time of Osiris was called Zep Tepi, and Bauval and Hancock both call it Zep Tepi in later works, as do several other authors. However, Gilbert mistakenly reverses the phrase to "Tep Zepi" throughout this book.

Gilbert's interpretation of biblical prophecy, using the fig tree parable (see chapter 5) and the beginning of the modern state of Israel in 1948, indicates that all prophecies would be accomplished by 2018. This also happens to be, according to Jenkins, the year that the winter solstice Sun will definitely clear the galactic equator, having spent thirty-six years crossing it in the other star gate.[52] But as we have already seen in chapter 5, if we take the UN resolution that created the state of Israel in 1947, the fig tree parable indicates that 2016 is to be the accomplishment of the prophecies. This is the last year of the galactic alignment, if 1998 is taken as the exact halfway point in the process. Gilbert calls the Silver Gate the "Gate of Hades," which suggests that there may be a few years of hell coming up before the opening of heaven's gate.

There are some remarkable parallels here to the work of Weidner and Bridges, with a twelve-year countdown from summer solstice 2000 to 2012 as opposed to a ten-year countdown from the fall equinox 2002 to 2012. Both approaches involve a ceremony at the Great Pyramid; both ceremonies connect to the start of the descent of the New Jerusalem[53] of the book of Revelation; and both mention the necessity of preparing our bardo body (as in the Tibetan system) or our sahu body (as in the Egyptian system) for 2012. One thing that Gilbert missed is that just above the hand of Orion is situated the galactic anticenter, as we saw at the end of chapter 7.

THE ORION PROPHECY

In November 2001, Belgian authors Patrick Geryl and Gino Ratinckx published *The Orion Prophecy,* a book that follows in the same vein as Gilbert and Cotterell's *The Mayan Prophecies,* and is also influenced by Gilbert and Bauval's *The Orion Mystery.* As in *The Mayan Prophecies,* Geryl and Ratinckx say that a sunspot megacycle is due in 2012 and will cause a global catastrophe. They say they have found evidence in the Egyptian Book of the Dead that in 9792 BC a very similar catastrophe occurred (resulting in the sinking of Atlantis), and that an astro-

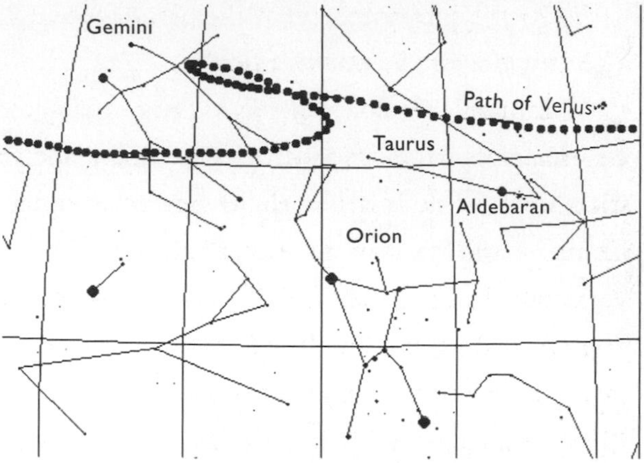

Fig. 8.23. The path of Venus during its retrograde loop in 9792 BC, according to Geryl and Rattinckx, is actually the loop that occurred in 9696 BC, according to SkyGlobe. Shown on CyberSky software—http://cybersky.com.

nomical sign of the imminent return of the catastrophe is a retrograde loop of Venus that happens above Orion and near Gemini. However, I checked and rechecked the authors' claims using astronomical software, and they just don't add up.

There are twelve retrograde loops of Venus that occur over Orion between 1924 and 2012 (see color plate 13), and of these, the best match to the 9792 BC position shown by Geryl and Ratinckx (in which Venus slows and changes its apparent reverse direction to resume forward motion above the hand of Orion) happened in 1940. You can check them on the shareware program SkyGlobe. I wrote a detailed review

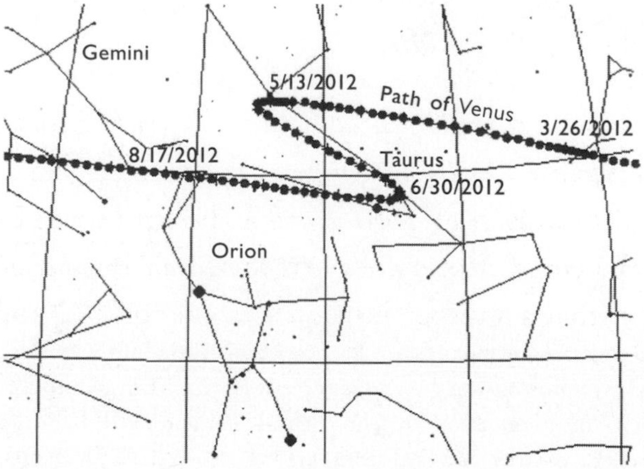

Fig. 8.24. The path of Venus during its retrograde loop in 2012 AD is actually a perfect match for a loop that occurred in 9389 BC, according to SkyGlobe. Shown on CyberSky software—http://cybersky.com.

of the book and posted it online at Diagnosis 2012,[54] but the authors failed to address the points raised.

Not many astrology software programs go back as far as 9792 BC, but SkyGlobe does the job nicely. I had not tried SkyGlobe when I wrote the review, but now that I have checked it out, I have found that the shape and position of the 2012 loop agrees with the one shown by Geryl and Ratinckx, but the 9792 BC loop differs considerably. The loop they show for 9792 BC looks most like the one that SkyGlobe shows for 9696 BC, ninety-six years later. This further enlarges a long list of problems with their theory. In fact, the Venus retrograde loop of 9792 BC, as shown on SkyGlobe, has an exact match in the year 1609 AD. After much searching with the software, I found that the exact match for the 2012 loop actually happened in 9389 BC. (The loop starts on February 12, 9389 BC.)

We are told by the authors that if a star map featuring Orion is laid onto a map of the Pyramids of Giza, with the belt stars aligned with the three pyramids, then other stars align with other pyramids.* In particular, they suggest that Aldebaran, in the star group called the Hyades (in

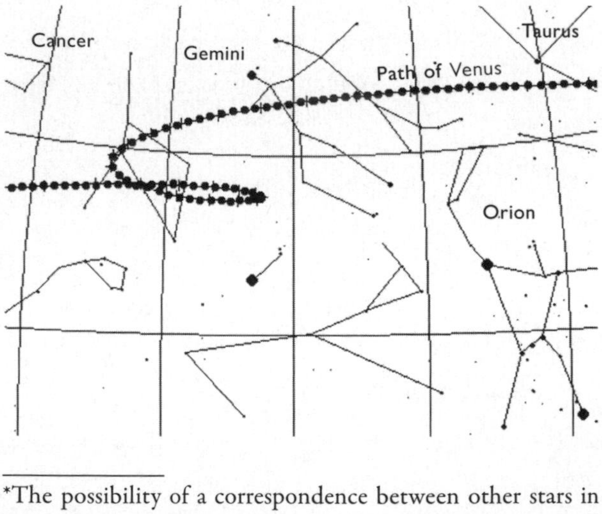

Fig. 8.25. The actual Venus loop pattern in 9792 BC according to SkyGlobe, which has an exact match in 1609 AD. Shown on CyberSky software—http://cybersky.com.

*The possibility of a correspondence between other stars in Orion (in addition to just those in Orion's Belt) to the pyramids on the ground is something that Bauval and Gilbert suggestsed in *The Orion Mystery* (pp. 296–300), but Bauval later admitted that only the correlation of the pyramids to Orion's Belt is convincing.

the Taurus constellation), would overlay the pyramid of Amenemhet III at Hawara. Since astronomers call the Hyades "the Labyrinth," this could indicate the location of the long-lost massive astronomical building also called the Labyrinth. They were convinced that it urgently needed to be found, since it contained knowledge of the events that led to the Atlantean flood.

This was quite easy to check up on using the map in Mark Lehner's book *The Complete Pyramids,* and I expected a slight error. However, when I aligned Orion's Belt to the Pyramids of Giza, I found that while Hawara lies about fifteen miles in a south-southeasterly direction from Giza, Aldebaran overlaid an area just over a mile southwest of Giza. This is a massive error. However, although *The Orion Prophecy* is seriously flawed, it did bring up some interesting points.

At the start of the book, Geryl was about to look for an answer in the zodiacs of the Temple of Dendera in Egypt. The rectangular zodiac particularly interested him, since he thought it contained coded information about the catastrophe that ended Atlantis, and how to calculate when the cycle would recur. It was at this point that Ratinckx showed him a rare copy of Frenchman Albert Slosman's translation of the

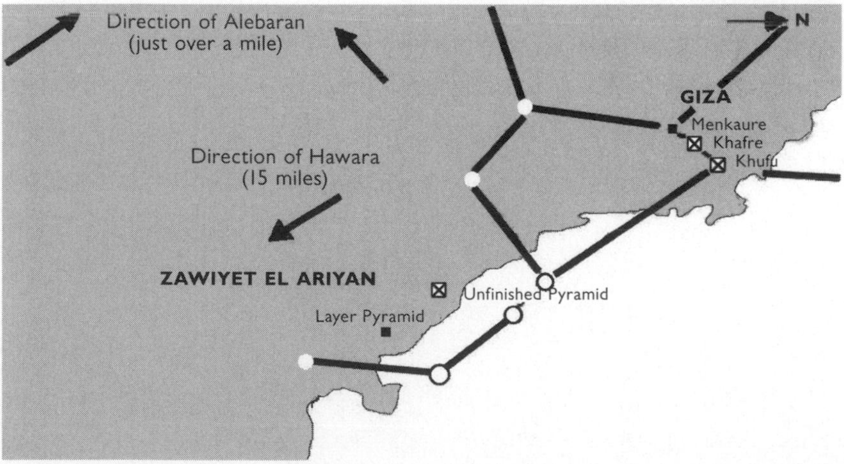

Fig. 8.26. When the celestial belt of Orion is lined up with the terrestrial one at Giza, Egypt, the celestial labyrinth is nowhere near the terrestrial one. Adapted from a map in *The Complete Pyramids,* by Mark Lehner.

Egyptian Book of the Dead (*Le Livre de l'Au-dela de la Vie*). Geryl was so amazed by what he read that he forgot all about the Dendera zodiac. This is a shame, since David Pratt has published information on the zodiac (see his Poleshifts: Theosophy and Science Contrasted website)[55] that finds evidence of three previous world catastrophes encoded (just as Slosman had discovered by decoding the Egyptian Book of the Dead). As we have seen, John Lash also found 2012 encoded in the round zodiac of Dendera. Pratt also quotes S. A. Mackey (a shoemaker and astro-mythologist) as saying that the design of the Labyrinth showed three pole shifts.

9
THE GEOMAGNETIC FIELD

MAGNETIC FIELD REVERSAL

Geologist Gregg Braden has pointed out that Earth's magnetic field has dropped 38 percent in the past two thousand years (equivalent to an average of 1.9 percent per century), and that the rate of decline has risen to an average of 6 percent over the last hundred years.[1] In other words, the global magnetic field is dropping at an accelerating rate toward a zero point, at which time the polarity of Earth will reverse.[2] Geological surveys of the Mid-Atlantic Ridge show that this has happened many times before in the history of Earth. Braden (at the time that he wrote *Awakening to Zero Point: The Collective Initiation,* which was first published in 1993) says

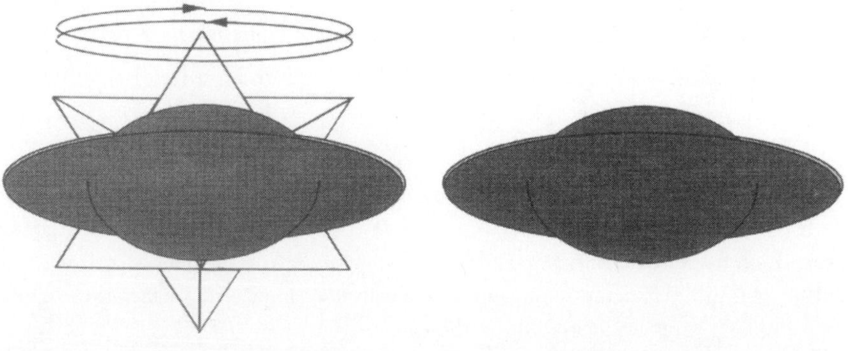

Fig. 9.1. The Mer-Ka-Ba (or Merkabah), with counterrotating tetrahedrons, appears as a UFO according to Gregg Braden. From *Awakening to Zero Point: The Collective Initiation,* by Gregg Braden.

that the declining magnetic field is causing the Schumann resonance in the ionosphere to increase, and that it is currently resonating at between 8.6 and 9 hertz and will level out at 13 hertz—the next number in the Fibonacci series.* The whole process is connected to the Maya calendar, says Braden, the largest increase having been recorded in 1987, which is when the Harmonic Convergence occurred. He says the last polar reversal was 13,200 years ago—close to half a precession cycle.†

To adapt, says Braden, we must allow our bodies to alter their resonant frequency to match that of Earth. The process of adaptation is blocked by negative emotion, so we must become compassionate and generate unconditional love, he says. Shortly before Earth shift, or zero point, some people may achieve resonance in advance of the mass of humanity, and their Mer-Ka-Ba fields—counterposed tetrahedronal fields—will counterrotate and merge to form a saucer-shaped light body, carrying those people into the fourth dimension.[3] This is the Rapture, according to Braden. He also claims[4] that many sacred sites were designed for initiation, including the Great Pyramid,‡ and are places with a low magnetic field, allowing an easier alteration of frequency, so we should visit them.[5]

Some information originating in a completely different field may help illuminate the low magnetic field–spiritual body connection. Bob Monroe, a pioneer of techniques to induce out-of-body experiences (OBEs), states that when an OBE is occurring, the body's polarity, or electromagnetic field, reverses itself. This polarity reversal has been witnessed by consciousness researcher Michael Hutchison[6] while

*The situation regarding the Schumann resonance is harder to answer. Scientists in the United States and Western Europe are mostly saying it hasn't changed (Braden got his information from Russia and Norway), yet the climatology department of Arizona State University is using changing Schumann resonance values as an accurate measure of global temperatures, so the jury is still out on this one.

†There is disagreement as to whether this "Gothenburg flip" was just a local field change or a full reversal; see note 10 in chapter 8.

‡See note 5 on page 434 for more on the geomagnetic interactions of the Great Pyramid.

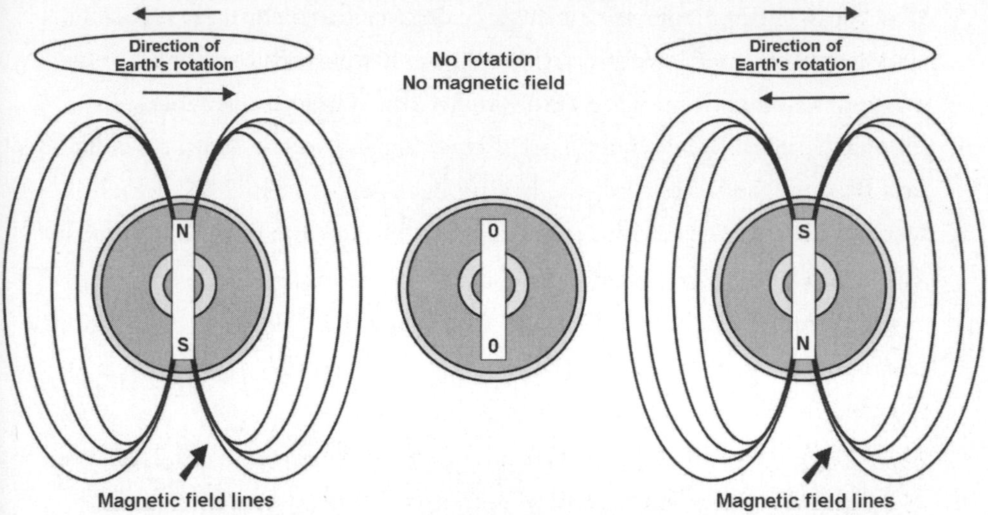

Fig. 9.2. According to Braden, Earth's rotation slows until it stops, when the geomagnetic field drops to zero. Then the rotation starts in the opposite direction, with a reversed geomagnetic field. Adapted from *Awakening to Zero Point: The Collective Initiation,* by Gregg Braden.

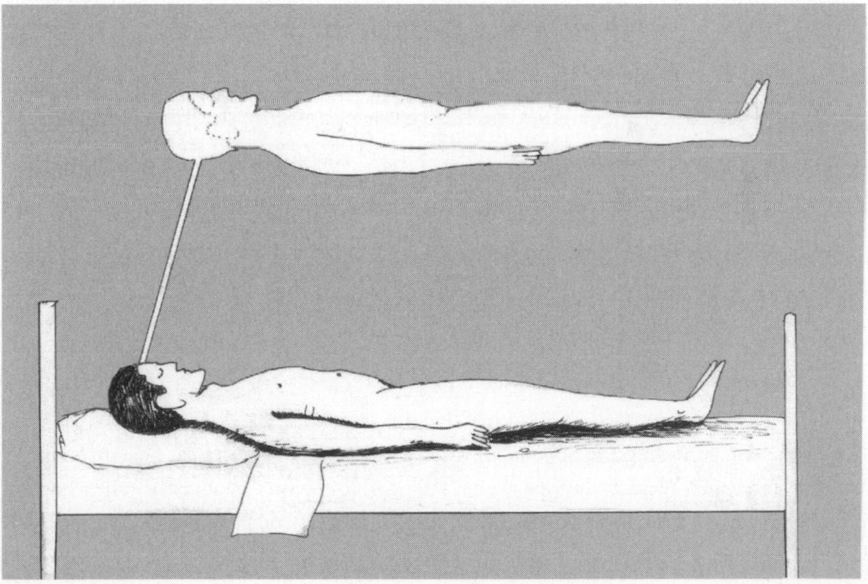

Fig. 9.3. In some accounts, the astral body is connected by a cord from its head to the physical body's head. Adapted from W. G. Watts in *The Projection of the Astral Body,* by Sylvan Muldoon and Hereward Carrington.

observing experiments at the Monroe Institute in Virginia. However, this begs the question: If Earth's magnetic field reverses, would this trigger the mass of humanity to have an out-of-body-type experience?

ONE-BAKTUN GEODYNAMO

In 1996 it was discovered that Earth's core revolves faster than its crust. According to the "Outline of Historical Geology," published online by biogeologist Ellin Beltz:

> The speed of the core's independent spin is between 0.4 and 1.8 degrees a year, which works out to a full lap, relative to a fixed point on the surface, about every 400 years. The leading suggestion for the cause of the independent motion is an interaction between magnetic fields generated by fluids moving in the Earth's outer core, which is molten iron, and the inner, [which is] solid crystal. The implications of differential rotation between the core and the surface are hot topics at professional meetings and in various physical and geophysical journals. The iron crystal in the core is longer from north to south, but it does not exactly match Earth's poles. As a result of the core's orientation and rotation, "magnetic north" moves over time. The physical pole of the planet also wobbles over time, a process called "precession of the equinoxes."[7]

The length of a baktun (of which thirteen make up the cycle that terminates in 2012) is 144,000 days, which is 400 tuns (or 360-day "years")—about 394 tropical years. Could there be a connection to the four-hundred-year core rotation period? Several sources, including Gregg Braden, have suggested that the end point of the Maya may be connected with an imminent magnetic pole shift; Ellin Beltz says that the four-hundred-year differential rotation of the core causes the magnetic poles to wander (polar wander is now known to be accelerating). Other sources have connected it to geomagnetic reversal, and some scientists have started saying we are due for geomagnetic reversal soon.

IMMINENT REVERSAL

On April 17, 2002, Telegraph.co.uk (website of Telegraph Group Ltd., which publishes the London *Daily Telegraph* newspaper) reported that the previous week a paper had been published in the scientific journal *Nature* concerning new discoveries about Earth's magnetic field:

> Dr Gauthier Hulot of the Institut de Physique du Globe de Paris and colleagues at the Danish Space Research Institute in Copenhagen used magnetic field measurements taken by the Danish Oersted satellites and compared them with similar readings made in 1980 by the American Magsat satellite . . . The results, published in *Nature,* confirmed that the Earth's magnetic field is getting weaker. If it continues to weaken at its current rate, the dipole field will have vanished in 2,000 years. But Dr Hulot and his team also found a large area of "reversed magnetic flux"—where the magnetic field runs counter to the rest of the world's field—below South Africa and the Southern Ocean. Normally, lines of field move from the south to the north. But in this "reversed flux" area, the magnetic field lines loop backwards and head south. Under this area, the columns of moving liquid iron in the core may be rotating a little differently than they are in the rest of the core, locally weakening and reversing the magnetic field.[8]

The same article pointed out that computer simulations (performed by Dr. Gary Glatzmaier of the University of California, Santa Cruz, and Dr. Paul Roberts of UCLA) of the magnetic field performed in the mid-1990s had already predicted not only that a magnetic field reversal would be preceded by a decrease in magnetic field intensity, but also that areas of reversed magnetic flux would appear prior to reversal.

In November 2002, the online version of the newspaper the *Observer,* reporting on Dr. Hulot's results, added that

> the magnetic field seems to be disappearing most alarmingly near the poles, a clear sign that a flip may soon take place . . . Using satellite measurements of field variations over the past twenty years, Hulot plotted the currents of molten iron that generate Earth's magnetism deep underground, and spotted huge whorls near the poles . . . Hulot believes these vortices rotate in a direction that reinforces a reverse magnetic field, and as they grow and proliferate, these eddies will weaken the dominant field: the first steps toward a new polarity, he says.[9]

The article also pointed out that although ancient lava beds show that some reversals lasted thousands of years, some researchers say "some flips may have lasted only a few weeks."[10]

The Canadian government funds a research group called Geolab, which tracks the movement of Earth's north magnetic pole. Previous measurements showed that between 1831 and 1904 it did not move much, but during the seventy years after that, it started a slow movement northward. Then it started to accelerate, so that in March 2003 the BBC reported that Dr. Larry Newitt of Geolab said it is currently moving four times faster than it was during the seventy-year wander. In the same report, David Kerridge, of the British Geological Survey, when asked about the geomagnetic field, is quoted as saying that "there is strong evidence that the field is decreasing by about 5% per century."[11]

So here we have some strong confirmation that Gregg Braden's prediction of a geomagnetic reversal is quite likely to happen; the articles cited above confirm many of his points, including the fact that Earth's magnetic field has been declining for the past two thousand years.

A 100,000-YEAR CYCLE

We have already mentioned ice cores in chapter 7, in the discussion of Dr. Paul LaViolette's galactic superwave. The first ice cores were taken from Camp Century, Greenland, and Byrd Station, Antarctica, in the early 1960s. LaViolette undertook the first ice-core study to look for cosmic dust levels in 1981–82, using the Greenland sample. The results were supported by a later beryllium-10 study of the Greenland cores, and of the later Vostok (Antarctica) cores. LaViolette interpreted the resulting graphs as indicating a 26,000-year period between cosmic ray

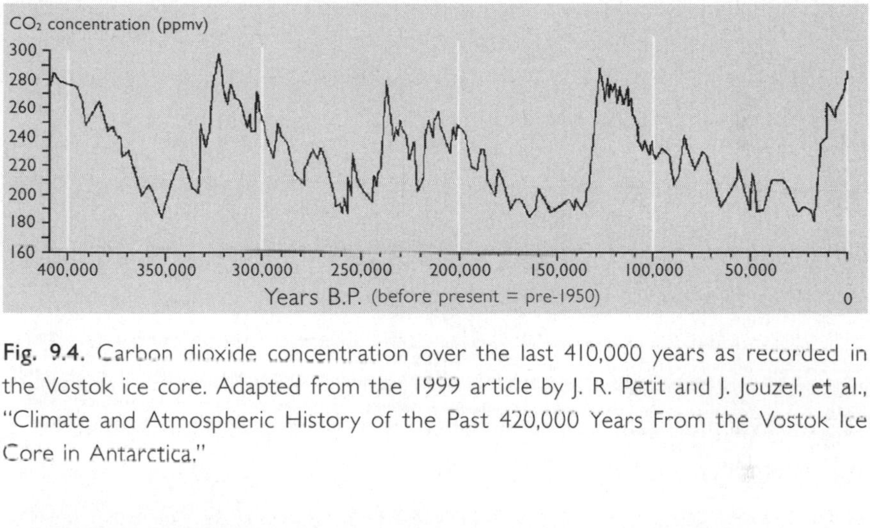

Fig. 9.4. Carbon dioxide concentration over the last 410,000 years as recorded in the Vostok ice core. Adapted from the 1999 article by J. R. Petit and J. Jouzel, et al., "Climate and Atmospheric History of the Past 420,000 Years From the Vostok Ice Core in Antarctica."

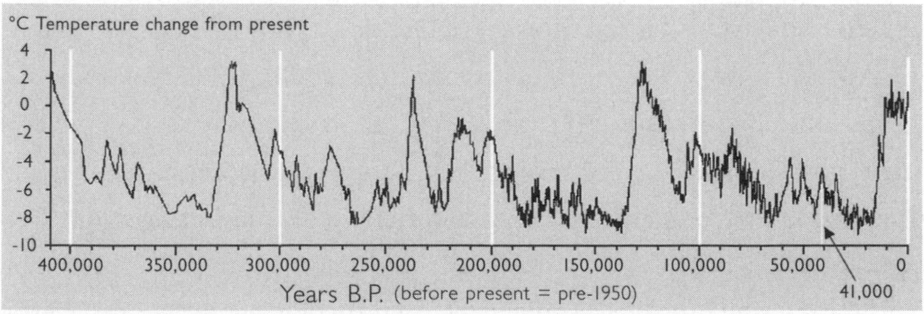

Fig. 9.5. Temperature change over the last 410,000 years as recorded in the Vostok ice core. Adapted from the 1999 article by J. R. Petit and J. Jouzel, et al., "Climate and Atmospheric History of the Past 420,000 Years From the Vostok Ice Core in Antarctica."

and cosmic dust peaks, with an occasional "13,000 year half-cycle recurrence interval."[12]

In 1999, a paper on the Vostok ice cores was published in *Nature*[13] showing graphs of temperature levels, methane levels, carbon dioxide levels, marine oxygen isotope levels, and solar output levels. They all show an approximate 100,000-year cycle with a high degree of correlation. Thus each of these 100,000-year cycles would equal about four 26,000-year precession cycles.

In the March 29, 2002, journal *Science,* the National Institute of Advanced Industrial Science and Technology (AIST) announced a discovery of a 100,000-year cycle in Earth's magnetic field:

> Using a high-sensitivity magnetometer to measure remnant magnetization of marine sediments, we obtained a continuous record of variations in strength and inclination of magnetic field that have occurred in the past 2.3 million years. Frequency analysis of this record has led to the discovery of a long-term periodic component of 100,000 years . . . The 100,000-year period of variation corresponds with variation in the eccentricity of Earth's orbit and the period of the glacial/interglacial cycles. Therefore, magnetic field changes must be understood within the context of variations in the Earth system.[14]

Then, two months later, on June 6, 2002, in a paper called "Variations in Solar Magnetic Activity during the Last 200,000 Years: Is There a Sun-Climate Connection?"[15] by Mukul Sharma, assistant professor of earth sciences at Dartmouth College, it was announced that he had just discovered a 100,000-year solar activity cycle. He apparently did a combined study of the ice-core results and the marine-sediment results mentioned above, and concluded that the solar magnetic field had a 100,000-year cycle linked to Earth's cooling and warming cycles—the ice ages.

This has led to renewed discussion of the controversial

Milankovitch theory, discussed in greater detail in the following chapter, which says that "the cyclical variations in the Earth's orbit around the Sun result in the Earth receiving varying amounts of solar radiation that, in turn, control the climate. This explanation is under dispute because the variations of the solar energy in relation to the changes in orbit are very small."[16] The fact that it was already known that Earth has a 100,000-year cycle of eccentricity (deviation from a circular orbit) means that these findings are probably more than just coincidence.

10
ICE CYCLES

THE CELESTIAL ICE CLOCK

A book by Dr. William A. Gaspar, *The Celestial Clock,* claims that the 5,125-year thirteen-baktun cycle is exactly one-eighth of a 41,000-year cycle of conjunctions of Neptune and Pluto—a period that is also the sum of a 23,000-year "truncated precession cycle" and an 18,000-year "solar magnetic reversal" cycle. Gaspar is a medical doctor, practicing in New Mexico, whose interest in Native American spirituality led him to research into long-term climate cycles.

The 23,000-year period is a "dominant ice volume collapse cycle," and every 11,500 and 23,000 years there is a major polar axis shift, causing severe earthquakes and flooding, according to Gaspar. Earth goes through warming and cooling cycles, which are a result of increased heat and carbon dioxide but independent from humankind. The "black hole" at the center of the galaxy exerts a cyclic gravitational pull every 23,000 years, when "the North Pole Axis aligns with the center of the Milky Way galaxy,"[1] resulting in friction between the magma and Earth's crust, which in turn produces heat and melts the polar ice caps. This can also happen at the half period of 11,500 years, when the axis "may be shifted back into position at every 11,500 years at its 'half cycle' alignment."[2]

Gaspar has taken on a very difficult job here, trying to unify the complex cycles of the Milankovitch theory with the 18,000-year solar reversal cycle of the Sun's warped neutral sheet as calculated by Maurice Cotterell; astrological cycles; Richard Noone's ice theory (in his book, *5/5/2000: Ice, The Ultimate Disaster*), which predicted an ice age following the May 5, 2000, "planetary alignment"; and the catastrophe theory of Derek S. Allan and J. Bernard Delair. He also brought in John Major Jenkins's essential interpretation of Maya myth and calendrics in the theory of galactic alignment.

In Allan and Delair's book *When the Earth Nearly Died,* a scholarly attempt was made to date Noah's flood and find a plausible cause. The massive amount of geological evidence that they supplied leads to the conclusion that the flood occurred 11,500 years ago.* Gaspar's theory is quite hard to follow, but he certainly raises some noteworthy points. It is remarkable, though, that he doesn't mention directly that the 41,000-year cycle is also the period known as the variation of the obliquity of the ecliptic, or the variation of axis tilt cycle.

*This differs by about 1,200 years from the estimate of astrophysicist Dr. Paul LaViolette of 12,700 years, in his galactic superwave theory as presented in *Earth Under Fire,* and by 1,700 years from geologist Gregg Braden's magnetic reversal theory in *Awakening to Zero Point*. However, Allan and Delair concluded that the flood was caused by the close flyby of a large planetoid. Unlike Sitchin, however, they don't suppose it to be on its way around an elliptical orbit, but instead consider it to have continued onward to collide with the Sun. *Uriel's Machine,* by Christopher Knight and Robert Lomas, concludes that Earth was hit 9,640 years ago (7640 BC) by seven comet fragments, causing massive tidal waves, and that this was the date of Noah's flood. Andrew Collins has criticized Knight and Lomas (claiming that they failed to reference him as one of their sources), but Collins's own conclusion in his book *Gateway to Atlantis* is that the comet impact occurred 10,600–10,500 years ago (8600–8500 BC). Hapgood's model, according to Graham Hancock, puts the flood (caused by crustal displacement) at 14,500 BC to 12,500 BC, whereas Hancock himself goes for 10,500 BC (though since ice cores have now shown Antarctica to have been fully glaciated for much longer, Hancock no longer suggests Antarctica was formerly Atlantis, and in *The Mars Mystery* he suggests that crustal displacement/the Flood was triggered by comet or asteroid impact.

MILANKOVITCH CYCLE

There are several websites that give a clear summary of the Milankovitch theory, first postulated in the 1940s, which is at the heart of Gaspar's book, and there you will find descriptions of a minimum of three long-term cycles that affect the climate. There is a good but simple explanation called the Milankovitch cycles and glaciation, which depicts a 100,000-year eccentricity cycle, a 41,000-year axis tilt cycle, and a 23,000-year precession cycle (precession is usually said to be 25,000–26,000 years—we shall return to this point). Since most of Earth's landmasses are in the northern hemisphere, the combined effects of the three varying cycles can trigger glaciation: "At times when northern hemisphere summers are coolest (farthest from the Sun due to precession and greatest orbital eccentricity) and winters are warmest (minimum tilt), snow can accumulate on and cover broad areas of northern America and Europe."[3] However, "at

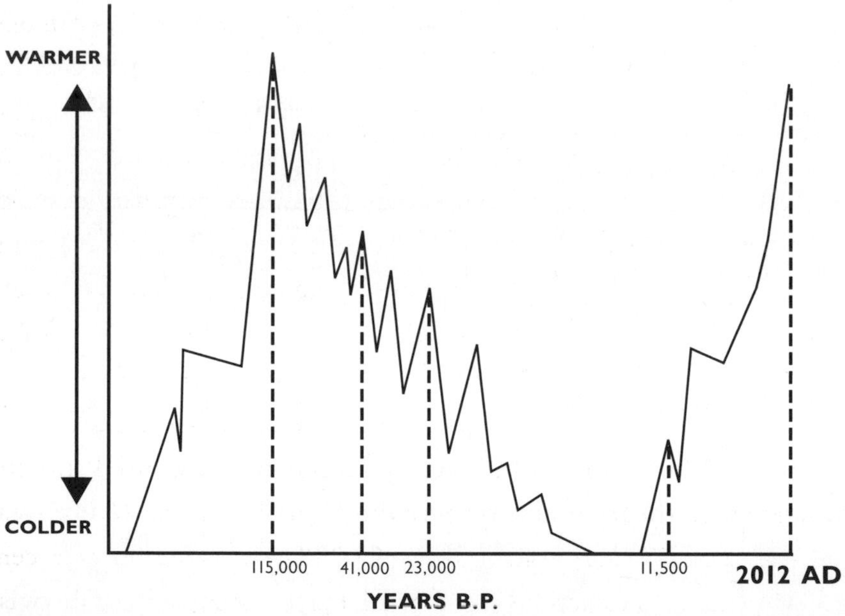

Fig. 10.1. Gaspar's take on the Milankovitch cycle was summed up in this distorted graph. After Joan Hultman in *The Celestial Clock,* by Dr. William A. Gaspar.

present, only precession is in the glacial mode, with tilt and eccentricity not favorable to glaciation."[4]

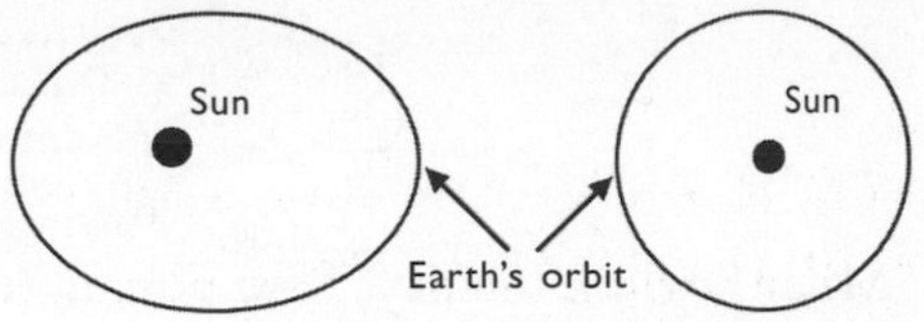

Fig. 10.2. Earth's 96,000–100,000-year cycle of eccentricity

On the U.S. National Oceanic and Atmospheric Administration (NOAA) palaeoclimatology website[5] there is a similar explanation, except that precession is given as 22,000 years. A more thorough explanation is given on the U.S. Naval Observatory site,[6] which mentions the eccentricity, axis-tilt, and precession cycles, plus a one-year cycle and a 21,000-year cycle. The basic climate cycle is, of course, the yearly cycle of the seasons. This is caused by the 23.4 to 23.5-degree tilt of Earth's axis, and on one day each year, the northern hemisphere is at its coldest, when tilted away from the Sun. This is winter solstice. Earth's orbit varies in eccentricity every 21,000 years, and it just so happens that perihelion (closest orbital point to the Sun) is currently very close to winter solstice. "It turns out that the proximity of the two dates is a coincidence of the particular century we live in."[7] (The 100,000-year cycle modifies the 21,000-year one.)

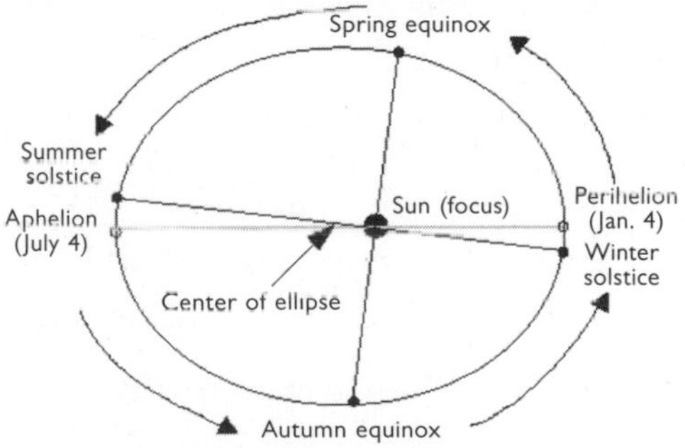

Fig. 10.3. The Earth's variation of eccentricity cycle. The elliptical orbit rotates in a counterclockwise direction once every 21,000 years.

The 22,000- or 23,000-year precession cycle used in the Milankovitch theory results from the combination of the 26,000-year precession of the equinoxes and the 21,000-year variation of eccentricity cycle. In precession, the axis of Earth rotates in a clockwise direction (if viewed from above the North Pole), and in the variation of eccentricity cycle, Earth's elliptical orbital path around the Sun

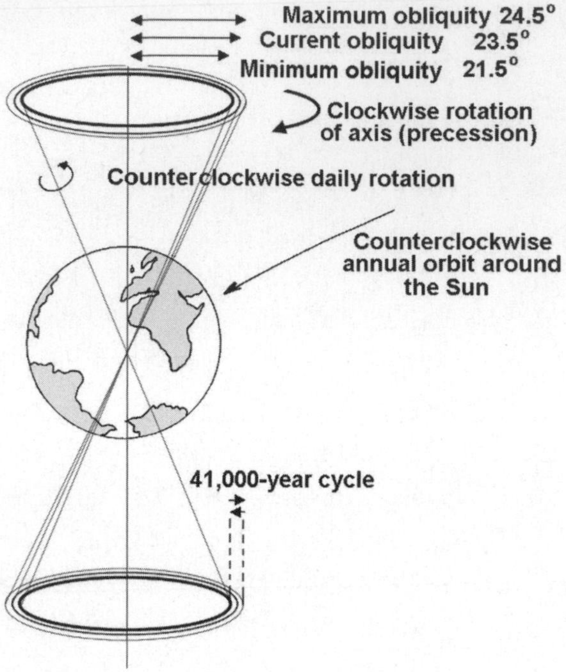

Fig. 10.4. Earth's 41,000-year axis-tilt cycle (or variation of obliquity)

rotates in a counterclockwise direction. Thus the combined clockwise and counterclockwise cycles mean that the proximity of Earth's North Pole to the Sun goes through one cycle every 22,000 to 23,000 years.[8] Gaspar speculates that the precession cycle is shorter than the usual 25,000- to 26,000-year figure due to sudden axis shifts, related to the changing magnetic field of the Sun. However, now that we understand that the 22,000- to 23,000-year cycle is not a precession cycle but the combined result of the precession and variation of eccentricity cycles, we can see that his explanation is unnecessary. The confusion would be less if the 22,000- to 23,000-year cycle had its own name, such as "the Sun–North Pole proximity cycle."

Gaspar's graph of global warming shows a peak 115,000 years ago (see fig. 10.1), which is made up of 5 × 23,000-year cycles. The 41,000-year variation of obliquity (tilt) cycle is also clearly visible, and is com-

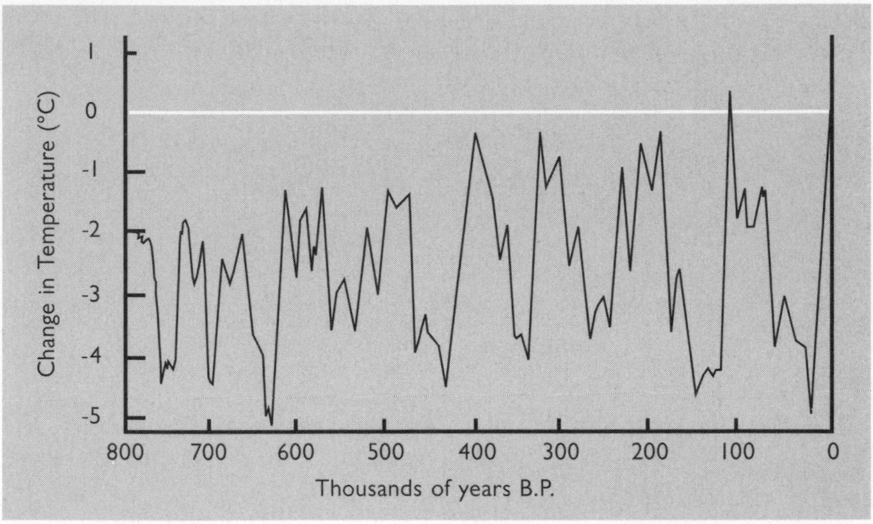

Fig. 10.5. Data from Greenland and Antarctica ice cores, ocean, lake, and bog sediments data, and tree-ring data are combined in this graph. The last section is the basis for Gaspar's distorted rendition (see fig. 10.1). After the University of Washington Department of Atmospheric Sciences (UWDAS).

posed of a 23,000-year cycle plus an 18,000-year cycle (however, this 18,000-year cycle is the one proposed by Maurice Cotterell in *The Mayan Prophecies*, and we have seen that this theory is flawed). At the peak of warming, the ice caps melt, causing flooding and constant rain, which washes the carbon dioxide out of the atmosphere into the sea, leading to sudden cooling and glaciation. We are currently at a warming peak of the same magnitude as that of 115,000 years ago, preceding the onset of the last ice age, and the 41,000-year cycle, which is clearly visible in ice-core studies, is exactly divisible by the 5,125-year thirteen-baktun cycle that ends in 2012.

Gaspar's key graph connecting 2012 to the ice theory (see fig. 10.1) differs from graphs depicting the results of the Vostok (Antarctica) ice-core studies (see chapter 9, figs. 9.4 and 9.5), which he also includes in his book. His graph is also distorted, with the dated peaks at disproportionate distances to one another. The source from which Gaspar generated his key graph is shown in *The Celestial Clock* on page 36 (see fig. 10.5). After a search, I found the origin of the graph on the website of the University of Washington's Department of Atmospheric Sciences.[9] The site explains that the graph was generated using a combination of palaeoclimatic data from Greenland and Vostok ice cores, plus ocean sediment cores, cross-referenced with insects and pollen found in lake sediments and bogs, and tree-ring data. It shows the 100,000-year warming cycle even clearer than the Vostok ice-core graphs.

Gaspar is not alone in his conclusion of a connection between Milankovitch cycles and 2012. I was amazed to find online a paper by E. C. Njau, chair of the physics department at the University of Dar es Salaam, in Tanzania, that studies global warming and Milankovitch cycles, and predicts an imminent climate change between 1997 and 2012.[10]

11

O.O.O.O.O

FLOOD AT 3113 BC

Egyptologist David Rohl has ruffled a few feathers in the Egyptology and archaeology communities with his revised chronology of Egyptian, Old Testament, and Mesopotamian timelines. Rohl is currently completing a doctoral thesis at University College, London, and is Britain's most high-profile Egyptologist, following his 1995 three-part TV documentary, *Pharaohs and Kings: A Biblical Quest,* and his two books, *A Test of Time* and *Legend.*

In an article called "Mountain of the Ark,"[1] published in the London *Daily Express* in 1999, Rohl reexamines the evidence for the biblical Flood. The story of the Flood originated in a Sumerian legend called the *Epic of Gilgamesh,* in which the original Noah was called Utnapishtim. Rohl shows that the eleven-foot-deep layer of silt found between 1928 and 1934 by Sir Leonard Woolley at Ur (where Abraham originated, probably bringing the legend with him) can be dated to around 3100 BC. Rohl also reasons that later conclusions about the discovery were wrong, since other, later-dated silt deposits were found to be the result of local flooding. Then he supplies some evidence that the biblical dating was also faulty, and that when rectified it supports his new archaeological date for the Flood:

> The Greek and Aramaic versions of the Old Testament (both of which are older than the earliest surviving Hebrew copy of the Bible)

> suggest that the Flood took place up to eight hundred years prior to the date calculable from the Massoretic (Hebrew) text (c. 2300 BC) from which the Latin and English translations of the Old Testament derive. A date of around 3100 BC is therefore quite possible if we take into consideration these earlier sources. It is also reassuring to note that the Maya tradition places the Great Flood in Mesoamerica at precisely 3113 BC, according to their recently deciphered calendar.[2]

Thus we have evidence that the Sumerian (biblical) Flood was contemporary with the start of the thirteen-baktun cycle. Most of the conflicting Mesoamerican accounts that remain regarding previous eras include an era that ended in flood. One of the Aztec versions (the Leyenda de los Soles, the "Legend of the Suns"—see chapter 1) says that it was the Fourth Sun that ended in flood (we are now in the Fifth Sun). If the five eras of the Aztecs are each 5,200 years in length, as suggested by Gordon Brotherston,[3] and this is a remnant of the original 5,200 tuns (about 5,125 years), then this means that the previous era, which ended in 3114 BC, ended in a flood. However, most of the Mexican accounts, and the Popol vuh of the Maya, indicate that it was the first era that ended in a flood.

COMET EXPLOSION: 3114 BC

David Furlong, author of *The Keys to the Temple,* has pointed out that around 3100 BC,[4] not only did the first phase of Stonehenge appear, but also the great stone circles of Castlerigg, in Cumbria; Stennes Stones, in the Orkneys; and Newgrange, in Ireland. At the same time, the building of long barrows ceased. Dynastic Egypt was also founded at this time, while the Maya, in a third area of the world, dated the beginning of the thirteen-baktun-cycle of their Long Count calendar from 3114 BC. Geoclimatologists have found a period of climatic change around this time, called the Piora Oscillation, which resulted in a "sulphate deposit" in the Greenland ice cores. This suggests volcanic eruption or cometary impact. There is evidence of flooding in Mesopotamia (as Rohl has pointed out), and evidence of ancient flooding also exists on the present-day Navajo reservation in the United States. A rise in sea level also occurred at this time on the coast of Brittany and at the Dead Sea in Israel, plus there was a sudden drying up in the area of the Sahara. Furlong then goes on to connect these events to the sinking of Atlantis, bringing the proposed date forward 6,400 years from Plato's 9500 BC to 3100 BC.

In an online article, "Astronomical Aspects of Mankind's Past and Present,"[5] climatologist Timo Niroma of Finland has independently observed several anomalies around 3100 BC. There was a methane peak (caused by fires); a cold spell, according to the tree-ring record of ancient bristlecone pines; and the coastal menhirs in Brittany suddenly appeared. Niroma also lists several of the events spotted by Furlong. Niroma suggests that a "huge meteorite swarm," possibly from the "break-up of a great comet in the inner parts of the solar system," caused tsunamis and an atmospheric dust blanket. Two small craters have been discovered from this period. Niroma points out that phase one of Stonehenge was purely astronomical; the later phases included ritual uses (though some of the fifty-six "Aubrey holes" were redug later to receive human cremations). The wave of stone-circle building, he says, was for the purpose of studying celestial phenomena and to try to predict cycles. He goes on to make the "outrageous suggestion" that the meteorites responsible were the Taurids, and, in particular, their companion, Comet Encke. The path of Encke has been tracked back to an Earth intersection about five thousand years ago. However, further investigation showed that another comet, Oljato, had a trajectory that was virtually the same as that of Encke about 9,500 years ago, and this seems a more likely candidate than Encke for the 3100 BC episode. The final suggestion is that a comet or asteroid one hundred kilometers across disintegrated in 9500 BC into three main fragments—Oljato, Swift-Tuttle, and another. The third fragment is what rained down on Earth in 3114 BC.

Support for this connection between the Long Count and comets comes from a book titled *Exodus to Arthur: Catastrophic Encounters with Comets,* by Mike Baillie. Baillie is a dendrochronologist (tree-rings and climate studies), and establishes correlations between tree-ring anomalies and comets, taking history and mythology into account. Comets are repeatedly referred to as flaming serpents and dragons throughout history. Gods and demigods such as Lug, Merlin, Phaethon, Typhon, and Quetzalcoatl, the angels Michael and Gabriel, and even Jesus have comet connections. When Baillie asked atmospheric physicist Gerry McCormac what would happen if a comet came very close to Earth, he replied: "If it came within the Earth's magnetosphere it would probably be spectacular . . . the sky would go purple or green, particles from the comet would spiral down the lines of force, and it is likely that you would have amazing auroral displays and coloured streamers."[6] Baillie replied, "Oh . . . you mean like Quetzalcoatl, the Central American sky serpent with his feather arrangement represented as flames of fire?"[7] There is accumulated evidence that a comet this close would produce an audible hissing sound, like a snake. Comets can often appear to be brilliant white, and were once known as "bearded stars." Is this why the myth of Quetzalcoatl (like that of Jesus) speaks of a white-bearded man who promised to return one day?

The ancient Egyptian Book of the Celestial Cow describes the myth of Ra getting angry with his subjects and sending his eye out to punish them, in the form of Hathor/Sekhmet/Amit, which, says E. P. Grondine on the Cambridge Conference website,[8] represents the fragments of Comet Encke that fell to Earth in 3114 BC. He also cites the work of Dr. Lonnie Thompson's team from Ohio State University, whose recent ice cores from Peru and earlier ones from Mt. Kilimanjaro in Tanzania show "a climate collapse circa 3200 BC, with dating determined by radio carbon dating."[9]

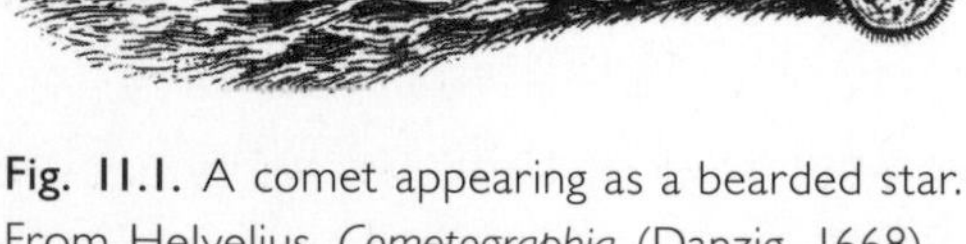

Fig. 11.1. A comet appearing as a bearded star. From Helvelius, *Cometographia* (Danzig, 1668).

THE EYES OF RA AND HORUS

Fig. 11.2. The pharaonic headdress with the uraeus snake—the (right) Eye of Ra.

The myth of the Right Eye of Ra says that in the first universe (World) made by Ra, the people began to plot against him, and so he hurled his divine right eye at them, in the form of the goddess Sekhmet. Murry Hope says that this eye was a "bright and burning object"[10] and may represent a comet or planetissimal that fell to Earth, causing devastation. Ra then departed for the sky and created the current world. The Eye of Ra became the uraeus snake at the center of the pharaonic headdresses, and we have already discussed the snake–comet connection.

There is another myth of Ra that may be pertinent to this account. Ra's wife, Nut, offended him by cohabiting with her brother, Geb. Ra then forbade Nut to give birth on any of the 360 days of the year. She appealed to Thoth, lord of time, who then played a game of senet (an Egyptian board game, similar to draughts) with the Moon and won "a seventy-second part of her light." One seventy-second of a part of 360 is five (360/72 = 5), and so Thoth won five days on which Nut gave birth to the five "epagomenal neters"—Osiris, Horus the elder, Set, Isis, and Nephthys. Murry Hope says this means that "a change in the Earth's orbit involving the Moon, which precipitated a change in the Earth's axis, was responsible for the five extra days we now have in our calendars."[11]

Since all five of these epagomenal gods, or neters, have associations with Sirius, Hope sees Sirius as playing some part in the drama. The presence of a right eye implies the involvement of a left eye, and the left eye was the lunar eye, or Eye of Horus, which, like Ra, was depicted with the head of a hawk.

The Eye of Horus—the Oudja, or Utchat, was a left, or lunar, eye, and a *hekat,* which was a unit of volume used for the measurement of

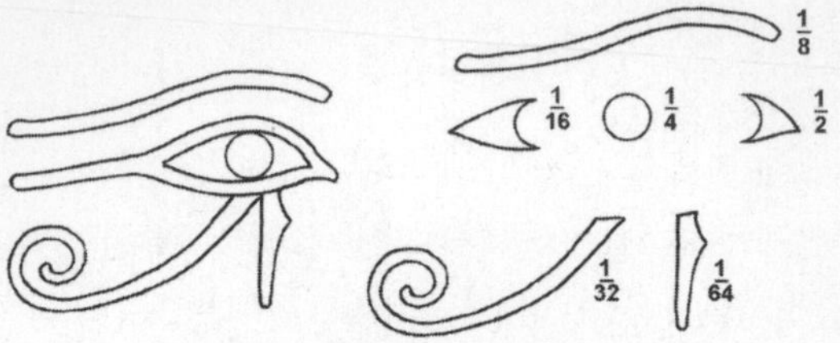

Fig. 11.3. The left eye of Ra, a.k.a. the Eye of Horus, and its component fractions of volume and time

grain. It was also seen as a measurement of time and volume. It is made up of several components, each of which represents a different fraction; there is a quarter, half, eighth, sixteenth, thirty-second, and sixty-fourth. When they are all added together, they total sixty-three sixty-fourths. In music, this is very close to the ratio of the Pythagorean comma, which is the difference in ratio between octaves and fifths, and amounts to about 1.0136, while sixty-three sixty-fourths is about 1.0158. Pythagoras was primarily a mystic, and the ratio has been seen to represent the difference between the imperfect mortal state and the perfected immortal state. It is also close to the ratio between solar and lunar years (1.0307),* which is the ratio between the lunar left eye and the solar right eye. However, while the Jewish Sadducees kept time by the Sun and the Pharisees kept a lunar calendar, the Pharisian lunar calendar was not 354 days in length, as in twelve lunar months, but was actually 360 days—the same as the tun of the Maya. The ratio between the solar year and a 360-day year (1.0146) is much closer to the Pythagorean comma and to the Eye of Horus ratio. In fact, it lies halfway between the two.

So, encoded in the myths of the solar and lunar eyes we have a suggestion of a past astronomical event that "altered the length of the year." Since our right eye is connected to our left rational hemisphere and our left eye is connected to our right intuitive hemisphere, there was a "fall" from an intuitive right-brain state of consciousness to a rational left-brain state. The galactic alignment process signals the activation of the pineal eye (Eye of Horus in a triangle) that lies between the hemispheres and will allow whole-brain consciousness. We shall return to this idea in part 4.

---

*The ratio between solar and lunar years (1.0307) is where a sidereal year of 365.256 days is divided by a lunar year of 354.3672 days, made up of twelve synodic lunar months of 29.5306 days each.

Plate 1. The Aztec Sunstone (see chapter 1 and the following page).

Plate 2. The Palo Volador of the Maya is a ritual that dates from preconquest days, in which a tall wooden pole represents the Mayan world tree—Earth's rotational axis. Four *ángeles* (hanging by one leg) spin to the ground, representing the journey of the Hero Twins of the Popol vuh as they descend to the underworld, Xibalba, to battle with the Lords of Death. The four men spin through thirteen revolutions each, a total of fifty-two, representing the fifty-two-year Calendar Round. This is when the New Fire ceremony was performed by the Toltecs and Aztecs in order to stop the world from ending, but fifty-two is also symbolic of the 5,200-tun thirteen-baktun cycle that ends in 2012 (see chapter 1).

Plate 3. The Aztec Sunstone with color restored, by Roberto Sieck Flandes. It is possible that the original design was devised as a working calendar, with moving parts, recording the passing of eras, including the Long Count. Theoretically it could record a unique date combination for each day over a period as long as 10,250 years (see chapter 1).

Plate 5 (above). The symbolism of the Star in the tarot, especially this version, circa 1943, by Lady Frieda Harris, is suggestive of the Blue Star (of the Blue Star Kachina Prophecy). The card is ruled by Aquarius (the coming age), and was painted according to a design by Aleister Crowley, head of the Argentum Astrum (AA), the Order of the Silver Star, which is Sirius, according to Crowley's executor, Kenneth Grant (see chapter 2).

Plate 4 (above). Painting of the three cosmic ages, according to artist Joachim de Fiore (ca. 1130–1202). The first age is the Age of the Father, at the bottom of the picture; the second is the Age of the Son; and the third is the Age of the Holy Ghost, at the top. These can also be called the ages of Law, the Word, and Spirit (see chapter 2).

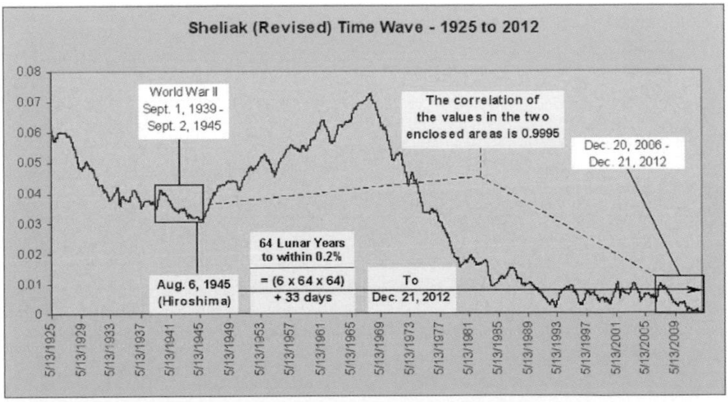

Plate 6. Zyzygyz's diagram showing how the 1939–1945 and 2006–2012 periods relate to each other on the timewave (see chapter 4).

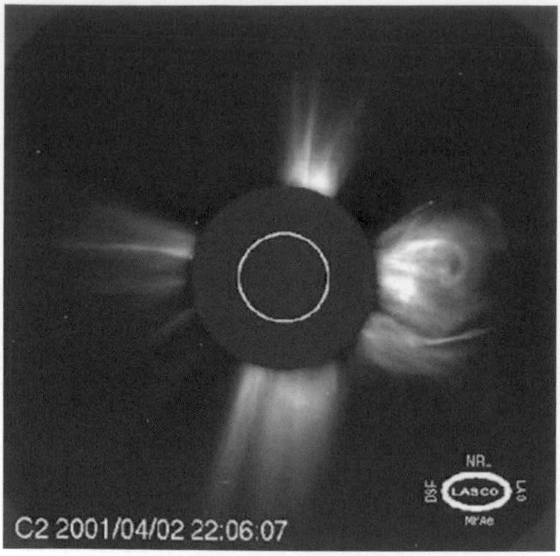

Plate 7. On April 2, 2001, this record-breaking X–22 solar flare occurred. Photo courtesy of LASCO/NASA.

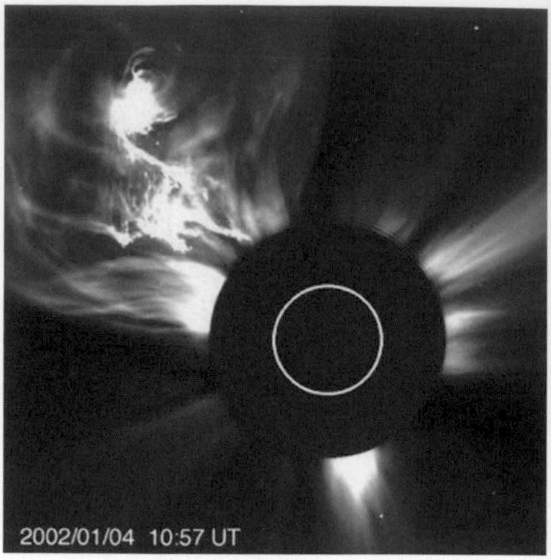

Plate 8. On January 4, 2002, the Sun provided the most complex CME (coronal mass ejection) ever recorded. Photo courtesy of NASA.

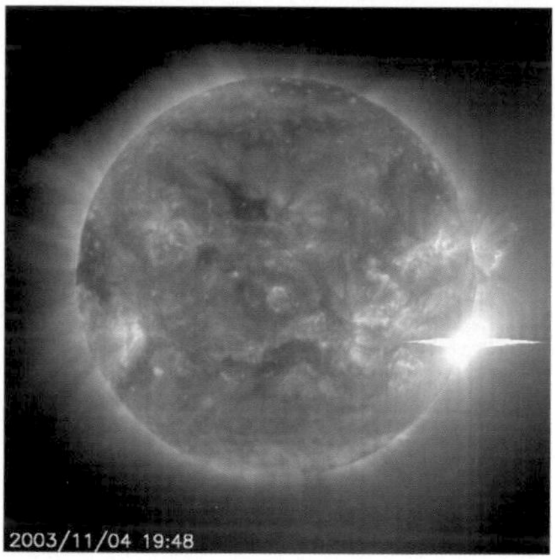

Plate 9 (left). This flare was released on November 4, 2003, and is now known to have been an X–45—more than twice as large as any previously recorded flare (see chapter 20). Photo courtesy of NASA.

Plate 10 (below). Kev Peacock's diagram showing the upward trend of earthquakes, geomagnetic activity, and solar activity, and the clear correlation among peaks on all three graphs.

See chapter 6 for details on the solar activity shown on this page.

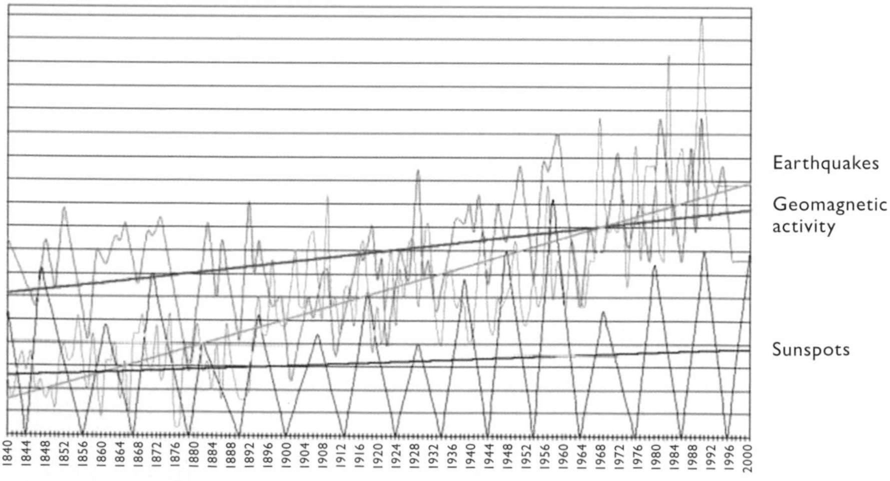

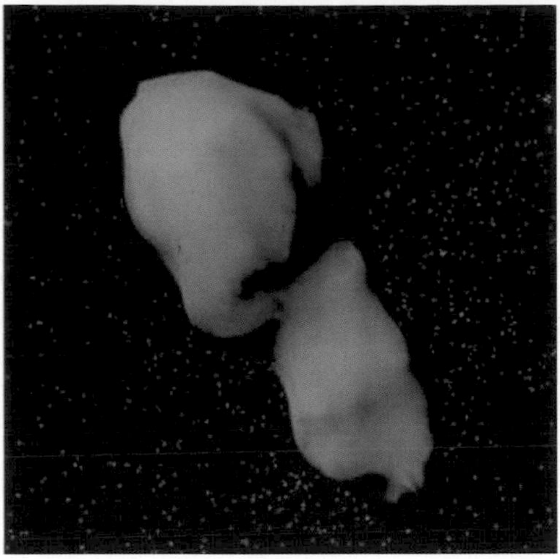

Plate 11. Computer model of the asteroid Toutatis, another celestial object some believe threatens our planet (see chapter 18). Image courtesy of Scott Hudson and Steven Ostro, JPL and NASA.

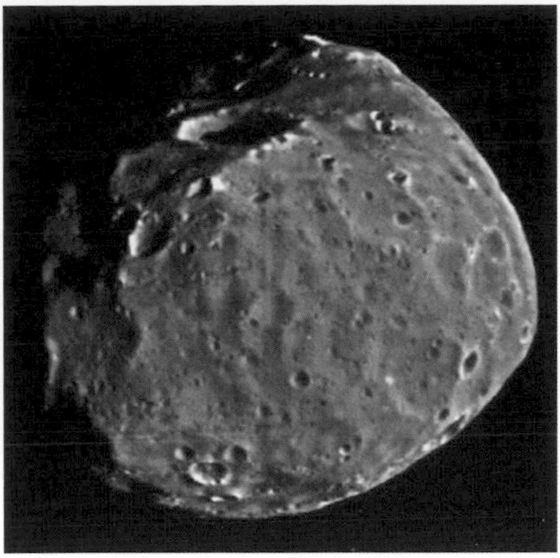

Plate 12. Mars's moon Phobos. Some have predicted this may be dragged from its orbit and onto a collision course with Earth in 2012 (see chapter 7). Photo courtesy of NASA.

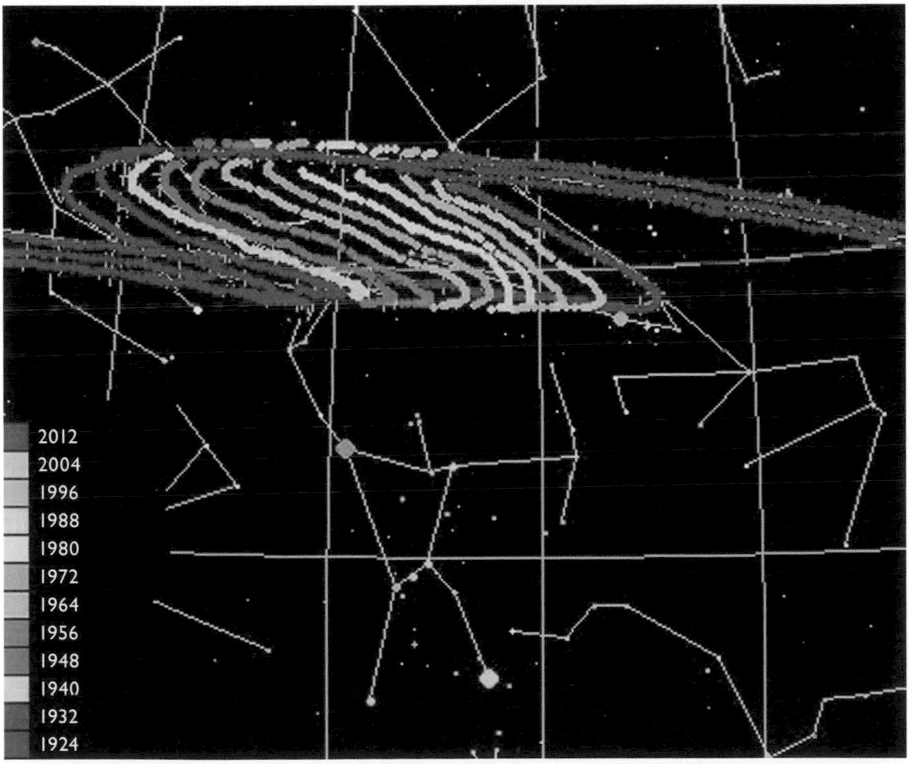

Plate 13. Some hold that a retrograde loop of Venus occurring above Orion and near Gemini will herald a solar megacycle catastrophe (see chapter 8). Here, the twelve retrograde loops of Venus between 1924 and 2012 are shown on CyberSky software, http://cybersky.com.

Plate 14. The Pyramid of Kukulcan at Chichén Itzá, with Moon rising (see chapter 8). Photo: Lee Anderson

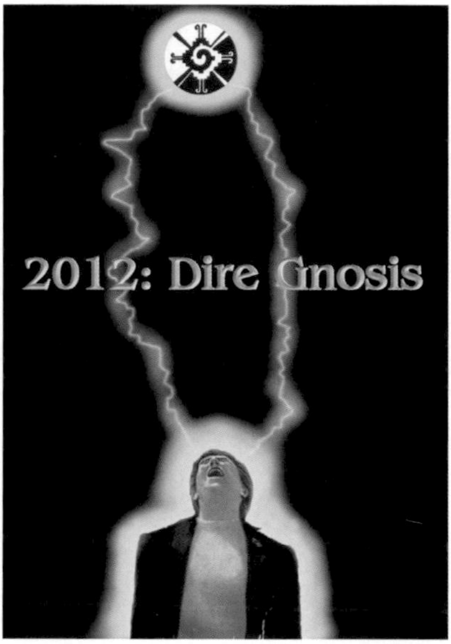

Plate 15. *Anamnesis*—intense mental download from Galactic Center—the Diagnosis 2012, or Dire Gnosis, website logo. See www.diagnosis2012.co.uk—the most comprehensive website on 2012 predictions.

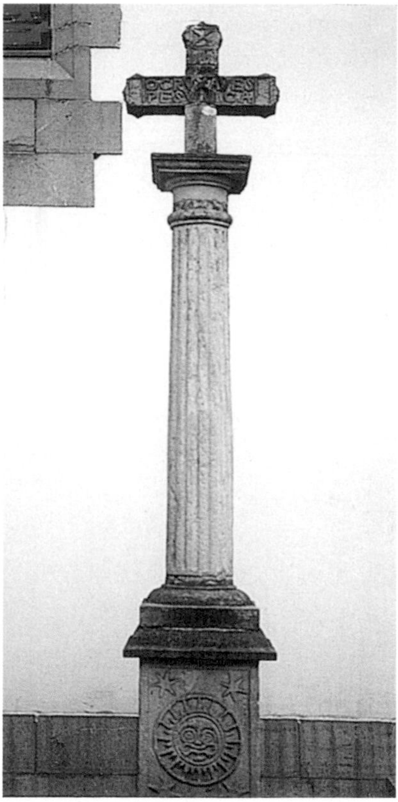

Plate 16 (above right). The Cross of Hendaye. Does this monument, built around 1680, encode a warning of "an imminent trial by fire for the northern hemisphere"? (See chapter 8.) Image from *The Mysteries of the Great Cross of Hendaye*, by Weidner and Bridges (2003), used by permission. Photo © Darlene, 2004

Plate 17. Examples of expulsion cavities found by the author in the East Kennett grid crop circle (below), indicating that the stems have been rapidly heated. Photo: Melissa White, www.fairlyte.co.uk/

Plate 18 (above). Crop circle at Huish, Wiltshire, July 20, 2003, demonstrating the Mayan thirteen-baktun cycle as depicted by José Argüelles. Photo: Bert Janssen

Plate 19 (left). The 20 × 20 grid crop circle at East Kennett, Wiltshire, July 2, 2000. This can be interpreted as a calendrical grid representing three cycles from the Mayan Long Count system. Photo: Francine Blake

Plate 20 (right). Crop circle at Milk Hill, Wiltshire, August 12, 2001. It has 409 circles and is 800 feet in diameter—but do the thirteen main circles in each arm represent the Mayan thirteen-baktun cycle? Photo: Janet Ossebaard

See chapter 16 for details of all the crop circles on this page.

Plate 21. This photograph was taken by Roz Savage at the annual festival of Snow Star, or Qoyllur Riti, held just below the glacier on the sacred snow peak of Mt. Ausangate, in Peru. There were 30,000 pilgrims in attendance. The festival celebrates the emergence of "the shining ones who will lead Peru and the world into a new era of peace" (see chapter 21). Roz Savage's website: www.rozsavage.com.

Plate 22. A Bronze Age calendar priest-king and queen standing in front of Newgrange megalithic passage tomb, in the Boyne Valley, Ireland. The golden hats and collars are real and are recent archaeological finds from Switzerland and Germany. Covered in thousands of Sun and Moon symbols, the hats record ancient time cycles. They are around three thousand years old (see chapter 22).

KEENE ON COMETS

In 2002, visionary novelist Joel J. Keene published a book called *Cosmic Locusts,* connecting a comet impact to 2012 following a series of personal synchronicities. Keene's birth date is on the cusp of Scorpio and Sagittarius, which is the celestial position of Galactic Center, and, even more accurately, the intersection point of the galactic equator and the ecliptic. When he heard that the thirteen-baktun cycle of the Maya was designed to terminate when the winter solstice Sun converges on this very point in 2012, and when he concurrently found that the biblical book of Joel seems to describe Earth's future passage through the tail of a comet, he became aware of a clustering of more synchronicities around himself. He started to write it all down in a fictional format, because it was all so incredible. But Keene found that there were even more synchronicities.

Fig. 11.4. A comet appearing as a sword. From Helvelius, *Cometographia* (Danzig, 1668).

He had written of a 2012 convergence of a burst of energy from Galactic Center and the return of the shattered remains of a comet, only to find that astrophysicist Paul LaViolette had already said that we are overdue for a burst of energy from Galactic Center, and tentatively connected it with the end point of the Maya. Then he found that dendrochronologist Mike Baillie had discovered evidence that an ancient shattered comet became associated with the mythology surrounding the birth of Venus (Quetzalcoatl), which is also associated with the start of the thirteen-baktun cycle. Finally, he found that the series of prophecies of the end times given to the children at Fatima by an apparition of the Virgin Mary spoke of the "birth of the planet Venus" being responsible for the forthcoming ascension of mankind, as well as mentioning the flaming sword of St. Michael (according to *Fatima Prophecy,* by Ray Stanford). The coincidences are continuing to the present.[12]

12

FRINGE SCIENCE

NAKED SINGULARITY

A 1999 essay titled "Awakening to the Omega Point,"[1] by UFO researcher Bob Buck, notes that the planets are losing their magnetic fields: Jupiter's red spot changed its direction of rotation in 1986–87, and the Sun is losing its polarity. In addition, Buck says that there has been a shifting of the spectrum in the auroras and rainbows, like a redshift, but in the other direction—a kind of violet shift. A powerful magnetic field is the cause, says Buck (the auroras or light displays are caused by the interaction of the solar wind and Earth's magnetic field). Buck describes the unknown magnetic field that we are approaching as the "naked singularity," and says that it will probably upset the Oort cloud (a spherical cloud of material including billions of comets, believed by some astronomers to surround the solar system at a distance of one light-year) and cause more comets to divert from their usual orbits. He thinks that every 13,000 years, Earth goes through one of these naked singularities, which he sees as a merging of dimensions, but also thinks a twenty-three-degree axis shift of Earth, back to its position perpendicular to the ecliptic, may be possible. He predicted that the encounter would happen sometime between 2003 and 2012.

In an update, Buck says that since reading a book titled *The New Gravity,* by amateur physicist Kenneth G. Salem, he now thinks that the speed of light is accelerating at a rate of 2.8 angstroms per second, and will continue to do so until it reaches a critical point, when a "phase change" will occur, shifting the whole universe to a higher frequency. Professional physicists have recently verified some of Salem's findings. Buck quotes an article by investigative journalist and filmmaker Linda Moulton Howe telling of a recent discovery that one of the heavier atomic particles, the muon, is not behaving the way it has been for the last thirty years. Buck predicts that other, less heavy particles will change next, and so on, until photons and other weightless particles are affected and the phase shift occurs (by 2012). He says that this will affect the whole universe, not just this solar system.

SHIFT OF THE AGES

Visionary David Wilcock's online book *The Shift of the Ages*[2] (originally called *Convergence,* and having been briefly called *Convergence II,* it currently carries the subtitle *Scientific Proof for Ascension,* with the alternative subtitle *Convergence Volume 1*) is an attempt to find the science and philosophy behind the thirteen-baktun cycle, using Cotterell's sunspot theory as a key element in the puzzle. Wilcock originally concluded that the end point should be 2039, following engineer and ufologist Maurice Chatelain's reinterpretation of the Maya calendar, which was based on the conjunctions of Jupiter and Saturn. But as Wilcock explained to me via e-mail, "Now I know that the 2012 date fits in much better, though the Jupiter/Saturn conjunctions are definitely still important."[3]

Shift of the Ages covers many subjects—Earth changes, superstring theory, sacred geometry, the Philadelphia Experiment, Bruce Cathie's Earth grids, Richard Hoagland's planetary geometry observations—and it forms an introduction to the more scientifically orientated follow-up, *The Science of Oneness*[4] (previously called *Convergence III,* it now carries the subtitle *Convergence Volume 2*). However, Wilcock believes himself to be a reincarnation of Edgar Cayce (the American clairvoyant who died in 1945 and whose photo shows similar facial features to those of Wilcock), and has been channeling Ra.

The Ra material was first channeled by Carla Rueckert and is the source of much of the cosmological information in Wilcock's books. What makes Wilcock different from most channelers, though, is that he manages to unearth obscure scientific papers that confirm many of the points that were channeled from Ra. But, says Wilcock, it turns out that Ra also indicated the importance of 2012: "Without any reference from either side to the Mayan Calendar, Ra independently stated that the cycle change would come between 2011 and 2013."[5] Accordingly, Wilcock's website, www.Divinecosmos.com, states that its objective is to scientifically prove that "a dimensional shift will occur on Earth on or before 2012."[6]

The follow-up to *The Science of Oneness* leaves channeling behind and is called *The Divine Cosmos*[7] (now carrying the subtitle *Convergence III*). It is a bold attempt to rewrite the current scientific paradigms using cutting-edge international scientific research findings (however, it is advisable to first read *The Science of Oneness* as an introduction). *The Divine Cosmos* covers such subjects as ether theory, zero-point energy, the amazing work of several Russian scientists, Earth changes, solar developments and changes on other planets in the solar system, tornados and vortexes, and plasmatic phenomena (a plasma is an electrically charged gas, or, as the *Concise Oxford Dictionary* puts it, "a gas of positive ions and free electrons with an approximately equal positive and negative charge").

In *The Divine Cosmos,* particularly in chapter 8, and even more so in the article that was originally a supplement to the book (but is now listed separately), Wilcock claims that he has found the "'smoking gun' to prove that an energetic transformation process is culminating in December 2012."[8] To do this, he cites the remarkable work of a number of Russian scientists.

SOLAR SYSTEM CHANGES

Dr. Alexey Dmitriev is a professor of geology and mineralogy, and the chief scientific member of the United Institute of Geology, Geophysics, and Mineralogy, within the Siberian department of the Russian Academy of Sciences, in Novosibirsk, Siberia. In a paper titled "Planetophysical

State of the Earth and Life,"[9] Dmitriev discusses recent changes on the Sun, Earth, and the other planets of the solar system. He notes the increase in the amount of solar activity since the Maunder Minimum (when sunspots were very scarce) in the late 1600s (see chapter 6, fig. 6.2), and the increased frequency of X-ray flares.

David Wilcock expands this subject in *The Divine Cosmos,* giving accounts of recent solar storms, X-ray flares that caused power-grid collapses and alteration to Earth's magnetic field, coronal mass ejections, and the increasing speed of the solar winds. The records are constantly being broken as the Sun gets ever more active, lending support to the idea that we are approaching the end of a solar megacycle.

Dmitriev points out Earth's magnetic anomalies, its weakening field, and the accelerating movement of its magnetic poles as evidence that a "geomagnetic inversion" is taking place. He mentions the increasing temperature and climate changes that are currently causing concern, and then shows that extreme weather and seismic phenomena such as floods, hurricanes, earthquakes, and landslides are also increasing, having collectively risen by 410 percent from 1963 to 1993 (Wilcock says the figure is now around 600 percent).

Wilcock supports these claims (his source is Michael Mandeville) with evidence that volcanic activity has increased by 500 percent since 1875. However, official sources in Western Europe and the United States say that increased reports of earthquakes can be explained by advances in seismic detection technology and the fact that there are more monitoring stations around the world, while the apparent rise in the number of volcanic eruptions is due to the population explosion, with more people living near volcanoes and thus reporting eruptions via an improved communication network.

The other planets in the solar system are also showing changes, says Dmitriev. There has been a change in the atmosphere, a brightening, and the appearance of dark and light spots on Venus, plus "a sharp decrease of sulphur-containing gases in its atmosphere" and a sodium (or natrium) atmosphere or "tail" forming around our Moon. The atmospheric density

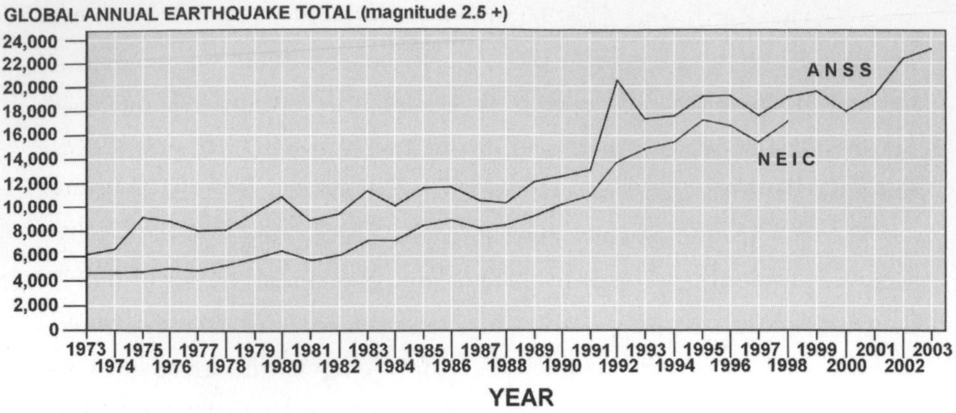

Fig. 12.1. The lower graphed line (NEIC) is generated from the figures gathered in April 1999 by Michael Mandeville when he did a laborious catalog search on the National Earthquake Information Center database, searching for global earthquakes of magnitude 2.5-plus from 1973 to 1999. The upper line (ANSS) is generated from a similarly laborious search performed by the author in 2004 through the Advanced National Seismic System catalog database. The information is more complete and up-to-date on the upper graph line, supporting Mandeville's conclusion.

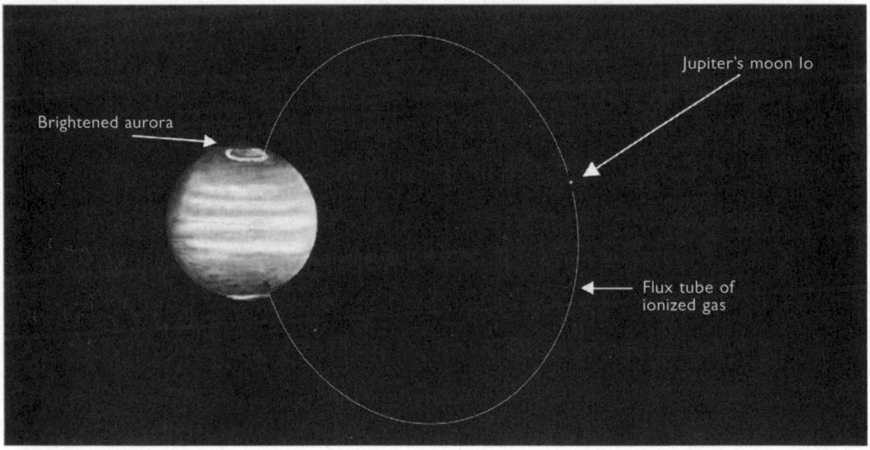

Fig. 12.2. Artist's impression of the flux tube going from Jupiter to its moon Io.

of Mars was found to be twice as thick as its former level, causing damage to NASA's Mars Global Surveyor satellite in 1997. There has also been warming on Mars, causing the melting of the Martian ice caps, plus a "cloudy growth in the equator area and an unusual growth of ozone concentration." Dmitriev says Jupiter's magnetic field intensity has doubled, and the planet has developed a visible one-million-ampere flux tube of ionized gases from its magnetic poles to the moon Io. He also says that Jupiter has had an "appearance of radiation belt brightening" and "large auroral anomalies." However, these are thought to be an aftermath of the collision with Comet Shoemaker-Levy-9 in 1994. On Saturn, the brightness has increased, along with the appearance of polar auroras. Uranus and Neptune have both had recent polar shifts, and the magnetosphere of Uranus has made a sudden leap in intensity, while Neptune has had "a change in light intensity and light spot dynamics." Pluto has also had "a growth of dark spots" (first photographed by the Hubble Space Telescope in 1994 and released to the public in 1996). In fact, in July 2003 the BBC announced[10] that astronomers have been baffled to find that Pluto's atmosphere has thickened over the last fourteen years (deduced by calculating the dimming of background stars as Pluto passes in front of them).

Wilcock supports Dmitriev's conclusions by adding that a new, thin atmosphere on Mercury has recently been observed.[11] He has also updated the Russian planetary observations (since Dmitriev's essay was written in 1997, and updated in 1998), showing that these changes are ongoing.[12]

Dmitriev says that there is a root cause for all these Earth changes, solar changes, and changes on the other planets: they can all be attributed to the fact that the solar system is moving into a new energetic area.

MAGNETIZED BAND OF PLASMA

The Russians have been studying the heliosphere boundary since the 1960s and have observed that the thickness of the "shock wave" at the leading edge has increased tenfold (1,000 percent) since then. It has gone from about four astronomical units (AU) to about forty AU. This is because the solar system is moving through a magnetized band of plasma that is sticking to the edge of the heliosphere, "which has led to a plasma overdraft around the Solar System, and then to its breakthrough into interplanetary domains."[13]

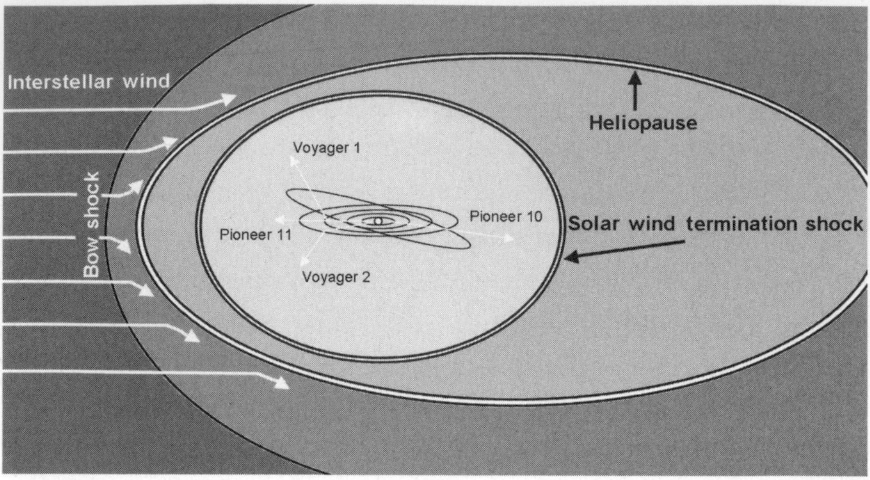

Fig. 12.3. A cross-section of the heliosphere showing U.S. mission trajectories that sent back some of this information. As the solar system moves through interstellar space, the solar wind continues out past all the planets, until it slows to a stop. This boundary is called the termination shock. This is comparable to the hull of a ship moving through the sea. The interstellar wind is like the sea, and there is a layer surrounding the termination shock that is comparable to a bow wave. The whole solar system, including termination shock and surrounding shock wave, is called the heliosphere, and the edge is called the heliopause. The leading edge of the heliosphere is called the bow shock.

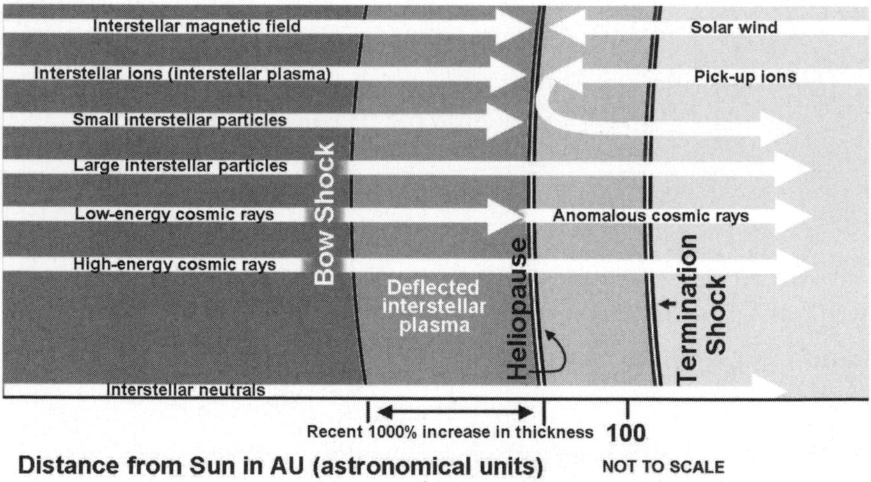

Fig. 12.4. A close-up of the heliosphere boundary, showing the incoming interstellar plasma buildup (the thickening layer of deflected plasma) and how other particles interact with the boundary.

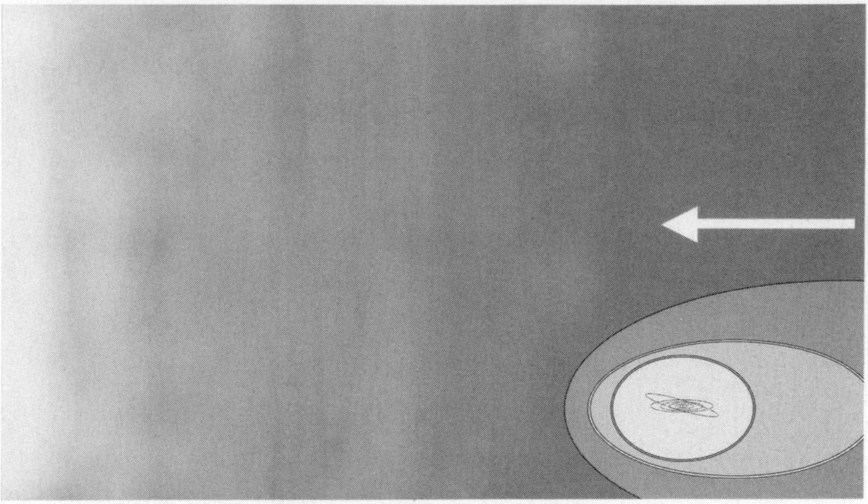

Fig. 12.5. The heliosphere travels into the magnetized plasma band.

It is this plasma, seeping into the solar system, that is causing the alterations, says Dmitriev, and the solar changes are reinforcing the changes on the planets. On Earth, there is a buildup of new plasma in the ionosphere (about sixty to a hundred miles above Earth) that is affecting our weather, and the solar changes are adding to these, causing the geomagnetic reversal. Dmitriev has written a paper on plasmatic light phenomena that he calls "vacuum domains," such as tornado lights[14] (lights that occur in the vacuum at the center of a tornado; the term *vacuum domains* also includes phenomena such as ball lightning, poltergeist lights, earthquake lights, and the elves and sprites in the ionosphere that we shall discuss later in this book). He concludes that these vacuum domains are on the increase, and cannot be explained by modern branches of physics. A version of the ether model (nowadays called zero-point energy) can be used to try and understand them:

> Because these non-homogeneous vacuum domain objects display not-of-this-physical-world characteristics such as "liquid light" and "non-Newtonian movement," it is difficult not to describe their manifestations as being "interworld processes" . . . Such disturbances cause and

> create energy and matter transfer processes between the ether media and our three-dimensional world . . . Hundreds of thousands of these natural self-luminous formations are exerting an increasing influence upon Earth's geophysic [*sic*] fields and biosphere. We suggest that the presence of these formations is the mainstream precedent to the transformation of Earth; an Earth which becomes more and more subject to the transitional physical processes which exist within the borderland between the physical vacuum and our material world.[15]

Dmitriev sees the solar system changes as soon culminating in an evolutionary change that involves interaction with something beyond the three-dimensional world. Wilcock has extended and supplemented Dmitriev's theory, suggesting that the magnetized plasma bands are part of a galactic version of a known solar-system structure called the Parker spiral (fig. 12.6). This is a series of magnetic plasma lines that spiral out from the Sun across the solar system in curved paths, radiating outward like a pinwheel in slow motion. Wilcock proposes that the area that the solar system is entering is a macroscopic version of this model, with the lines radiating from Galactic Center.

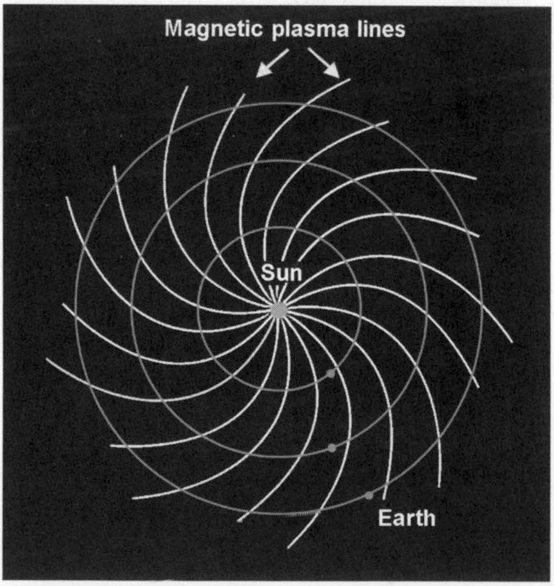

Fig. 12.6. The Parker spiral—magnetic plasma lines in the solar system.

Other scientists have interpreted the solar system changes as the approach of Planet X,[16] or as the effects of the hypernova explosion of the star Eta Carinae.[17] But these theories are much less convincing than Dmitriev's. Wilcock claims that at the climax in 2012, we are due for a "matter transmutation," and thinks our entry into the plasma beam is through ripples spaced out by phi, the golden ratio. Also known as the divine proportion (and what some people such as Cotterell call the Golden Mean), phi is defined as the ratio that results when a line is divided into two sections so that the ratio of the smaller segment to the larger segment is the same as the ratio between the larger segment and the whole line. It can be calculated to many decimal places, but is approximately 1.6180339. It is a ratio that appears throughout nature: for example, in the proportions of the human body, animal bodies, plants, and DNA. The orbital periods of some of the planets in our solar system are related by phi, and the rings of Saturn are spaced by phi.

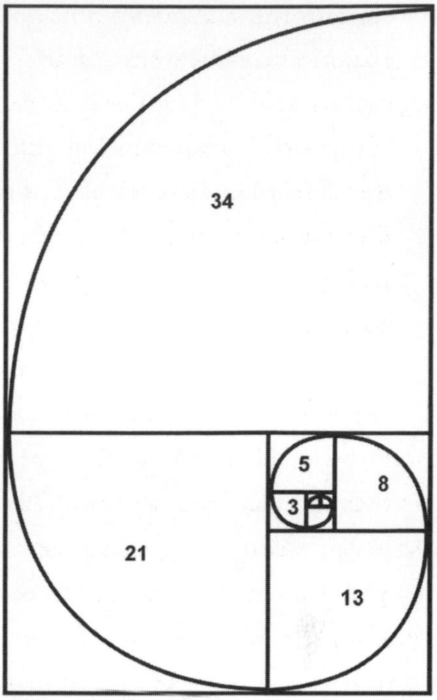

Fig. 12.7. The phi spiral generated by the Fibonacci series.

In the twelfth century, Leonardo of Pisa (known as Fibonacci) introduced to Europe a number series related to phi that starts with a 0 and 1 then continues by adding the two previous numbers to get the next one (0, 1, 1, 2, 3, 5, 8, 13, etc.). The ratio of any two successive numbers in the series gets nearer to phi the further along in the series it is. The ratio can be used to generate a spiral, and that phi spiral is found in places as diverse as pinecones, pineapples, the shell of a nautilus, and the spiral form of a galaxy. The golden ratio also occurs in the musical scale and in classical works of art.[18] Wilcock says,

> I believe that we are moving into a new zone of aetheric energy, and that such zones have "ripples" surrounding them in either direction that are spaced out by the phi ratio. The closer that we get to the new zone (actually like a spiral-cut piece of pie when seen from an overhead view of the galaxy, similar to the structure of the Parker Spiral magnetic field in the solar system), the more rapidly we move through these threshold points. Each one of them creates a sudden spike in the level of ambient aetheric energy moving through our Sun and planetary systems. This in turn creates a sudden spike in the pressure of luminous plasma energy at the Earth's core. These pressure surges are released as earthquakes . . . The pressure surges also are related to the sudden increases in consciousness, since torsion waves are the carriers of consciousness.[19]

The "smoking gun" article (originally a supplement to *The Divine Cosmos*) that proves that an energetic transformation process is culminating in December 2012 is titled "The Ultimate Secret of the Mayan Calendar—An Imploding Cycle of Energy Increase, Culminating in 2012–2013 AD."[20] It attempts to explain, in laymen's terms, a lengthy paper by Sergey Smelyakov and Yuri Karpenko titled "The Auric Time Scale and the Mayan Factor: Demography, Seismicity, and History of Great Revelations in the Light of the Solar-Planetary Synchronism."[21]

THE AURIC TIME SCALE

Sergey Smelyakov is a professor of mathematics at Kharkov University, Ukraine, with an interest in astrology. He has found that the golden ratio (which he calls the golden section), expressed by phi, governs the movements of the asteroid belt and the cycles of solar activity, as well as all the planetary orbits in the solar system, including comets. He has also found that phi governs the stock market and population growth. By applying the same phi harmonics to the thirteen-baktun cycle, he has, according to Wilcock, found a phi spiral that seems to spiral in on itself on December 21, 2012—a kind of time implosion, or final "point of bifurcation."

I read through the Ukrainian sixty-four-page, equation-drenched paper, and found that the authors had based their theory on a start-date of August 6, 3113 BC, taken as a Gregorian historical date as given by José Argüelles in *The Mayan Factor,* when in fact the start-date 0.0.0.0.0 was actually August 11, 3114 BC, as we now know. As a result, the authors had shifted the end-date forward to 2013. I contacted Sergey Smelyakov,[22] wondering whether or not the theory would survive such news, and was relieved when he told me[23] that "several numerical conclusions might be slightly changed," but that the theory was not seriously damaged by the modification. This meant that Wilcock's smoking gun was still smoking—or would be after reloading.

Smelyakov told me he had realized there was a discrepancy in Argüelles's *The Mayan Factor,* since the given start- and end-dates (3113 BC and 2012 AD) did not allow for the 1,872,000 days to elapse between them, being a year short. He had correctly guessed that the error was due to the so-called missing year zero, which has to be accounted for when counting years from BC to AD or vice versa (remember, there was no year zero, since the year following the hypothetical day of Christ's birth was 1 AD and the year directly before it was 1 BC). However, Smelyakov assumed that Argüelles's start-date was a Gregorian historical year, and that the end-date was a year early, so he moved the end-date forward a year to 2013. Actually, he placed the end point at November 19, 2013, because of a calculation using the Julian day number, but confused it with the Julian date. I brought this matter to Smelyakov's attention in my first e-mail to him, referring to several online Julian date and day calculators that he could use to check the calculation, showing that the true end-date is December 21, 2012.*

In another communication, Smelyakov said that the revised figures for the second version of his hypothesis "the Auric Time Scale," rather than detracting from the theory, enhanced it: "If we take the year of 3114 BC for the starting point, even more intriguing results follow; e.g., the separating epoch of 1997 comes to 1996—the year of comet Hale-Bopp."[24]

This was looking good, but just then I discovered something else

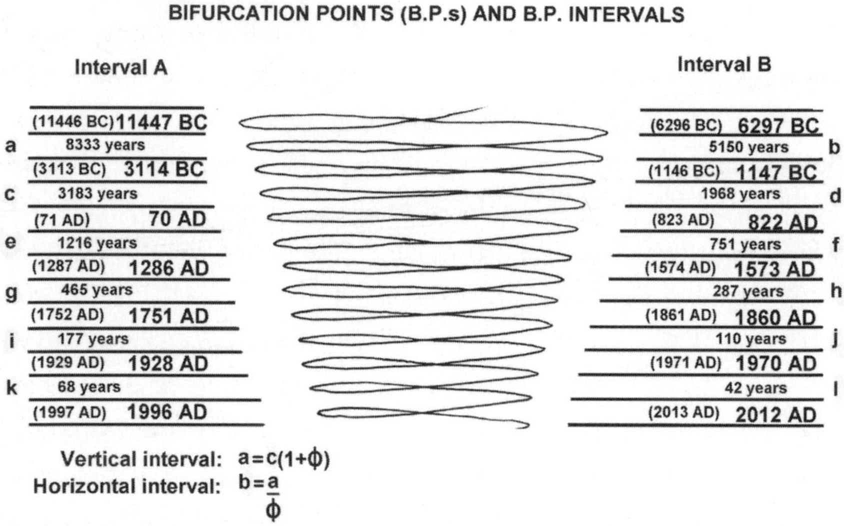

Fig. 12.8. An explanation of how the bifurcation points are generated for the Auric Time Scale (*auric* comes from the chemical abbreviation for gold—*au,* after the golden ratio).

*The Julian calendar preceded the Gregorian calendar and, as an alternative to back-calculating Gregorian dates, archaeologists often use Julian day numbers, avoiding the "year zero" error, since the Julian calendar's zero point is in 4714 BC (Gregorian equivalent). The Julian day numbers correspond to a Julian date, which is different from the Gregorian. For example, the start-date for the thirteen-baktun cycle 0.0.0.0.0 is August 11, 3114 BC (in True Count correlation), back-calculated in the Gregorian calendar, but in the Julian calendar it was September 6, on Julian day number 584283. To find the end-date from this Julian day number, just add 1,872,000 (the number of days in the thirteen-baktun cycle) to 584283 and you arrive at JD number 2456283. Then use the online converters or a table to find the Gregorian equivalent (December 21, 2012). See www.diagnosis2012.co.uk/mlink.htm for online Julian date and day calculators.

about the Auric Time Scale, and it did not bode well for the theory. The scale is a series of "bifurcation points" that are generated by "phi harmonics" and form a time spiral. Smelyakov had calculated the two bifurcation points that occurred in 6296 BC and 11,446 BC, before the start of the thirteen-baktun-cycle. The further back in time you go, the farther apart the bifurcation points get. However, at the other end of the spiral, the last two points he gave were 1997 and 2013, so theoretically the bifurcation points could be calculated on past his original 2013 termination point all the way to the center of the spiral, where they would converge on one point in time. I continued the calculation forward in time, and found the phi spiral whorls into itself, with time speeding up to infinity, on June 11, 2037 AD. This was an approximate calculation, and alternative calculations resulted in this end-of the-whorl point being projected anywhere between May 2036 and November 2038, depending on the criteria used in the calculation (such as recalculating from 11,447 BC; from 3114 BC; projecting from December 2012; and depending also on how many decimal places were used). However, the main point here is that the center of Smelyakov's spiral is over twenty years after the end of the thirteen-baktun cycle.

I wrote to Smelyakov to tell him about my findings, and he replied, saying that he wasn't trying to say that 2012 was the end of the world, just that the thirteen-baktun cycle was resonating to the Auric Time Scale.[25] Even so, he said, I had made a valid point:

"Your suggestion as to forward continuation of this series might also be significant. Frankly speaking, I did not think of it . . . Now, I am surprised that this idea had not come to me."[26]

David Wilcock was certainly under the impression that the spiral whorls in on itself in 2012, as he says in his "smoking gun" essay that explains the Auric Time Scale, entitled "The Ultimate Secret of the Mayan Calendar":

> Here, it is important to remember that the entire cycle decreases to a certain final moment at the end of the cycle. Even though

> the phi-based series F can theoretically extend infinitely in either direction, when you are dealing with a series that is exponentially decreasing in duration, the cycle does converge on one end-point, simply representing smaller and smaller intervals of time in that moment. This would be the same as how the central point of the spiral on a seashell is theoretically infinite in its imploding "recursiveness," yet you nevertheless can clearly define where the end of the seashell's spiral is.[27]

In October 2004, over two years after my e-mail communications with Smelyakov, he sent me his modified theory. However, the modification only extended to altering the dates of bifurcation points, when the thirteen-baktun start point was corrected by shifting it a year, to 3114 BC (Gregorian historical). The continuation of the spiral to its implosion point was not addressed, but there are indications that the implosion point should have been incorporated into the modified Auric Time Scale theory, as we shall see.

CRITICAL MASS

Pierre Teilhard de Chardin was a paleontologist and also a Jesuit priest. His study of the development of early hominids and the increasing size of their brain cavities led him to a conclusion that was unpopular both with the church and with other paleontologists.

He developed the law of Complexity-Consciousness, which indicates that "the greater the complexity of brain and nerve apparatus, the higher the form of consciousness." He saw the primary function of his religion, Christianity, as a mechanism to prepare humanity for a profound evolutionary event in its future that he called the "Omega Point."

Teilhard described Earth as a lithosphere (solid crust) covered by a hydrosphere (watery layer) and then an atmosphere, or gaseous envelope. Within the hydrosphere and atmosphere is the biosphere, or layer of living organisms, and he thought that the purpose of evolution was to develop a self-reflective consciousness that could build up a noosphere, or mind layer around the planet. Humankind is the species that has started the process, and it will culminate in a direct telepathic communication between people, which would be the birth of the ultrahuman. This would finally cause the noosphere to become an awakened noosphere, or "Christosphere," an event that will happen "at the apex of the cone of spacetime"—that is, the Omega Point.

In *The Awakening Earth,* consciousness researcher Peter Russell has built on Teilhard's ideas in saying that as world population approaches the critical number of ten thousand million (10^{10}), we could suddenly find ourselves functioning on a new level, as cells in a "global brain." Since there are 10^{10} atoms in a neuron and 10^{10} neurons in the human cortex, Russell theorizes that when the numbers of living units approach the critical mass, they start to behave like parts of a larger organism. World population is currently well above 10^9 (one thousand million), at about six billion (six thousand million, or 6×10^9), so we are definitely now in the ballpark.

In *The White Hole in Time,* Russell analyzes the acceleration of human knowledge, at first comparing it to an ever-steepening curve, but then he observes that the evolution and history that have led to the present time can be seen as a spiral, and that "paradoxically, although such spirals go on forever, they have a finite length."[28] Immediately after this, he

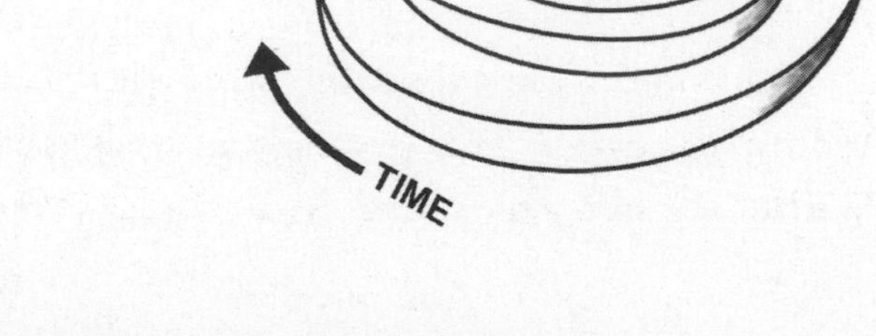

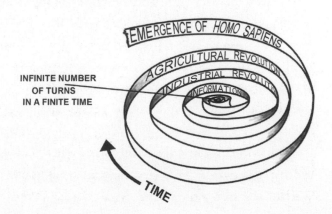

Fig. 12.9. The white hole in time is at the center of a spiral converging on a point in time that seems to be around 2012 AD. After *The White Hole in Time: Our Future Evolution and the Meaning of Now,* by Peter Russell.

discusses the similarity between Terence McKenna's Timewave Zero[29] and the spiral of time, with ever-shortening epochs and accelerating change, converging on a point of maximum novelty. He calls this "compression," and notes the coincidence between the end of the thirteen-baktun cycle and the end of McKenna's timewave, seeing 2012 as the center of the spiral and a likely consciousness shift point. He compares the process to the evolution of a star that becomes more compressed until it becomes a singularity, or black hole. This is what lies at the center of the spiral—what Russell calls the "white hole in time." As we approach it, "we will be experiencing an ever-accelerating spiral of inner awakening."[30]

INFORMATION DOUBLING

Philosopher, futurist, and psychonaut Robert Anton Wilson, in his 1982 book *Right Where You Are Sitting Now,* included an analysis of history he calls the "Jumping Jesus Phenomenon."[31] In a 1973 study by French economist Georges Anderla, of the Organization for Economic Cooperation and Development (OECD), a basic unit was used to represent the scientific knowledge of humanity in the year 1 AD. Calculations were then made to determine how long it took for that information to double. Tongue in cheek, Wilson renamed the unit "one jesus," after a certain "celebrated philosopher" who was born that year. Then he charted the time it took for the knowledge to double, to the level of two "jesuses"(2 *j*), and again, how long it took to double again, to 4 *j,* and so on, illustrating his account with historical examples.

So starting at the estimated beginning of *Homo sapiens,* it took until 1 AD to arrive at the basic unit of 1 *j;* that is about 40,000 years if we take 40,000 BC as the beginning of the species. It took 1,500 years for the next doubling; then 250 years for the next; then 150 years; then fifty, ten, seven, six. That takes us to the date Anderla's study was completed, in 1973. In 1982, when Wilson published *Right Where You Are Sitting Now,* he estimated that the next doubling happened in 1978–79, and that in 1982 it doubled again to a level of 512 *j.*

By 1990, when Wilson wrote another article on information doubling, "Gaia: The Trajectories of Her Evolution (part 4),"[32] he said that information was then doubling every eighteen months. Further investigation had caused him to modify the earlier figures slightly, so that it took 40,000 years to reach the first unit of knowledge, which was now placed in the Bronze Age. A further three thousand years then led to the first doubling at the time of the Roman Empire, around 1 AD. The name of the knowledge unit was thus no longer relevant, though Wilson didn't bother to replace it, just referring to them as "doublings" (I would have changed a "jesus" to a "merlin" if it was up to me). The other times of doubling remain as they were in Anderla's study.

Wilson shows the acceleration process by comparing the revolution in agriculture, which took thousands of years to transform society, to the Industrial Revolution, which happened about ten times faster, to the Computer Revolution, which is happening ten times faster again. He goes on to say that it seems to be "a fractal process, with some quantum jumping included." He then points out the similarities to the fractal timewave of McKenna, and its acceleration of novelty:

> [McKenna] predicts that by 2012 we will pass through major quantum jumps in information/power every day, at the beginning of the year, and eventually every second, in the winter, climaxing when the evolutionary breakthroughs start coming in nanoseconds. I have never quite succeeded in explaining to myself why I find that scenario unbelievable. I do find it lovely, mathematically; I find it intuitively appealing; I admit it fits most of the data of my own survey of history, but . . . I evidently have my own conservative streak. I cannot imagine socio-evolutionary quantum leaps accelerating to the rate of 1,000,000 per second.[33]

So even though a graph in Wilson's 1983 book, *Prometheus Rising,*[34] shows the information doubling as an asymptotic curve exhibiting an exponential growth that goes vertical just after the year 2000, it is only

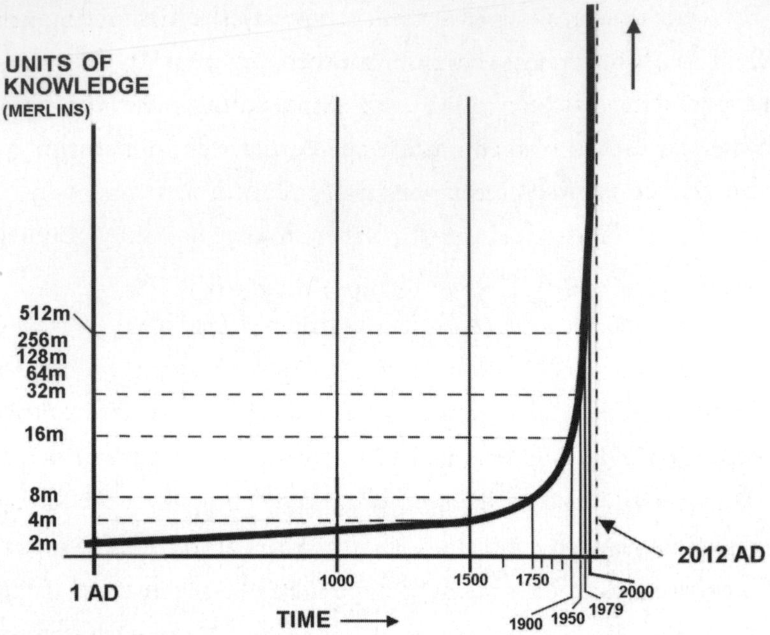

Fig. 12.10. Information-doubling time. Adapted from *Prometheus Rising,* by Robert Anton Wilson.

Wilson's "conservative streak" that stops him from saying any more. But he agrees with Peter Russell, in *The Awakening Earth,* that Gaia (a name used for Earth as a living being, originally an ancient Greek Earth goddess) "has eighteen of the nineteen traits of a living organism, as formulated in General Systems Theory. The one and only trait of a living creature missing in Gaia, Russell points out, consists of *ability to reproduce . . .*"[35]

Wilson thus sees future space colonies as seeds that will lead to the reproduction of Gaia, when Earth will exist as a living being. However, if we think of human children, they cannot reproduce until they have finished developing; until then they only have the potential to reproduce.

Although she gives no references, near-death researcher and visionary author P. M. H. Atwater claims, in her book *Future Memory,* that information doubling has now been scientifically projected to 2012: "It has been estimated by some scientists that by the first six months of 2012, global information could double each day. During the second six months of that year, they claim that global information could double each hour. If these scientists are right, come 2013, the doubling of global information could well be each and every second."[36]

QUANTUM AWAKENING

On December 27, 1978, Ken Carey was in bed with a cold and high temperature when he experienced what sounds at first like an auditory hallucination. When he realized the importance of what he was hearing, he grabbed a typewriter and began to transcribe what he heard. He kept on typing for eleven days, well after his cold had passed. He felt as if "something enormous" was looking through his eyes, which he interpreted variously as angelic, as extraterrestrial, or even as "informational cells in a Galactic organism." He is convinced that it was not channeling, since there was no trance, loss of consciousness, or voice change.

The information he received reads like a schedule for Teilhard's awakening noosphere, described earlier in this chapter. He was told that there was a series of three waves of transmissions of telepathic awareness. The first had taken place between 1967 and 1969; the second was happening then, between 1977 and 1979; and the third wave would be between 1987 and 1989, which was verified ten years later when Carey found himself receiving another set of transmissions. The transmissions, as he said later in his *Starseed* book series, were to help prepare humankind for the "first unified movement of the awakened planetary organism."[37] Carey says that this will be a "birth," when "the Star Maker will consciously awaken in all systems capable of sustaining universal awareness," and that "the Earth has a due date sometime during the second decade of the twenty-first century."[38]

Even the nineteenth trait of a living organism, the ability to reproduce, is foreseen in the millennium after "Quantum Awakening."[39] The transmissions also seem to confirm the connection to the critical mass of world population, since levels at the moment of quantum awakening are predicted to be around eight billion (8×10^9),[40] and afterward will level out at around eleven billion.[41] There is also a description of time cycles that is very reminiscent of Timewave Zero: "The curves of every cycle since the beginning of time—cycles short and cycles long—will crest together in a single moment soon after the turn of the Third Millennium AD."[42]

As with timewave theory, the event is not just confined to our solar system, but in this version it will coincide with the time when "the universe reaches a point of maximum expansion,"[43] and we presumably wouldn't know that was about to happen because the farther we look into the universe, the further into the past we have to also look (because of the time light takes to travel).

Carey says that the "return of the gods" was pinpointed by the Maya at winter solstice 2011 AD,[44] but as we know, the end of the thirteen-baktun cycle corresponds to winter solstice 2012 in the Gregorian calendar. So is this Carey's conscious mind clouding the transmission with misinformation he may have obtained from Michael Coe or Frank Waters?*

If the transmission is pure, then how did this error creep in?

It is now agreed among most Mayanists that the thirteen-baktun cycle will end in December 2012, though some say December 23 instead of 21. Yet Carey's transmissions repeat the 2011 date throughout all three books in the series, which was written over a period of nine years.[45] If the awakening event was actually due in 2011, and the Maya had been a year off, then the transmission should have been worded differently. Unfortunately, this makes me suspicious of the transmissions, and I notice that the first *Starseed* book was copyrighted in the same year that Russell's *The Awakening Earth*

*It seems that Frank Waters, in *Mexico Mystique,* quoted the December 24, 2011, date from Michael Coe's book *The Maya* (1966), but in later works Coe used the Lounsbury correlation that fixes the thirteen-baktun end point (13.0.0.0.0) as December 23, 2012.

was published—1982. However, Carey's description of the Quantum Awakening process is still a fascinating possibility.

> There will be a great shift then, a single moment of Quantum Awakening. In this moment, the smallest interval of time measured in these dimensions—the interval that occurs in every atom between each of its billions of oscillations per second—will be lengthened unto infinity. An interval of non-time will expand. Through that expansion eternity will flow. Some will experience this moment as minutes or hours, others as a lifetime. Still others will experience this flash of non-time as a succession of many lives, and some few will, in this moment, know the Nagual itself, the great nameless Presence that exists before and after all these worlds.
>
> In the expanse of the non-time interval, human beings will have all the time they require to realize, experience, and remember full consciousness of their eternal spirits, and to recall the origin of their individuality in the primordial fields of being. All will have ample time to recharge their form identity and its biological projection with the awareness of who they are, why they have individualized, and why they have chosen to associate with the planet's human expression. Each one will have the choice to return to biological form or to remain in the fields of disincarnate awareness.
>
> Those who choose to return to human form will do so, fully aware of who they are. No longer will they be but partially incarnate; they will resume biological residence with the full memory and consciousness of their eternal natures, sharing the creative capacities of the Star Maker, whose reflective cells they will then know themselves to be.[46]

PART 3

Beyond the Veil

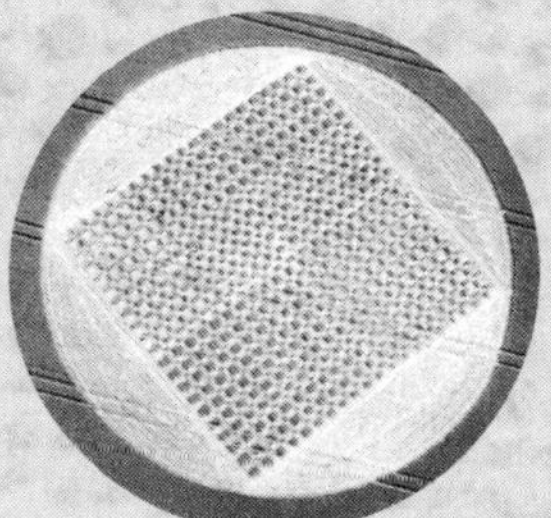

The fragment of the Paris Papyrus suggests that the original light which shines in is the light of the Sun. But what if there is a special kind of light, such as the glow of an exploding Galactic Center, which triggers this transformation, whether we are prepared or not? In addition to the physical disaster, we could also be faced with a spiritual or psychological disaster in which the transformational process is triggered automatically, creating havoc among those who are not prepared for the astro-alchemical transmutation.

Jay Weidner and Vincent Bridges,
A Monument to the End of Time

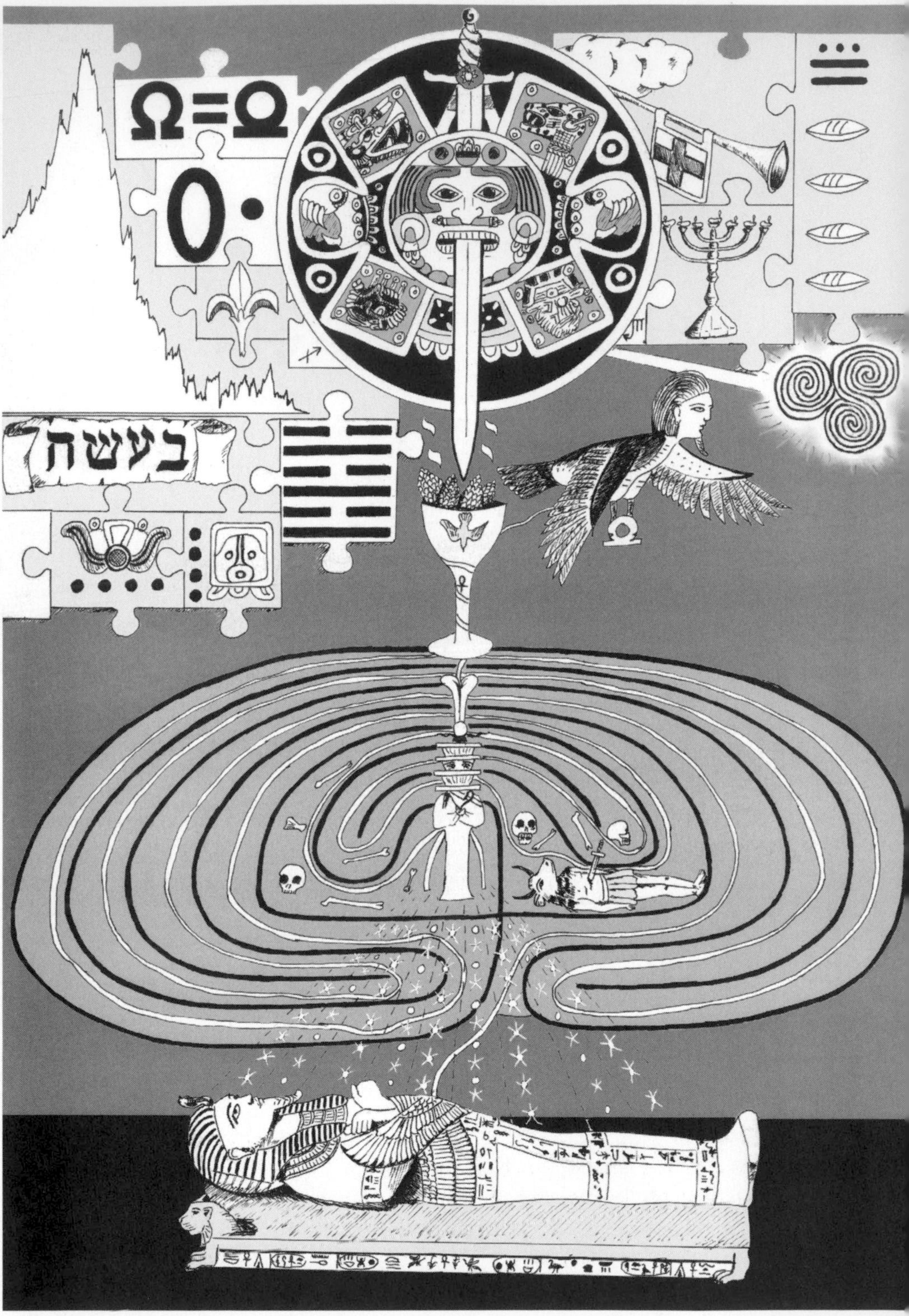
Ω=Ω
0
בעשח

13
SHAMANISM

HOLY FLESH

The Spanish historian Bernardino de Sahagún, a Franciscan missionary to the Aztec people of Mexico, recorded that the Aztecs consumed certain intoxicating mushrooms. In the Nahuatl language these mushrooms were known as *teonanácatl,* which means "flesh of the gods." The Maya shamans and kings are known to have used a mushroom called *k'aizalah okox,* or the "lost judgment mushroom," to produce visions and to travel to the underworld, Xibalba. They also used the venom of a poisonous toad called *Bufo alvarius,* which was probably dried and then smoked.* The mushrooms are varieties of *Psilocybe,* probably *Psilocybe mexicana* or *Psilocybe cubensis,* and contain the hallucinogens psilocin and psilocybin, which are both tryptamines (4-hydroxy-DMT and 4-phosphloroxy-DMT). The toad venom contains 5-methoxy-DMT, and bufotenine.†

Fig. 13.1. Mushroom use, depicted in the Codex Vindobonensis of the Mixtecs.

*This is found today in the Sonoran desert of Mexico and should not be confused with the Colorado cane toad, *Bufo marines,* or *Bufo americanus,* the venom from which only contains bufotenine. Mexico "represents without a doubt the world's richest area in diversity and use of hallucinogens in aboriginal societies." Devereux, *The Long Trip,* 105.

†While bufotenine is also a tryptamine (5-hydroxy-DMT), its hallucinogenic effectiveness is disputed (see www.erowid.org).

It was *Psilocybe cubensis* (also called *Stropharia cubensis*) that the McKenna brothers ingested during their trip through the Amazon, combined with the smoking of the bark of another hallucinogenic plant called *Banisteriopsis caapi* (better known as ayahuasca—note that the term also applies to a hallucinogenic brew that contains the vine plus additional psychoactive plants). This combination, which triggered communication with an "alien insectoid anthropologist" being, resulted in the I Ching–based Timewave Zero theory. In fact, Terence McKenna later stated that when pure DMT (*N-N*-dimethyltryptamine) is smoked, it facilitates communication with extradimensional life-forms that he calls "self-transforming machine-elves," who told him that the "laws of physics will be transformed around 2012."[1]

I recently found another user of the sacred fungus, whose experience seemed to verify what McKenna said:

> Our Awakening . . . My partner and I took some Magic Mushrooms, and what should have been a fun trip turned into a vision of the future. It woke us up and pulled us out of the Old World for once and for good . . . We also received a date, like a programmed ending point, or beginning . . . 2012. This is the moment where time and therefore also space will fall apart, and we will remain with only our spirit (energy). And it isn't anything dramatic, because at that point you will realise the body is only a blockage, it cannot deal with the speed, and you will want to be released from it.[2]

In June 2004, I received more confirmation on the mushroom–2012 connection, when a Web surfer known as Unborn Mind wrote to tell me about the experience, a few weeks previously, of a friend of his. This person was inexperienced in the use of psychedelic mushrooms, and he ate four to five grams of them with the sole intention of becoming intoxicated: "He has no knowledge of religion or prophecy. What followed after[ward] was described by him [as] 'the ripping away of everything as we knew it' and he was very adamant about the year this would happen (you guessed it, 2012). I just think it is very interesting coming from someone with absolutely no knowledge of what a Mayan even is."[3]

THE SPIRIT MOLECULE

So the Maya shamans ingested psychedelic mushrooms and toad venom containing methylated tryptamines. As we know, they (or the Olmecs or Izapans) also devised a calendar consisting of 260 katuns that terminates in 2012 AD. Terence and Dennis McKenna, while in the Colombian rain forest, ingested the very same methylated-tryptamine-containing mushrooms that the Maya had used, plus they smoked the beta-carboline-containing ayahuasca bark. They then decoded the twenty-six-level temporal hierarchy of Timewave Zero from the I Ching and found that the termination point was in 2012 AD.

The pineal gland is a small gland in the center of the brain, the

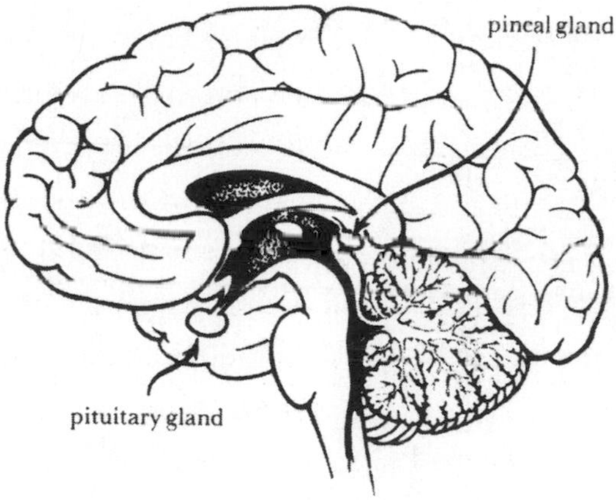

Right: Fig. 13.2. The pineal gland is situated at the center of the brain. From *The Opening of the Third Eye,* by Dr. Douglas Baker.

Below: Fig. 13.3. A comparison of tryptamine molecules that occur in plants. Adapted from *The Invisible Landscape: Mind, Hallucinogens and the I Ching,* by Terence and Dennis McKenna.

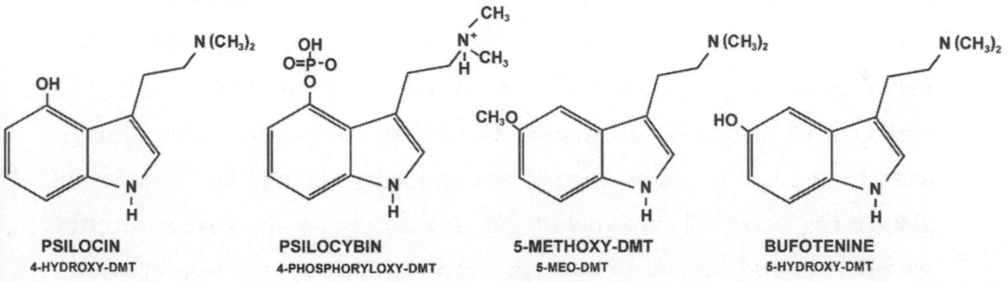

function of which we are only just beginning to understand. Descartes said it was the "seat of the soul"; others call it "the third eye," but it now seems that among its many functions, it can, under certain circumstances, produce both types of hallucinogen mentioned above—methylated tryptamines and beta-carbolines.

We have known since 1967 that a hormone found in the pineal gland[4] is actually identical to 6-MTHH,[5] which is a hallucinogen found in *Virola* and related species, and is sometimes used in the ayahuasca brew. It is a beta-carboline synthesized from melatonin, and at least two other beta-carbolines have also now been reported as being synthesized in the pineal gland.[6] One of these is known as pinoline, first identified in 1961,[7] and we will have more to say about it later. In addition, I have just discovered that there is also evidence that the pineal gland produces 5-meo-DMT,[8] and we now have new evidence that it also produces DMT itself.

Between the 1960s and 1970s, it was found that the hallucinogen DMT (*N-N*-dimethyltryptamine) was synthesized in human lungs, blood, and brain (known as endogenous DMT, or DMT made inside the body, as opposed to exogenous DMT, which is taken into the body from an external source). But in 1990, Dr. Rick Strassman (then associate professor of psychiatry at the University of New Mexico School of Medicine, in Albuquerque) succeeded in getting funding for a five-year study of DMT, and found substantial evidence for "an essential function in DMT production and assimilation"[9] in the pineal gland. His book, *DMT: The Spirit Molecule,* which summarizes the five-year study, was published in 2001 and makes for fascinating reading.

Strassman finds evidence that the pineal gland manufactures DMT in certain circumstances, and these include birth, death, near-death experiences (NDEs), and other mystical states, such as close-encounter incidents, as detailed in the work of Harvard psychiatrist John Mack, who worked with over a hundred abductees before his untimely death under the wheels of a drunk driver in 2004. Graphologist Caroline Taylor wrote a review of *DMT: The Spirit Molecule* exclusively for the Diagnosis 2012 website, and has included a dialogue between Rick Strassman and herself, in which she asks his opinions regarding McKenna, the timewave, and John Mack's work. She summarizes her impressions of the book:

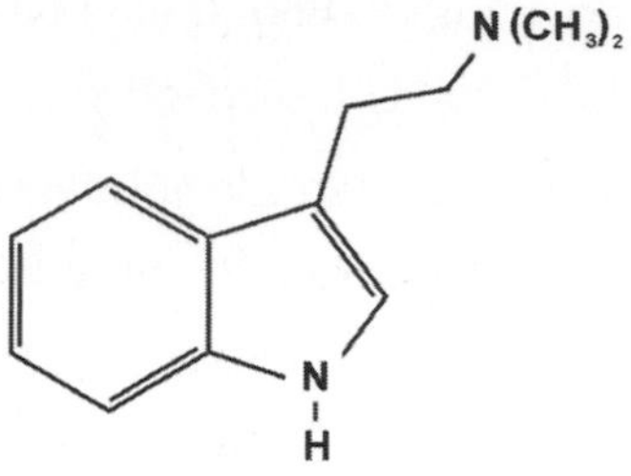

DIMETHYLTRYPTAMINE
DMT

Fig. 13.4. The "spirit molecule"—pure dimethyltryptamine, or DMT, apparently made naturally in the pineal gland as well as the lungs and the blood, according to Dr. Rick Strassman's findings. After *The Invisible Landscape: Mind, Hallucinogens and the I Ching*, by Terence and Dennis McKenna.

> The "spirit molecule" is, for Dr. Strassman, essentially a kind of chemical "vehicle," enabling disembodied consciousness to be carried to other realms of existence. If McKenna and the Maya Calendar creators were right, it may be that we will be needing this vehicle—en masse—in the near future. Rather than a redundant organ like the appendix (as it was once thought), the pineal gland may in fact be a FUTURE ORGAN. Upon this little pine-nut of an organ could rest a new model of brain and mind—indeed a whole new paradigm.[10]

At the end of her review, Taylor asks a very interesting question of Rick Strassman, and gets an equally interesting reply:

> **Caroline Taylor:** Put more specifically; do you feel that endogenous DMT could have a role to play in some future scenario, involving mass major trauma of some kind? [Do you agree] this would be followed, potentially, of course, by a sudden evolution of consciousness to a happier state, for some/all of our species . . . in other words, a biologically based eschaton? Such a scenario could be the result of some major environmental collapse, of course—or some harder-to-imagine one, involving the final timewave "concrescence" in 2012. In either case, do you agree with my feeling that endogenous DMT could, figuratively speaking, be waiting in the wings right now for its

star performance as a vehicle of mass transcendence, or purveyor of large amounts of human (and animal?) consciousness to some other dimension? (I hesitate to say "higher," for it could also be called "deeper," or "fuller.") In any case, the pineal gland (body) could be both the "port" of this vehicle, and possibly a vital "future" organ.

Rick Strassman: I have been nursing a theory along these lines for some time. Briefly, that the *N*-methylating enzyme responsible for DMT production turns on in everyone across the planet at the same time, thus ushering in the eschaton, messianic age, noncorporeal consciousness, or what have you. This could take place any number of ways; one way is a common cold virus gets a bit of DNA implanted into it that is inserted into all of us, which is programmed to turn on at some specific time, unleashing the *N*-methylating enzyme effects. This time would be an astrologically determined event, such as solar flare, particular constellation alignment, etc. Of course, along the lines of your interest in John Mack's work, that stimulus for enzyme activation could come from "them."[11]

THE PSI MOLECULE

In her book *Where Science and Magic Meet,* the parapsychologist Serena Roney-Dougal has studied the endocrine system and compared it to the yogic system of chakras. It has been known for many years that the seven power zones that lie along the spine between the perineum and cranium correspond to the positions of the glands in the endocrine system, but Roney-Dougal's study makes some important new connections. The central channel that connects the chakras is called the *sushumna,* and this is analogous to the central nervous system, while the other interweaving channels are called *ida* and *pingala,* which relate to the sympathetic and parasympathetic parts of the autonomic nervous system. A study of the definitions of the chakra functions, as given by Satyananda (Swami Saraswati Satyananda is a modern yoga teacher, of the Bihar School of Yoga, Bihar, India), shows a correlation with the

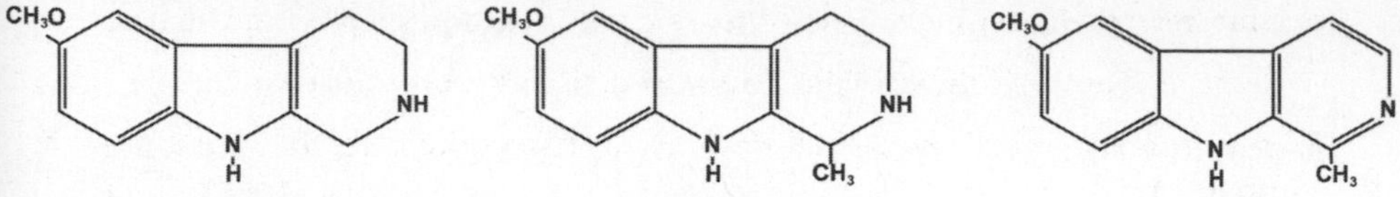

Fig. 13.5. Two beta-carboline molecules (on the left), which occur naturally in the pineal gland, compared with the harmine molecule that is found in the bark of yage/caapi/ayahuasca and in Syrian rue (*Peganum harmala*). Pinoline occurs in *Virola* and related species of plants, as well as in the pineal gland of humans.

biological functions of the related endocrine glands, and this includes the ajna, or third eye, since the pineal is like a vestigial eye and governs hormonal output according to how much light is received by the retinas of the eyes. It is responsive to the daily solar cycle, the monthly lunar cycle, the seasonal variation over a year, and even the 11.3-year (average) sunspot cycle (but in the case of the Moon and sunspot cycles, these are detected via the changing geomagnetic field). The ajna is a "command" chakra, just as the pineal is the biological clock that functions as an off switch for hormones and the pituitary is the on switch.

Roney-Dougal says: "Together with serotonin and melatonin in the pineal gland and retina, there is another class of compounds called beta-carbolines, which are produced by the pineal gland, our third eye, and which are chemically very similar to the harmala alkaloids, which are potent hallucinogens. The pineal gland is therefore involved in altering our state of consciousness to a potentially psi-conducive state."[12]

The pineal is involved in timing the onset of puberty, at which time it calcifies, and it has now been established that there is a relationship between poltergeist cases and puberty and adolescence. Roney-Dougal goes on to say that "young children have no sense of direction until about age 11, and then girls become really remarkable in their accuracy, while boys remain inept."[13]

It has been established by Indian researchers that a change in the geomagnetic field of about twenty nanoteslas (a tesla is a unit of magnetic flux density of one unit per square meter, and a nanotesla is a billionth of a tesla) is enough to "lower seizure thresholds and precipitate convulsions in people who suffer from epilepsy"[14] (in tribal societies, these people would be trained as shamans, and taught to control the condition and use it to advantage). But following solar storms or flares, the geomagnetic field can change by as much as 500 nanoteslas.[15] Roney-Dougal thinks that the calcification at puberty allows the pineal to become sensitive to electromagnetic fields. There is also evidence (after dowser Bill Lewis detected a spiral of energy around a standing stone that was later verified by a physicist from Imperial College, London, using a gauss meter) that water dowsers are responding to changes in the geomagnetic field.[16] As Paul Devereux, coordinator of the Dragon Project, which measures energies at ancient sites, puts it, "It is certain that many living organisms can sense variations in the geomagnetic field directly, and it is highly probable that human beings have such sensitivity as well."[17]

A correlation between ghosts and sudden increases in geomagnetic activity has now been established by the head of the Neuroscience Research Group at Laurentian University, Canada, Professor Michael Persinger, and the same for poltergeists,[18] while another correlation has been established between quiet geomagnetic activity and successful remote viewing and telepathy experiments.[19]

Knowing, as we do, about the changing geomagnetic field and its trend toward inversion, we could speculate that a sudden change of geomagnetic field on that kind of scale could be enough to trigger a pinoline-activated leap in psychic abilities. As the general electromagnetic field intensity drops (i.e., quiet geomagnetic activity), there may be a tendency toward increased telepathy and clairvoyance. But at the time of sudden jumps (i.e., sudden increases in geomagnetic activity) due to increasing solar storms and flares, there could be outbreaks of poltergeist activity and apparitions. In fact, harmine was actually called telepathine when it was first extracted from *B. caapi,* due to its effects; then it was found that the same alkaloid had already been extracted from Syrian rue (*Peganum harmala*) and named harmine. Another possibil-

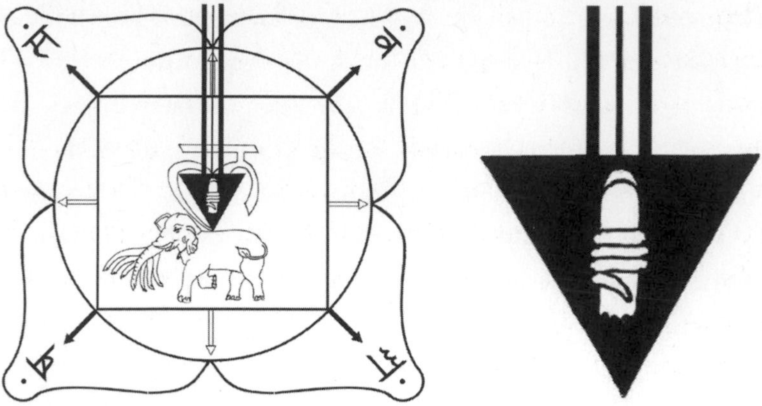

Fig. 13.6 The *muladhara* chakra (left), located at the base of the spine. In the center of the symbol for the base chakra (close-up on right), we can see the kundalini fire snake coiled three and a half times around the Shiva line of force that descends from above. Adapted from *Kundalini in the Physical World,* by Mary Scott.

ity is that the geomagnetic changes will cause kundalini, the fire snake, to involuntarily awaken from its slumber in the base chakras and rise up the spines of humans, hopefully causing enlightenment, or *samadhi*. This is a scenario predicted by several researchers, based on the legend of the return of Quetzalcoatl, the Feathered Serpent.

BICYCLE ACCIDENT

In 1985, a Dutch boy (with a Swiss-Norwegian lineage), Ananda Bosman, had a bicycle accident while living in England. It occurred when he was about fifteen years old, before the calcification of his pineal gland. The trauma triggered off a seven-year process in which the pineal eye was progressively opened, allowing an alternative neurochemical process to begin, in which endogenous (made in the body) pinoline and DMT were being synthesized and secreted into his system. As Ananda later wrote, "This increased especially during the night, when more melatonin was available to be synthesized into Pinoline, and hence allowing 5-meo-DMT."[20]

The altered biochemistry led to the boy having an out-of-body experience (OBE) in which he met "non-localised quantum selves." These beings were beyond extraterrestrial, or even ultraterrestrial, defining themselves as "InterUniversals" and collectively calling themselves "Emmanuel." One being identified himself as "Salvana." After the accident, Ananda witnessed multiple UFOs during a visit to England, and the following year one of the InterUniversals appeared to him while he was awake, with his eyes wide open, and Ananda later says he was "alerted by 'Emmanuel' in 1989 to the significance of 2012."[21]

Over the next few years, he was given all sorts of information, and developed several projects to help humanity prepare for the forthcoming Omega Point by developing a method to bring about "the translation of our biological body back into its vacuum hyperspatial Original Light Body form."[22] Part of this process involved a fourteen-day darkroom retreat, in which melatonin floods the system. Ananda writes, "Once there is a threshold of melatonin, then pinoline results. After some four to seven days, 5-meo-DMT awakens. One can see in three-dimensional Holon pictures, as the thoughts behind language."[23]

A holon is something that is simultaneously a whole and a part—something seen from a holistic perspective. In Ananda's words: "After some 10–12 days, one starts to see in infrared and ultraviolet, one can actually run across the room and touch a person, by seeing their heat patterns. The Holon images exteriorise and one is walking in a virtual reality . . . the DNA language, macro uploaded and interactive."[24]

The darkroom retreat is combined with a tryptophan-rich diet to give the pineal gland the raw ingredients it needs (tryptophan is destroyed by cooking). The diet must also avoid dairy products, alcohol, drugs, and many everyday foods that interfere with the process.*

---

*Special attention to diet is necessary because beta-carbolines inhibit a substance called MAO: monoamino-oxidase, which protects us from the effects of tyramine, which is in many foods and normally rendered harmless by MAO. This is why shamans usually go on a fast before taking ayahuasca. www.erowid.org/chemicals/maois/maois_info2.shtml.

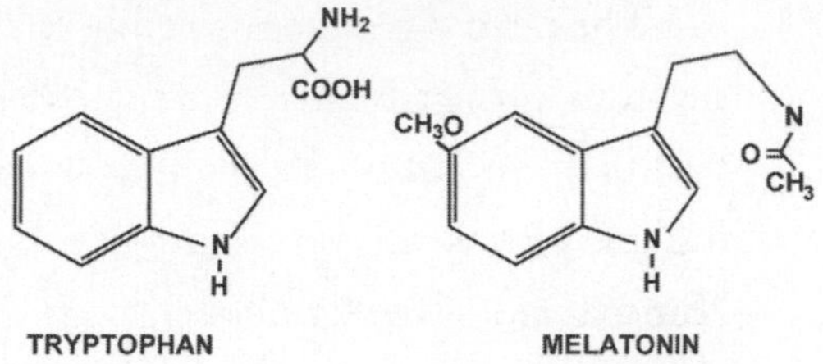

Fig. 13.7. Tryptophan is the raw material from which the pineal synthesizes melatonin and tryptamine. From tryptamine is synthesized DMT, and from melatonin comes pinoline and 6-MTHH. Adapted from *The Invisible Landscape: Mind, Hallucinogens, and the I Ching,* by Terence and Dennis McKenna.

Ananda reveals what may be news to Rick Strassman: Dr. Carl Callaway, of the University of Kuopio, Finland, already "conclusively demonstrated that the Near-Death Experience, Out-Of-Body Experience, and death itself (by blood plasma spinal tap readings), are based on pinoline, DMT, and 5-meo-DMT release."[25] Callaway's work was published in 1991 and 1993,[26] so it was being done at the same time that Strassman was doing his research. Callaway had also demonstrated that lucid dreams are induced by DMT, and this is something that will be relevant later.[27]

Ananda also claims that the gnostics and alchemists used a combination of Syrian rue (which contains harmine or telepathine) and acacia (acacia contains DMT) as a kind of Mediterranean ayahuasca, and he has produced a limited-edition book on this, *The Soma Conspiracy,* and one called *The Rig Veda,* in which he identifies the Rig Veda (or Vedas, a collection of ancient Hindu holy writings) as the source for the "Apocalypse code" of the book of Revelation.[28]

In 2002, seventeen years after his first contact with Emmanuel, Ananda was reading the Vedas (which some claim are as much as seven thousand years old) when he was "thunder-struck" to realize that they mention a deity, or *manu,* named Salvana—the same name as the one from the Emmanuel group who contacted him. Moreover, he was said to be "one of the central Manus from the seven Alpha-Omega Manu couples that overwatch and influence the species development" and, according to Ananda, is also "the Omega Manu of our cycle, into which our reality is emerging." Ananda claims that the Vedas also time this "emergence" as coinciding with the galactic alignment process, and he identifies Salvana with the noosphere: "The Vedas conveyed such an emergence to be astronomically marked by the precise alignments of our astrophysical back garden that accurately took place in 1998 and will again in 2012—an emergence we are amidst. The Omega Manu, in hadron physics jargon, acting as a kind of Macro Irreversible Coherent Noosphere of our Age."[29]

SYRIAN RUE

In the online *Diary of a Psychonaut,*[30] I found an account called "Time Stop," by someone known by the pseudonym Copehead, who had combined Syrian rue, mushrooms, and laughing gas (nitrous oxide).

> Let me add a little disclaimer here. I am not a follower of Terence McKenna. I do not subscribe to his theories, although I have read both *Food of the Gods* and *True Hallucinations,* as well as much of the material online. His ideas are definitely entertaining, but I have a hard time lending much credence to them. Having said that, here is the message I received, reconstructed from my messy notes as best as possible: "Time stops in 2012. Please be ready. The work must be finished. This is the moment all the great books of history have been written about. This is the birth of religion."[31]

Copehead's next account, "Mushroom Love," describes an experience following the ingestion of *Psilocybe cubensis.*

> Again I received the idea that something monumental was to happen in 2012. I was seeing images of McKenna, along with things like "2012, are YOU ready?" I was told that we make our own reality [and] that no one needed to be unhappy. When I asked if the 2012 thing were [*sic*] real, I was told (the voice was now present) again: "We make our own reality. It is as real as you want it to be." I knew suddenly that this could happen. I realized that 2012 is just a date and that I could transform my own consciousness by then if I

> wished to do so. I decided right there to do so, and to help as many others to do so as I could. It was a very religious feeling.[32]

In Copehead's third relevant account, "Monkeys in Space," the writer says:

> I had a vision of the universe, containing itself. I saw a series of calendars, clocks and other time measuring devices, the most perfect/accurate of which seemed to be the Mayan calendar. Again I experienced the "time stop" phenomenon as well as the mushroom voice. Again it told me many strange and wonderful things. Here are a few of those things: The mushroom, mankind, and several other species on this planet (including dolphins) are all collaborating on a colossal project, the ultimate nature of which was somewhat elusive to me. Most of humanity is completely unaware of their participation in this project, but the majority of the other players are fully aware of their own parts in it. This project will reach completion in the year 2012. It has something to do with time, consciousness, and perhaps leaving the planet or returning it to a pristine state.[33]

So what all this seems to mean is that tryptamines and beta-carbolines enable contact with some information source that is trying to get us to prepare for 2012. But does this happen only with these molecules?

SALVIA DIVINORUM

Salvia divinorum, diviner's mint, is widely thought to be the "divine plant" of pre-conquest Mexico, known in the Nahuatl language as *pipiltzintzintli.* The Mazatecs still use it today. I have encountered several people who have received information about 2012 following the use of this sacred plant. As I write, it is still legal, as far as I know, in most places except Australia. Charlie Sabatino, who lives in Mallorca, Spain, told me: "I took some strong drugs (*Salvia divinorum*) and 'heard' that I was to write a screenplay called 2012 (I'm a writer). I grabbed a pen and then started scribbling whatever came into my mind; it was like taking dictation."[34] The notes extended for six pages and included such subjects as the Mayan calendar, UFOs, standing stones, telepathy, visions, and dreams. Sabatino concluded, "Geoff, I was a really normal guy once. Italian butcher's son from the Bronx, N.Y. I didn't ask for any of this . . . I blame my wife."[35]

Someone called Damaeus also had a vision of 2012 with *Salvia:* "Concerning *Salvia divinorum* and 2012, I gather from my experiences with *Salvia* that none of us will ever see the destruction of the Earth. We will hear the calling before it ever happens and our consciousness/spirit/soul will simply float away, leaving our physical bodies to be destroyed with the worldwide earthquakes or whatever other disasters nature has in store."[36]

Another Web surfer, who calls himself Canul, offers the following description about his first experience with *Salvia:*

> In my first experience I felt and saw things of this reality that made me think of 2012. I had a perception of reality that is commonly described in Salvia divinorum literature: experiencing two

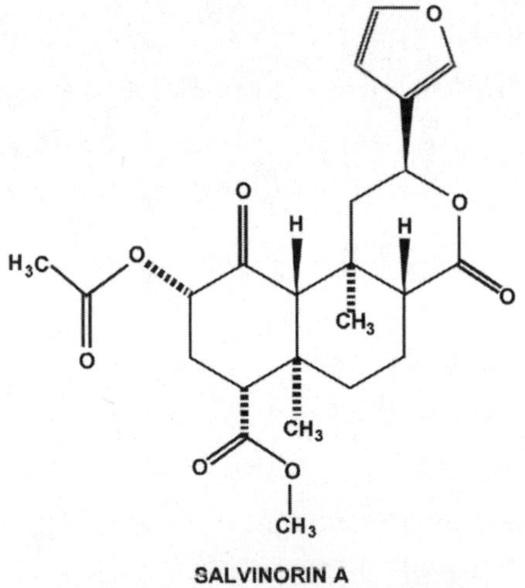

Fig. 13.8. The salvinorin A molecule

> different realities in parallel. One reality was the one that habit tells me [is everyday reality], the other one was 2012 nearby. I don't attribute this to the enthusiasm I have over 2012. It was authentic. I am reminded of the scene in the movie Artificial Intelligence, where the boy runs away with the other A. I. models from the "robot gatherers."[37]

A psychonaut named Shivasix says: "Salvia divinorum is a powerful plant helper sent to ease our transition at the end of time,"[38] which he equates with the end of the thirteen-baktun cycle in 2012.

And the final anecdote on 2012 information via exogenous molecules comes from Tomorrowlander and his girlfriend, who had a large dose of a rather dangerous cough mixture and touched their foreheads together (pineal area), at which point an audible angelic voice said, "In the year 2012, we will come . . ."[39]

14

OTHER ALTERED STATES

NEAR-DEATH EXPERIENCES

I have recently discovered several apparently independent accounts in which people undergoing a near-death experience (NDE) have been given access to information about 2012. The most detailed account I found is that of Cassandra Musgrave, whose experience happened in 1992.

> I feel that my near-death experience was a real gift. I feel it was a real blessing. It really awakened me. I was water skiing in northern California and when I fell down by some freak accident, the rope twisted around my left arm and dragged me behind the boat. And I found myself being pulled at a very rapid speed and unable to get any air. And my friends, who were goofing off, didn't stop the boat because they weren't paying any attention.[1]

As Cassandra began to drown, she entered into a dreamlike state that was the start of her NDE.

> All of a sudden, I was out of my body watching myself being pulled along and thinking "This is really incredible. This is really quite amazing." On an inner level, I was being pulled through a very dark tunnel.

> I didn't feel afraid at all. All of a sudden, I found myself coming into a wide open space with stars all around, like out in the universe.[2]

It was then that Cassandra received a prophetic vision.

> Basically, I saw that there was a twenty-year period from 1992 to 2012. Things will be greatly accelerated on Earth. All these things will be manifest by great Earth changes: earthquakes, floods, tidal waves, great winds. I also saw there were certain areas that would be particularly affected—the areas of the east coast [of the United States], which will be surprised regarding earthquakes. I remember very clearly Japan slipping into the ocean. I was shown there was going to be something akin to three days of darkness. I don't feel it is from a nuclear war. To me, it was more of a feeling of natural Earth disasters with smoke from volcanoes that would block the Sun. We are all going to be on a roller-coaster ride, and yet it is not forever. If we have darkness for three days, it will pass. We will always have the light.[3]

Ken Kalb's book *The Grand Catharsis* is a history of consciousness from 1987 to 2012, weaving together many bodies of knowledge to focus on the real issues of the new millennium. It was based on a lucid vision gained during a near-death experience in 1969. Dannion Brinkley, now a best-selling author, is an even more famous case of NDE, since his experience lasted a full twenty-eight minutes, the longest clinically documented NDE ever recorded up until that time (I have since heard of a case that occurred in 1982 that lasted over an hour and a half).* Brinkley's experiences during those twenty-eight minutes are related in great detail

*The NDE of Mellen-Thomas Benedict lasted for at least an hour and a half, verified as an hour and a half from the time the body was discovered by Benedict's hospital caretaker, using a stethoscope, blood-pressure readings, and heart-rate monitor. Whether this counts as "clinically documented" or not is unknown. See: www.near-death.com/experiences/reincarnation04.html.

in *Saved by the Light.* Mike Pettigrew, the founder of the Institute for Afterlife Research, provides a compact summary of Brinkley's book on his website: www.mikepettigrew.com/afterlife/html/dannion.html:

> On September 17, 1975, Dannion Brinkley was talking on the phone during a thunderstorm. A bolt of lightning hit the phone line, sending thousands of volts of electricity into his head and down his body. His heart stopped, and he died, but in the process, he had a near-death experience. When Brinkley revived in the morgue after 28 minutes of death, he had an incredible story to tell.

Fig. 14.1. *Entrance to the Celestial Paradise,* by Hieronymus Bosch. Many near-death experiencers describe traveling through a tunnel into a bright light.

> Brinkley found himself floating above his body, and he then went down a spiraling tunnel into a brilliant light, where he was approached by a glowing silver being that radiated love. He then reexperienced his life (which had been one full of aggression, including a spell in the Marines). He judged himself, feeling the pain he had caused others, and also the positive effects he had on others. He was taken to a crystalline city of light. He saw a group of twelve beings representing the signs of the zodiac, plus a thirteenth being that presided over the other twelve.

> Dannion was approached by each one of these beings with what looked like a small box. Each time this happened the box opened up

to reveal what looked like a small TV screen. Scenes showing events from the future appeared on the screen that would grow in size until Dannion found himself inside the scene. These were major world events that would take place in the future. There were over 300 of these scenes. One hundred and seventeen of these were major world events that Dannion was told would take place in the future (1975 to 2012). Dannion was immersed in events ranging from the nuclear accident at Chernobyl, the break-up of the former Soviet Union, the Gulf War, changes to our Earth's climate, breakthroughs in technology and many other things. Now [in 2002], about 100 of these very specific global events have already occurred.

After getting instructions for the building of special centers, Brinkley was sent back to his body, which had already been issued a death certificate, and was cold and covered in a sheet (and, in this version, about to be taken to the morgue). Following a lot of pain and paralysis, Brinkley found he had telepathic and precognitive powers. His predictions are said to be coming to pass with astounding accuracy—100 out of the 117, according to Mike Pettigrew in 2002. Regarding 2012, when asked by an interviewer, "How bad will it get, Dannion?" Brinkley replied:

Or how good will it get? I think it will get really bad geographically and topographically, but in spiritual consciousness it is going to grow in leaps and bounds. Oh, I wouldn't miss it for the world! This Earth is changing as the entire universe is changing, because between 2004 and 2014, more precisely between 2011 and 2012, we will experience the return of an energy system that existed here a long time ago. You can call it a spiritually uplifting consciousness, the Second Coming, or the birth of the Aquarian Age.[4]

In closing, Brinkley says, "This is my hope, rather than a prophetic prediction: as all the Earth changes, both subtle and massive, come to pass one by one, and most especially with the occurrence of the electromagnetic polar Earth shift (between 2012–2014), I hope we all will seize the opportunities for growth, which will undoubtedly present themselves to promote our inner development of spiritual virtues."[5]

P. M. H. Atwater underwent three NDEs and a kundalini experience (when the serpentine energy that lies dormant at the base of the spine is activated and moves up through the seven chakras, or power zones, on its way to the crown chakra). As a result, she became a researcher of near-death experiences and the author of eight books. In her research into child NDEs, she found that many of the children reported predictions of Earth changes, and she has correlated the timing of the culmination of these to December 2012, which she says will be not only the start of the Fifth World of the Maya but also the "final blossoming and fruition" of the "fifth root race"[6] in the Theosophical scheme. The fruits would be the first manifestation of the next evolutionary stage, the sixth root race (the Fifth World of the Maya is the Sixth Sun of the Aztecs). These forerunners of the next phase are known as the gifted and psychic Indigo children.*

An Australian woman named Jackie, who experienced two NDEs in 1979, says that she received an important message from "the angels," but couldn't quite remember what it was. Ever since then she has been looking for cues to jog her memory. While attending a Vancouver

*The concept of seven root races, which relates to the development of the seven chakras, was brought to the West via Madame Helena Blavatsky, and also used by Rudolf Steiner and Edgar Cayce. The base chakra was developed in the first root race, the Polarians; then came the Hyperboreans, then the Lemurians, the Atlanteans, and the Aryans (nothing to do with Nazi beliefs). Since each chakra relates to a color, the current outbreak of psychic children, known as Indigo children due to the color of their aura, is taken by many to be a sign of a coming transition to the sixth root race. Indigo relates to the color of the sixth chakra, the ajna, or pineal eye. Atwater originally put the start of the fifth root race around the time of Christ, whereas other sources put it much further back; her more recent work now concurs with this. Each root race has seven subraces, and we are said to now be in the fifth subrace. The implication is that we may simply be at the juncture of the fifth and sixth subraces. The next subrace is called "Nova men." See www.kheper.net/topics/Theosophy/root_races.html for more information. For more about psychic/Indigo children, see www.psychicchildren.co.uk.

NDE conference, she got the cue she needed, during a talk by P. M. H. Atwater. After the talk, Jackie said:

> It's all come back. It started yesterday in the experiencers' session. Then this morning—I already knew everything that woman said! In her talk she informed us that NDErs are, in effect, God's new chosen people. They're sent back to life with extraordinary powers to pave the way for the consciousness revolution that will transmute our planet, beginning in 2012.[7]

In 1991, at the age of thirty-five, singer-songwriter Pam Reynolds of Atlanta, Georgia, had one of the best-documented near-death experiences of NDE research, involving a vivid vision of 2012. Radio host Art Bell interviewed her on the "Coast to Coast" AM radio talk show on the night of December 6–7, 2001. The following summary was recorded on Linda's Coast to Coast weblog (by Linda): "This happened about ten years ago when she was dying from a brain tumor . . . She saw future events like Earth changes, a breakthrough in physics and a consciousness change near 2012. She didn't see the end of the world, just changes."[8]

OUT-OF-BODY EXPERIENCES

Professor Christopher Bache is director of Transformative Learning at the Institute of Noetic Sciences in Petaluma, California, and does research on self-induced out-of-body experiences. O. Frank Turner, author of *The Science of Spirit,* says of Bache, "In one of his OBEs, he was told by a number of points of light (he assumed they were spirits) that around 2012 the world would go through a global crisis unparalleled in human history (apocalypse?). They did not tell him what would happen. Because he is a scientific thinker, he began to look for what this crisis could be."[9] Bache has now concluded that it will be an environmental disaster largely caused by humanity. Coincidentally (or not), 2012 is the target date for the world to get its greenhouse gases

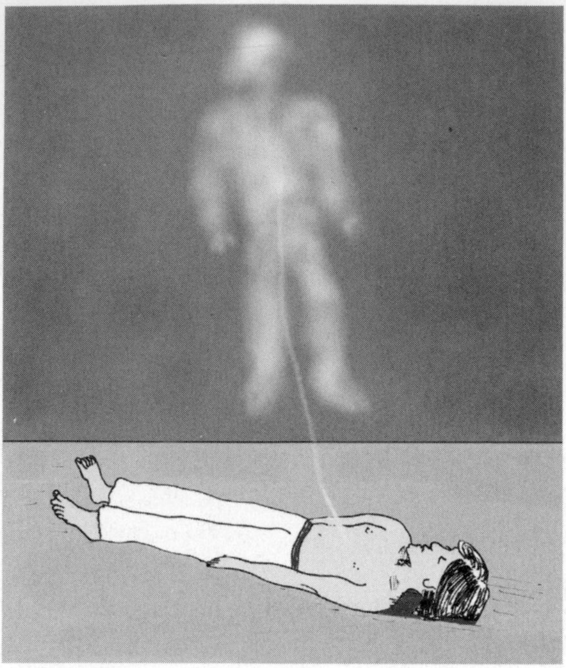

Fig. 14.2. Some out-of-body experiencers report that their astral body is attached to the physical body via a silver cord that connects to the solar plexus (though others report the head connection already depicted in chapter 9, fig. 9.3).

under control. This is known as the Kyoto Protocol, and infamously the United States has been dragging its heels, avoiding signing the treaty.

Native American elder J. Reuben Silverbird (a half Apache and half Cherokee—or half Navajo by some accounts) was born blind, and up to the age of five he traveled out of his body, at which time he miraculously regained his sight. At the age of seven (sometime before 1960), he says,

> Just two years after my wonderful miracle of sight, I had another, most horrifying vision that will remain with me until the day the Creator decides I am ready to go to the mountain and join my dear parents, Sun Bear and Thomas Banyanca. The experience conceived [*sic*] the year 2012 and my death and bringing back to life by a spiritual, almost Holy Stranger. This story and the one about my blindness is too long and will be a part of my book *Spiritual Healing Words of Wisdom.*"[10]

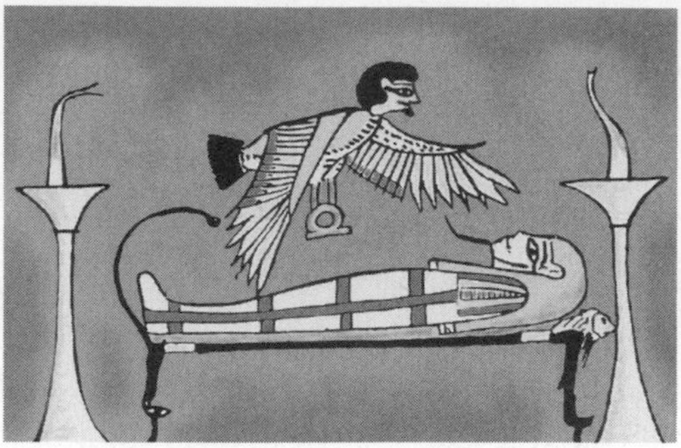

Fig. 14.3. The ba, or Egyptian conception of the astral body, visiting its mummy. It is holding the symbol of eternity. Adapted from the Papyrus of Ani.

LUCID DREAMS

When a dreamer becomes aware that he or she is dreaming, the whole dream becomes very realistic and the dreamer can then control the dream. This state is known as lucid dreaming. Lucid-dream researcher Stephen LaBerge, of the Lucidity Institute, says, "OBEs are actually variant interpretations of lucid dreams,"[11] and to support this position he cites a laboratory study.[12] However, other researchers, such as parapsychologist Keith Harary, disagree, citing other studies as evidence:

> The OBE and the lucid dream are not one and the same. Though the two are often confused with each other, laboratory studies show they are psychologically and physiologically distinct. For one thing, you don't have to be asleep to have an out-of-body experience. For another, the brainwave and eye movement patterns that have emerged from OBE studies in the laboratory are quite different from those associated with dreams.[13]

Since techniques have been developed to induce an OBE from within a lucid dream, researchers like Dean Walker believe "OBEs and LDs to be part of a continuum of experience with considerable overlap between them."[14] Whichever position we take, the fact is that we have already seen that OBEs are a source of 2012 prophecies. It is also a fact that there are some accounts of lucid dreams featuring 2012.

Nick Cumbo, a.k.a. Explora, runs a website called Dreampeace,[15] which contains a forum that "aims to bring together a large circle of dreamers, collaborating in mutual dreaming adventures." In 2002, Cumbo had a dream that included the following:

> Suddenly, my attention is drawn to a large object in the sky. Looking up, I am stunned to see the words "2012, it is their last superstar" printed on a large banner. The banner is huge, but seems to be moving away. It is attached to an amazing spaceship, a white and blue ship, with bright flashing lights, which are travelling away quite quickly. I am amazed by the reality of the ship and its message, and continue focusing on what the banner says. As I do this, the words change to "2012, it is their next superstar."[16]

Since then, in March–April 2003 and March–April 2004, the forum has attempted a collective induction of dreams about 2012.[17]

The imagery from another lucid dream, at a site called DreamImages, implies the advent of "electrical changes" and invisibility, just prior to a general "ascent on an escalator" that is labeled "2012–2025" in surroundings that look like New Zealand.[18]

In a precognitive dream that software designer and crop circle researcher Paul Vigay dates as "sometime between 1998 and 2012," and that "almost became a lucid dream," he reports witnessing a decloaking of extradimensional UFOs in the East Kennett (near Avebury, Wiltshire, UK) area.[19] For more 2012 dreams, see the Great Dreams website at www.greatdreams.com.[20]

MEDITATION

Holistic counselor, teacher, and author Gloria Karpinski gave a workshop in 2003 titled "Who Shall I Say Sent Me? Rediscovering Our Power before 2012," in which she explained the significance of the subtitle.

> "I was 'getting' that year in meditation long before I knew why," Gloria told the September workshop participants. Indeed, it was Gloria who first mentioned the year 2012 to us at the original Women and Wisdom group in New York, 25 years ago. "I had no idea that year was pinpointed in the Mayan Calendar as the end of a 25,000 year cycle," she explained. "The last time this kind of cycle ended was, of course, before recorded history. So we don't know what form it took then or will take this time."[21]

The implication is that it was before 1978 that Karpinski first picked up the significance of 2012 while meditating. In her book *Barefoot on Holy Ground: Twelve Lessons in Spiritual Craftsmanship,* she reveals that this was not just an isolated event, but one that recurred.

> The first few times that I received the number 2012 in a meditation as a significant date, I took it with a big dose of cosmic salt. Like the contents of Pandora's box, the prophecy pot has been overturned and the contents have poured out into the collective for anybody to pick up and interpret. I asked myself: Is this a truth I have been shown or have I simply picked up on an arbitrary date because so many people have energized it that it is accepted as truth? But I realized that this date had been showing up in meditations before I knew anyone else was getting it. Then, as I read various prophecies, learned the unusual astrological lineups of that year [Karpinski may be referring to galactic alignment, but see the discussion of the Great Conjunction in chapter 19 for other possibilities], and talked to others around the world who had received the same information, I realized that if I was off, then lots of us were off at the same time.[22]

She goes on to explain that she doesn't see 2012 as a date of catastrophe, but as the end of a period of purification and chaos.

PAST-LIFE RECALL

Graphic designer and author Randolph Weldon visited a regression hypnotist who took him so far into the past that he reexperienced memories from several different lifetimes. These previous lives included an Egyptian potter who had apparently lived through the disasters that occurred at the time Moses led his people out of Egypt. The potter witnessed the cyclone, earthquakes, and clouds of red dust that turned the rivers "red with blood," as described in the book of Exodus. Then came large hailstones, one of which crashed through the roof and killed his mother-in-law. With his remaining family, he joined the people stampeding out of the city and experienced a rain of burning oil. Weldon's regressed self observed that many Egyptians joined the fleeing Israelites, since Moses had threatened that his god would bring about the destruction of Egypt if the Israelites were not freed. The potter then witnessed the "pillar of cloud" that the people followed by day, that appeared as a "pillar of fire" by night, and the close pass of a "huge, dazzling, fuzzy yellow ball" that caused the waters of the Red Sea to part, while most of the refugees crossed to the other side. The potter and his family were at the back of the crowd, with the Egyptian chariots behind them, when a massive bolt of electricity flashed from the cometlike object to the ground and the waters crashed in and ended his life.

This vision precisely matched Immanuel Velikovsky's interpretation of the Exodus, which he put down to the "Birth of Venus," described in chapter 7. In Velikovsky's depiction, a cosmic collision—with a passing planet or star—caused a portion of Jupiter to be pulled out, which became a comet with a fifty-two-year period, and which eventually cooled and became the planet Venus. Velikovsky dated the event at around 1500 BC, though astronomers might argue Venus couldn't possibly be that young.

In his book *Doomsday 2012: A Survival Manual,* Randolph Weldon relates the aforementioned regression and its correlation to Velikovsky, and by combining the results of a similar past-life regression, he con-

Fig. 14.4. In Greek mythology, Vulcan cured Zeus (Jupiter) of his headaches by splitting his head open with an ax—and Pallas Athene (Venus) was born. *Atlanta Fugiens* by M. Maier, Oppenheim, 1618.

cludes that we are due for a repeat visit from the huge celestial object that triggered the Venus events, and that this visit is due in 2012.

Weldon cites Barbara Hand Clow's book *Eye of the Centaur,* in which she describes a past life as a Minoan priestess (in Knossos, on Crete) called Aspasia, who received a warning that a massive comet was coming that would cause Thera (a huge volcano, the remains of which are now the island of Santorini, in the Mediterranean Sea) to erupt. Aspasia was also informed that the comet returns every 3,500 years. Clow later recalled the experience of the passage of the comet and the resulting earthquake in which her life as Aspasia ended.

As also mentioned in chapter 7, Graham Phillips has already connected the events of the Exodus to the explosion of Thera, while Nigel Appleby has connected the giant planet Nibiru (usually said to be on an approximate 3,600-year orbit) to the same event, and predicted its return in 2012 (3,500 years later). But Weldon does not refer to Sitchin, Phillips, Appleby, or even Nibiru.

Amazingly, Weldon seems not to have read Zecharia Sitchin's *The Twelfth Planet,* but refers to another book, titled *The Genius of the Few,* by Christian and Barbara Joy O'Brien (Turnstone, 1985). They, like Sitchin, can translate Sumerian cuneiform script. The O'Briens come to many of the same conclusions as Sitchin. Weldon concluded, using Barbara Hand Clow's figure of a 3,500-year comet cycle, added to Velikovsky's date for the Venus incident, that a "Death Star" that contributed the matter that became the embryonic Venus (pulled out by Jupiter, rather than coming *from* Jupiter as Velikovsky argued) was expected to return around the turn of the millennium. It would be red in color, to fit in with the past-life memories, and would therefore be either a red-dwarf star, an infrared dwarf, or possibly a smaller, Jupiter-like planet. It couldn't be a red giant, as that would be the size of Earth's entire orbit. The minimum size of a red or infrared dwarf is between a tenth and a twelfth the size of the Sun, according to Weldon, but another possibility is a "proto-sun" like Jupiter (which some consider to be a failed star), but with a minimum scale of two times the size of Jupiter (or six hundred times the size of Earth). On page 173 of his book, Weldon says that the IRAS (Infrared Astronomical Satellite) should have detected any infrared dwarf out there when it scanned the entire sky in 1983, but it failed to pick anything up. He obviously hadn't seen the article in the *Washington Post* (see "Return of Nibiru" in chapter 7) that reported that the IRAS had actually picked up a planet the size of Jupiter, which may be part of our solar system.

It is ironic that Weldon used a past-life memory of Barbara Hand Clow to make his case for an imminent 2012 cataclysm. Clow has since written a book titled *Catastrophobia: The Truth behind Earth Changes in the Coming Age of Light,* which argues that the current obsession with imminent catastrophe is a kind of mental block in the unconscious mind of the species caused by a racial memory of a catastrophe in 9500 BC (as posited by Allan and Delair in *When the Earth Nearly Died*). She says 2012 will be a leap forward in consciousness.

The predictions of seers, including Cayce and Nostradamus, of cataclysmic events around the turn of the millennium added fuel to Weldon's fire, but the fact that they have not yet come true has led to Weldon's conclusion that the Maya and Hopis must have been correct about 2012 being the year of change, and the other predictions (Cayce expected it in 2000 AD and Nostradamus mentioned 1999 AD) are all about twelve years early.

Weldon says that the difference between the coming event and the previous visits of the Death Star is that this time we have over five hundred nuclear installations, millions of gallons of radioactive waste, and thousands of nuclear bombs that will all be unleashed in the mayhem, leading to a double catastrophe, the smattering of survivors all dying of radiation poisoning. That's the bad news.

The good news, says Weldon, is that life will continue in the astral planes and that it is up to us to use our remaining time to try and clear our karmic debts; otherwise we will end up on the lower astral, in the "House of Hopelessness" or the slightly better "Plane of Illusion," both of which the author claims to have visited during his nocturnal projections with his guides. He believes we should actually seek out adversity for ourselves in order to help balance any nasty behavior in this life or any previous ones.

It is intriguing that Weldon comes to the same conclusion as Appleby (though with more emphasis on the catastrophic), but using completely different sources. However, his principle evidence for the 3,500-year period is supplied by Barbara Hand Clow, who obviously doesn't agree regarding the recurrence of a catastrophe in 2012. It also turns out that Weldon's regression hypnotist is the same person who regressed Clow, Gregory Paxson; Paxson has reported that a third client of his also "recounted life as a young seer-priestess on that island, and died in the same apocalyptic disaster."

Taking these things into account, and including the possibility that Weldon's prior knowledge of Velikovsky and some of the millennium prophecies may have tainted his regression, we may be able to relax a little. There was actually nothing in Weldon's book to tie this event specifically to 2012, except that it is a supposed date of prophesied disaster that is close to the Cayce and Nostradamus predictions. Velikovsky's date of 1495 BC, plus Clow's 3,500 years, gave the year of catastrophe as 2006.

FUTURE LIFE PRE-CALL

In September 2003, the *Hindustan Times* reported that "an eminent film and TV personality, Smita Jaykar," underwent "future progression" under the guidance of Dr. Kondaveti Newton, and reported that only one-fourth of the world's population would survive to see 2012. The story ran: "'I could see the Earth enveloped in a golden light. Everything was in absolute peace and though the world was sparsely populated, all the inhabitants looked like highly enlightened beings,' says Smita while describing life on Earth as it would be by the year 2012. The Marathi actress was talking about her belief in the concept of past and future lifetimes."[23]

So, what would it be—earthquakes, volcanoes, floods, or the Third World War? The answer: "'Natural calamities yes, but the major part of destruction would be caused by a war, no doubt,' she says empathically. 'It was quite a few years back that Dr. Newton had carried me to the future and I couldn't quite believe what all [*sic*] I saw, but now I can see signs of it all coming true,' Smita adds."[24]

This again is not a good outlook, but 25 percent survival is certainly an improvement on Weldon's 0 percent.

15
UFOS AND ETS

UFO WAVES

A UFO "wave" (or "flap") simply means a concentration of reports of unidentified flying objects in a certain area at a certain time. The world's biggest-ever UFO wave started in Mexico in 1991 at the time of that year's total solar eclipse, and continues to this day. This has helped focus world attention on the Maya calendar and prophecies. A UFO wave over Arizona may also point to the Hopi prophecies, just as an Israeli wave highlights the biblical prophecies. In a similar way, during the total solar eclipse of 1999, the eclipse shadow passed over Turkey and appeared to trigger not only an earthquake but also a UFO wave that is continuing.

Following the 1991 wave in Mexico, a series of three videos was released[1] showing nearly two hundred video clips of UFOs over Mexico from 1991 to 1996, and stating that the 1991 eclipse had been predicted in the Dresden Codex (one of only three surviving codices, or books, written by the Maya). The editors of the video, Lee and Britt Elders, along with Mexican TV commentator Jaime Maussan, concluded that the 1991 eclipse was the start of the Sixth Sun, the event that is usually associated with 2012.

I have looked into this matter and found that errors were made. There is good reason to conclude that the original UFO was actually Venus, appearing during the darkness of the eclipse.[2] But that doesn't alter the fact that there has been a massive wave of UFOs in Mexico since 1991, as shown in the other videos and on Mexican TV. They are seen disappearing behind buildings, flying in groups, and buzzing (harassing) commercial and military aircraft.

However, the 1991 solar eclipse was not predicted in the Dresden Codex after all. I found Mayanist Mike Finley's website, The Real Maya Prophecies: Astronomy in the Inscriptions and Codices,[3] which explains that to use the Dresden Codex for eclipse prediction, the periods must be "recycled," or extrapolated, into the future. In other words, it may be possible to find the 1991 eclipse date by extrapolation, but the actual (Maya equivalent of July 1991) date is not recorded in the Dresden Codex. So this is one area in which the video editors have made misleading statements.

Another point made by Finley is that the accuracy of the charts fails after a few recyclings due to accumulated errors, and so a correction factor must be used over the thirty-eight recyclings necessary to bring the eclipse table forward to cover 1991. I wrote to Mike Finley to ask if he thought the 1991 eclipse could have been predicted using the GMT correlations. He took a detailed look at the problem and his reply is posted at the Diagnosis 2012 website[4] for those interested. He concluded that "the table isn't even close to the mark!!!"[5] Therefore there was no reason for believing that 1991 was the start of the Sixth Sun of the Aztecs (equal to the Fifth Sun, or World, of the Maya and the Hopis). This event is still scheduled for 2012.

Fig. 15.1. Two Aztec priests adjust a calendrical device. The National Stone, Museum of Anthropology, Mexico City.

CONTACT

Jon King, editor of the late, great *UFO Reality* magazine, had several contactee experiences between 1980 and 1984, in which he encountered humanoid ETs in a parallel dimension. When asked if he had been taken aboard an alien craft, King said, "Well, I wouldn't call it an 'alien craft.' But for lack of a different way to describe it, I suppose that's what I'm saying, yes."[6]

He was then asked if he had been abducted by aliens.

> No, I'm not saying that at all. I'm definitely not saying that. There was never any time I felt that I'd been abducted. Quite the opposite. It was as if the experience came in response to an inner need in me, a need to discover the truth about life. And this experience, once I'd processed it, satisfied that need. It was as if I called the experience to myself. Of course, I say all this with hindsight. At the time, it was very confusing to say the least. I went through hell; my life changed dramatically. From being a fairly successful musician, I virtually became a monk overnight! And all because of that experience.[7]

The beings King encountered were from "a different or parallel world rather than another star system,"[8] or, alternatively, "the next kingdom in nature, or the next dimension of reality."[9] They told him, telepathically, about a schedule that was unfolding between 1977 and 2013. The schedule is split into three phases, or "timegates." Phase one, from 1977 to 1989, is "contact." Phase two, from 1989 to 2001, is "landings/resonance shift." Phase three, from 2001 to 2013, is "decloaking/colonization." Each phase is designed to bring about "specific effects and changes in global consciousness."[10] In phase three, by the end of 2004, a "5D World Wide Web" should be in position, synchronized with Earth's geomagnetic grid, with "website stations" and "strategic free domains" at Earth's "grid cells." The

THE SCHEDULE

Schedule—Phase 1:
Primary Data Interactive/Contact
Timegate 1977–1989

Schedule—Phase 2:
Primary Landings/Resonance Shift
Timegate 1989–2001

Schedule—Phase 3:
Primary Decloaking/Colonization
Timegate 2001–2013

* * *

Primary Designated Landing Area
x-vis-Barbury Grid Coordinate
Wiltshire, England

Primary Earth Time Landing Coordinate:
Northern Hemisphere Sirius Alignment
[Stardate 13/20] circa July 17–Aug 17, 1998

5D World Wide Web
[Strategic Free Domain/Web Site Station]
Deployment Timegate: 1998–2004

4D–5D Migration Timegate: 2004–2010

5D Harmonic Interface Timegate: 2011

Domain HQ Decloaking Timegate: 2012

Earthshift: 2013

Fig. 15.2. The schedule that was received by Jon King during his contact experiences. Adapted from *UFO Reality,* October–November 1996.

D, as in "1D" through "5D," represents dimensions of consciousness, and refers to the following kingdoms: mineral, plant, animal, human, and alien (ultraterrestrial).

King was told that a large triangular area of southwest England, centerd at Barbury Castle hillfort, near Swindon, in Wiltshire, would be receiving "energy codings" (which sound remarkably like the crop circles that would soon proliferate in that area—see the next chapter) to prepare humanity for ET contact between 1998 and the dimensional shift, or "Earthshift," around 2013. I spoke to Jon King on July 29, 2000, and he told me that at the time he had his contact experiences, he had never heard anything about the Maya calendar, Terence McKenna, or 2012. That eliminates any possibility of subconscious interference and makes this an independent source of information on 2012 (although it seems that the later fictional TUMI column in *UFO Reality,* in which Jon King introduced the subject and presented the schedule, was influenced by J. J. Hurtak's *The Keys of Enoch*).[11]

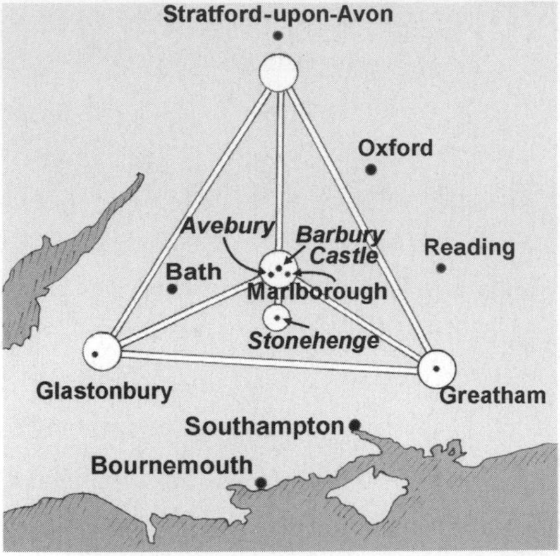

Fig. 15.3. Jon King's "Aquarian Triangle." Adapted from *UFO Reality*, April–May 1997.

ALIEN TELEPATHY

In the first issue of *UFO Reality*,[12] a UFO contactee named John Craig described the appearance of a UFO following his cry for help on May 16, 1992. The sighting lasted half an hour and was also witnessed by his family. He started writing a book with his wife (who is Japanese), to be titled "Merging Point," which "detailed the acceleration of evolving consciousness to a point in 2012 AD, when we would transmute our 3D consciousness to become truly multidimensional, galactic beings,"[13] but the book seems only to be available in Japanese. On May 16, 1993, the anniversary of the contact event, Craig received a telepathic message leading him to a deserted beach, where he witnessed another UFO, and he began to chant in an unknown language. The phrase "Echan Deravy" kept recurring, and he eventually realized that this was to be his new name. After this, the UFOs appeared "as if on command" in front of witnesses, and telepathic communication continued. Deravy stated:

> These were not objects in a 3D sense. They were "real apparitions," behind which some enormously evolved consciousness seemed to

> be at work. Which is why I now say they are not separate from us. They do not come from another planet, but rather from hyperspace itself—from the place all shamans come and go to.[14]

In a 1998 interview with *UFO Magazine* (UK version), Pam Hamilton described how, while she was growing up in Indiana, she experienced close encounters that began at a very early age, before she could walk. These earliest memories were with the Grays, which are the ET-type familiar to us from television and films. When she was three years old, in 1951, she had her first encounter with a tall, pale being with fair hair. Ufologists now know this race as the Nordics. Like the Grays, the Nordics communicated telepathically, and became Hamilton's constant companions. Unlike the Grays, there were identifiable males and females, "sometimes even materializing when she was with other people." This is a clear sign that the beings were not of the same density as our reality, and visible only to Hamilton at the time. According to *UFO Magazine*'s account,

> In subsequent contacts, the creature imparted specific dates into her mind, along with images associated with those dates. For example, she remembers him telling her that "2012 would be very important to me. I would be leading people through the mountains. What mountains? I have no idea. He said as I grew I would know certain things." Yet, in her conversations with other abductees, she's learned that her experiences are not unique. There is a sense among people who've been abducted that they have something inside of them that will "switch on" at some point.[15]

Since Pam was told by military personnel that they knew she had implants in her eye and ear, the implication is that perhaps these implants will be activated around 2012. However, since the military then said they had to "adjust" the implants so they could tell what the ETs were up to, perhaps they put them there in the first place.

THE ASSESSMENT

Robert O. Dean, retired command sergeant major (United States Army, Intelligence Field Operations), who had "Cosmic Top Secret security clearance," apparently the highest possible security clearance in NATO, was posted in Paris at Supreme Headquarters Allied Powers Europe (SHAPE) in 1964, when a secret document called "The Assessment" was published. "It was a military study that had been undertaken over a period of three years," he said, "and it concluded that although we were being monitored by extraterrestrial intelligences, there was no apparent extraterrestrial threat. It concluded that what was going on had been going on for a very long time, and that we are dealing with not one, but several extraterrestrial intelligences."[16]

Bob Dean retired in 1976, and at that time four types of extraterrestrials had been identified by the military.[17]

> But it was not simply that we were being visited by interplanetary visitors, or interstellar visitors, or even intergalactic visitors. When I retired in '76, many people in the military, government and science knew we were not really dealing with those issues. Rather, we were dealing with an intelligence that apparently had a multidimensional source. Now I'm not enough of a physicist to explain to you exactly what that means, or even what another dimension is. But I think there's an analogy to be found in death. I've always suspected that death is simply a journey to another dimension, another reality. I think we've come from there, we incarnate here, we live our lives, and we go back there. I've concluded that there are other levels of reality. And in a different way, the SHAPE study concluded something similar.[18]

Dean forsees an imminent "transcendent transformation" into another realm of reality that will be "so unbelievable that you can't even imagine it."[19] However, it seems to Dean that there is at least one ET race

Singular avistamiento de ufólogos
en el Centro Cívico de Lima

Fig. 15.4. Some of the speakers at the World UFO Congress in Peru in 2003. Back row, l–r: Geoff Stray, Salvador Freixedo, Haktan Akdogan. Middle row: Brian O'Leary, Daniel Munoz, Bob Dean. Front: Marcia Schafer.

that, along with governments and other powerful people, hopes to thwart the transformation, and that humanity has failed to make the transition many times in the past. At the end of his presentation at the first World UFO Congress in Lima, Peru, in August 2003, Dean mentioned that the end of the Maya thirteen-baktun cycle in 2012 could well be the crucial time.

In the late 1990s, Dean told Chris Carter, producer of *The X-Files* TV show: "You're not kidding me. Don't tell me that you're getting all these story ideas out of your own head. You're getting some input from someone on some of these ideas, because I know how accurate they are."[20]

THE X-FILES

The final episode of *The X-Files* was seen on TV in the United States in May 2002 and on British TV on March 16, 2003. The famous slogan started by *The X-Files,* "The Truth is out there," was finally explained. This particular show was called "The Truth," and in it, Fox Mulder finally discovers the ultimate secret: that the final alien invasion is scheduled for December 22, 2012, and that the Maya designed their calendar to end in 2012 for this very reason. However, as Bob Dean said, "When people talk about alien invasion . . . I mean . . . it's all gone on for so long now. If they had wanted to invade us, they would have done so a long time ago."[21]

Dr. Richard Boylan, a research behavioral scientist and cultural anthropologist, or, alternately, psychotherapist and clinical psychologist, has been saying for years that sources in the U.S. government have told him that they have agreed to join an "extraterrestrial federation of planets."

> This federation is made up of beings of all types, from Whitley Streiber's Greys to non-corporeal thought essences. From gasbag beings to intelligent insect races. This federation is a non-monetary, non-violent, telepathic coalition of beings from numerous planets. The federation has already greatly affected our physical and spiritual evolution throughout history and even before.[22]

They will supply us with all their technology, and train us in telepathic expertise, says Boylan, who adds, "The Maya knew this would be taking place. By the end of the Mayan calendar [in] 2012, Earth and all of Humanity will be wholly integrated into this extraterrestrial federation."[23]

16
CROP CIRCLES

BARBURY CASTLE

The first recorded cases of crop circles appear to go back hundreds of years, the cause sometimes attributed to fairies or demons, as with the "mowing devil" event of 1678 in Hertfordshire, England. Since the 1950s, crop circles have been connected with UFO appearances, and sometimes called UFO "nests."

The most famous of these appeared in the mid-1960s, in reed beds in Tully, Australia. From the mid-1960s to the early 1970s, there was a UFO flap in Warminster, Wiltshire, UK, that included several crop circles (still seen as UFO nests). From the late 1970s through the 1980s, they were thought by many to be a weather phenomenon, but they were getting more complex, appearing in pairs, then threes, then as quincunxes. Since then (with the possible exception of simple circular designs), an individual design is usually referred to as a crop formation, while they are often still collectively called crop circles, and researchers are still called crop circle researchers. They were also increasing in numbers. In 1990 they were strung out into long pictograms of connected glyphs, and reached record-breaking figures across England, with the biggest concentration in Wiltshire. In 1991, another jump in complexity occurred, and it clearly seemed as if the phenomenon was evolving. Then a popular newspaper ran an article in which two old men, Doug

Fig. 16.1. The Barbury Castle pictogram, July 17, 1991. Photo: Richard Wintle/Calyx Photo Services.

Bower and Dave Chorley, claimed to have made all the crop formations (now no longer just circles) as a joke. So the bubble burst, and nobody had to worry about reconstructing his or her "reality tunnel" anymore. That Bower and Chorley contradicted each other, changed their minds about which formations they had made, and failed to make any convincing demonstrations went unnoticed.[1]

Jon King says he was totally unaware of crop formations when he had his contactee experiences between 1980 and 1984 (see previous chapter). But the appearance of these designs in the fields certainly seems to fit his prediction of "energy codings," since crop formations cluster in the "Aquarian Triangle" (see fig. 15.3) which has actually been the center of crop-circle activity, though the focus is thought to be Avebury or Alton Barnes, rather than Barbury (three and a half and six miles away, respectively).

On July 17, 1991, a field at the base of the Barbury Castle hillfort was the site of a stunning triangular crop formation that looked like King's diagram of the Aquarian Triangle, and had appeared in the

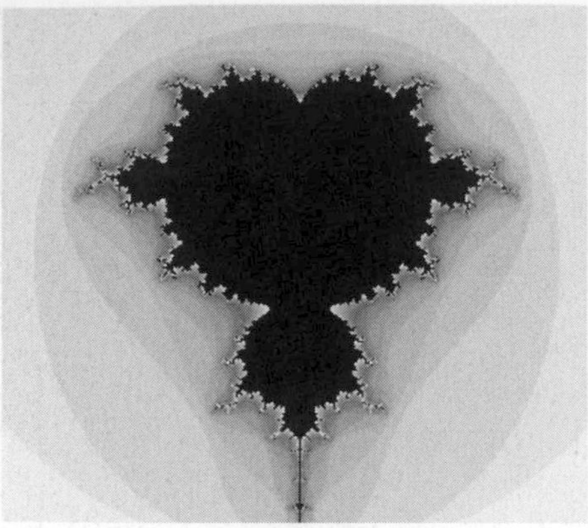

Fig. 16.2. The Mandelbrot set. A representation of this shape appeared as a crop formation in 1991. Produced by David E. Joyce's Explorer Applet—http://aleph0.clarku.edu/~djoyce/julia/explorer.html.

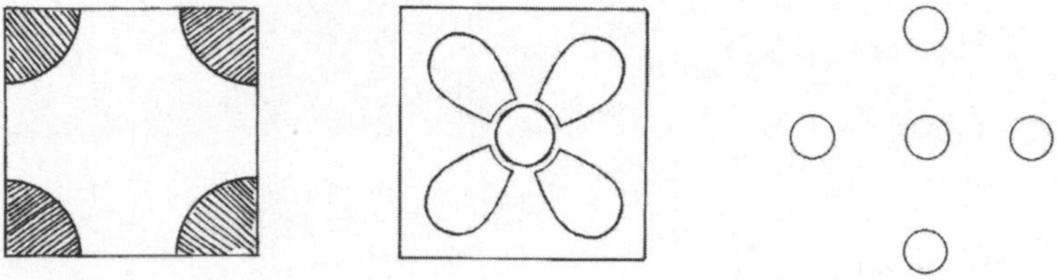

Fig. 16.3. Common Mexican forms of quincunx. After *Burning Water: Thought and Religion in Ancient Mexico,* by Laurette Séjourné.

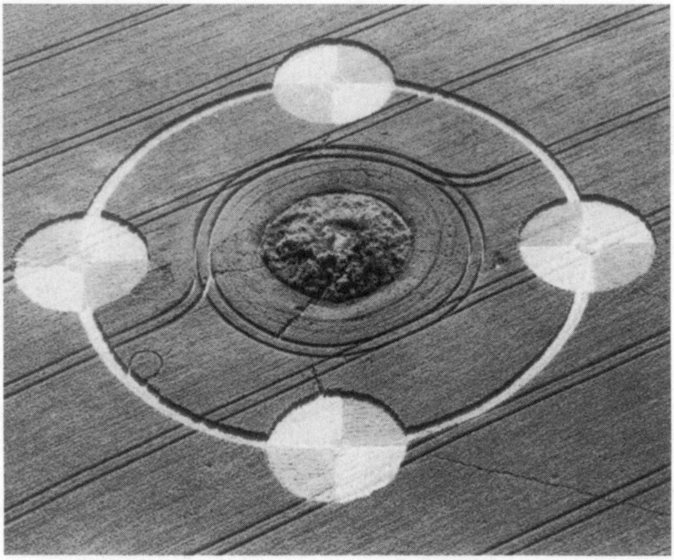

Fig. 16.4. A quincunx crop formation that appeared at Everleigh Ashes, Wiltshire, on July 19, 2000. The central circle is a Bronze Age round barrow. Photo: Andreas Müller

Fig. 16.5. Crop formation at Overton, Wiltshire, on July 13, 1996. Note the resemblance to parts of the Hunab Ku (below). Photo: Lucy Pringle

Fig. 16.6. Symbol adopted by many to represent Hunab Ku, from the Aztec Codex Magliabecchiano. From *Maya Cosmogenesis 2012,* by John Major Jenkins.

Fig. 16.7. The crop formations at Etchilhampton, Wiltshire, on August 1, 1997. The grid has a striking correlation with the Maya calendar, while its companion may represent the six magnetic fields of the Sun. Photo: Werner Anderhub

landscape right at the centerpoint of it. It was later described by crop-circle researcher Brian Grist as "an alchemical Mandelbrot Set," due to its similarity to alchemical diagrams that also represent chaos and order.[2] The Mandelbrot set is a fractal—the pictorial representation of a complex equation from Chaos Theory, in which the whole picture is contained within itself at varying degrees of magnification.

Within a month, on August 12, 1991, the Mandelbrot set itself appeared at a field near Ickleton, in Cambridgeshire. The original fractal discovered by Benoît Mandelbrot, it describes a system with a hierarchy of repeating patterns. So these two formations (at Barbury and Ickleton) imply the significance of 2012 via Jon King, since his schedule predicts that "Domain HQ" (the 5D mothership that forms the main strategic free domain and hub of the World Wide Web described in the previous chapter) will be situated at Barbury Castle, and is due to de-cloak in 2012, just prior to "Earthshift." In addition, they remind us that McKenna's Timewave Zero is also a fractal (originally called a modular wave hierarchy in 1975, before Mandelbrot had coined the term *fractal*) that terminates right at the end of 2012. There is even a rumor that McKenna claimed that the timewave is a Mandelbrot fractal,[3] but I have found no corroboration either that McKenna said this or that the timewave is a Mandelbrot set.

QUINCUNXES

A *quincunx* is defined as "an arrangement of five things at the corners and center of a square."[4] Apart from simple circles, the quincunx is the most enduring crop formation design, some versions of which (for example, the Maltese Cross, in Morestead, Hampshire, in 1997) are almost identical to some of the Mexican forms.[5] In his book *The Secret History of Crop Circles,* Terry Wilson reveals that the first-known appearance of the quincunx was in the 1950s, in Heytesbury, Wiltshire, about the time of the first recorded crop circle in Mexico (in maize). Mexico also had its own quincunx crop formation, in

Milbaheta, in 1977. In *Mexico Mystique,* Frank Waters describes the meaning of this figure:

> It symbolizes the four previous Worlds, or Suns, and the present Fifth Sun as their unifying center. But reconciliation of the bi-polar opposites during this era must take place within the human heart. So the quincunx also symbolizes Quetzalcoatl, who by penitence and self-sacrifice made a final ascent out of matter into spirit, and was transformed into Venus.[6]

The quincunx can be seen at the center of the Aztec Sunstone (see chapter 1), where the four previous Suns surround the Fifth Sun, in which we are living now, usually interpreted as the Sun of Movement, indicating that the cycle's end will be marked by an earthquake. Some researchers think the five Suns represent a series of five thirteen-baktun cycles, of 5,200 tuns each, adding up to 26,000 tuns (one full precession of the equinoxes), meaning that 2012 will be the end of the last of the series, and the end of a full cycle of 25,000–26,000 years. The Aztecs had forgotten about the Long Count and its thirteen-baktun cycle, but this possibility implies there may be some forgotten knowledge hidden in the Sunstone.[7]

CALENDRICAL CROP CODES

The crop formation found at Overton, Wiltshire, on July 13, 1996 (see fig. 16.5), suggests a symbol often referred to as Hunab Ku, originating in the Aztec Codex Magliabechiano. It incorporates the *G* symbol, which represents visual Galactic Center to the Maya, according to Yucatán daykeeper Hunbatz Men.[8] If this interpretation is correct, then the attached crescent—a symbol interpreted in the later 1999 formations as the eclipsed Sun—may represent the eclipse of Galactic Center by the winter solstice Sun, the event that John Major Jenkins has argued corresponds to the end of the thirteen-baktun cycle on December 21,

2012. In later photos of the formation, the small outlying satellite circle in the foreground has a tail and looks more like a comet,[9] suggesting the return of Quetzalcoatl, but in the earlier images, such as here, the tail is missing, so perhaps that was just a path left by a visitor.

On August 8–9, 1997, at Etchilhampton, Wiltshire, two formations appeared next to each other. One was in the form of a grid, 30 × 26 squares, and the other was a six-pointed whirling star. Crop-circle researcher Michael Glickman noticed that the grid was laid down boustrophedon style, or "as the ox ploughs" (in other words, the direction of lay alternates from line to line). He also pointed out that as there are twenty-six weeks in six months, and these are multiplied by thirty in the grid, the formation was signifying a time fifteen years in the future, which would be the year 2012.

Upon further study, I found other 2012 indicators in the Etchilhampton formations. The tzolkin, or 260-day Maya Sacred Calendar, is a grid of 13 × 20 squares, and the 780-square grid is divisible by exactly three tzolkins. In fact, a 780-day cycle was important to the Maya (in the Dresden Codex there are tables giving multiples of 780 days) because it is the synodic orbital period of Mars (the synodic periods of Mars were actually specifically demonstrated in an earlier formation that appeared at Titchbourne, Hampshire, in 1995). The number of units could also be taken as 781, if one includes the large rectangle that contains the grid. The number 781 is the same number of "bits" as are in Maurice Cotterell's 187-year sunspot cycle, which Cotterell has related to the thirteen-baktun cycle ending in 2012 (although he actually failed to do this convincingly). The six-pointed whirling star accompanying the grid could then represent the six rotating magnetic fields of the Sun (two polar and four equatorial), which cause the sunspot cycle.

SOLAR PREDICTION

In the November–December 1998 edition of the journal *Sussex Circular* (*SC*),[10] amateur astronomer Jack Sullivan analyzed a crop formation that appeared at West Stowell, Wiltshire, on July 23, 1994. The formation included a collection of standing clumps of cereal crop combined with three rings and a crescent. Sullivan reasoned that the clumps probably represented visible stars and the rings were planets, while the crescent was the Moon. Briefly, he found an exact match for the clumps in the constellation Cetus, which lies near Pisces on the ecliptic (or path of Sun and planets). Using astronomical software, he found the formation was depicting a conjunction of Saturn, Mars, and Jupiter in the Cetus constellation. The date of the conjunction was April 6–7, 2000, which was six years in the future. Then Sullivan checked the position of the Moon, and found to his amazement that at midnight not only was it in the exact position shown in the formation, it was also four days after the new moon, and would therefore appear as a crescent, as depicted. Jack Sullivan found that two other formations from 1994 also showed the same conjunction of April 2000.

Andy Thomas, then editor of the *SC* (who now runs the www.swirled news.com website and is also author of the crop-circle bible *Vital Signs*), was driving through Wiltshire around midnight on the night of April 7, 2000, near the very spot of the original formation. He noticed that the sky had strange red patterns in it, and wondered why, having forgotten that this was the very time predicted by the crop formation six years previously. It was only during the following day that he realized this fact, and also heard what had caused the phenomenon he had observed. This is how Thomas reported it in the 2002 edition of *Vital Signs:* "On 6–7th April 2000, one of the largest solar storms in a century erupted on the Sun, flooding the Earth's ionosphere with particles and creating the widest sighting of the aurora borealis across the UK and parts of Europe for many years, usually being restricted to more northern regions (hence its more common name of 'northern lights'). I was lucky enough to witness this effect as a strange shifting curtain of red light above Alton Barnes."[11]

The aurora was seen as far south as North Carolina. Here we have a strong suggestion that crop formations may be linked to solar changes in particular, as well as being able to predict—incredibly—astronomical events and changes in the geomagnetic field.

VENUS TRANSIT PREDICTION?

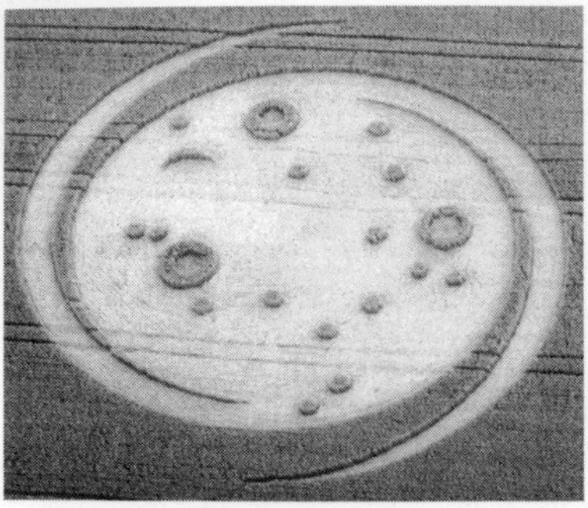

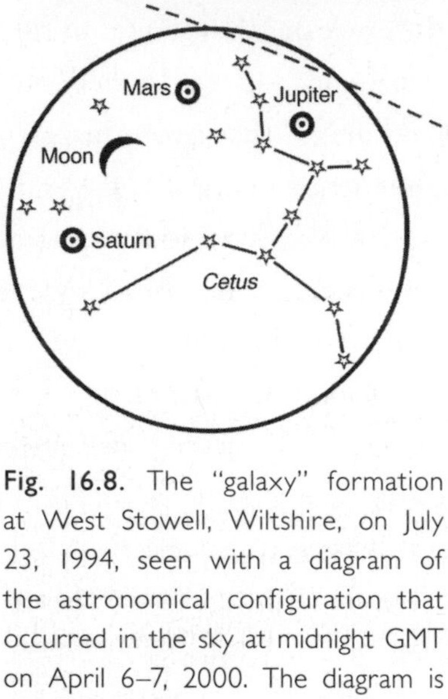

Fig. 16.8. The "galaxy" formation at West Stowell, Wiltshire, on July 23, 1994, seen with a diagram of the astronomical configuration that occurred in the sky at midnight GMT on April 6–7, 2000. The diagram is shown here upside down from normal protocol (the dotted line is the ecliptic plane) for clearer comparison with the crop-circle image above. Photo: Werner Anderhub. Diagram adapted from Andy Thomas's *Vital Signs.*

Jack Sullivan made another significant discovery concerning the Missing Earth formation that appeared at Longwood Warren, Hampshire, on June 26, 1995. At the time it appeared, the design was widely interpreted as having catastrophic implications for our planet. It seemed to show the Sun and the orbiting inner planets of the solar system as far as the asteroid belt, but although the orbit of Earth was depicted, unlike the other inner planets (marked as small rings), our own planet was absent from the diagram of the solar system. In 1995, Jazz Rasool said that the formation depicted the position of the planets at the time the formation appeared. However, this was true only if the solar system was viewed from "below," whereas the astronomical protocol is to view it from "above." It did not explain why Earth was missing, either.

In *SC,* issue 84 (May–June 1999), Jack Sullivan pointed out that the formation was a much more accurate rendition of the solar system from above on January 16, 1998. This showed the inferior conjunction (during the retrograde loop) of Venus, when Venus is directly between Earth and the Sun, in a plane view of the solar system. The formation was also an accurate rendition of the positions of the inner planets on June 4, 2004. This, again, was during the retrograde loop of Venus and only four days before the Venus transit, where Venus passed directly between Earth and the Sun (in horizontal and vertical planes).

Thus the Missing Earth formation had a double connection to Venus and its 584-day (or 583.92-day) retrograde loop cycle. The Maya followed this loop cycle closely, and it was tracked and recorded in the Dresden Codex over a period of 312 vague years (haabs are also known as vague years because they don't include leap days). This is equivalent to three Venus Rounds of 104 haabs each (and three Venus Rounds is equivalent to one Mars Round, or 146 Mars cycles of 780 days each); each Venus Round consists of sixty-five of the 584-day cycles. As if to confirm this connection, we find sixty-five "asteroids" depicted in the outer belt of the formation as small circles of varying sizes. The Venus Round is also equivalent to 146 tzolkins, or thirteen Venus pentagrams (eight haabs each), and each of the Venus pentagrams is equivalent to 146 uinals.

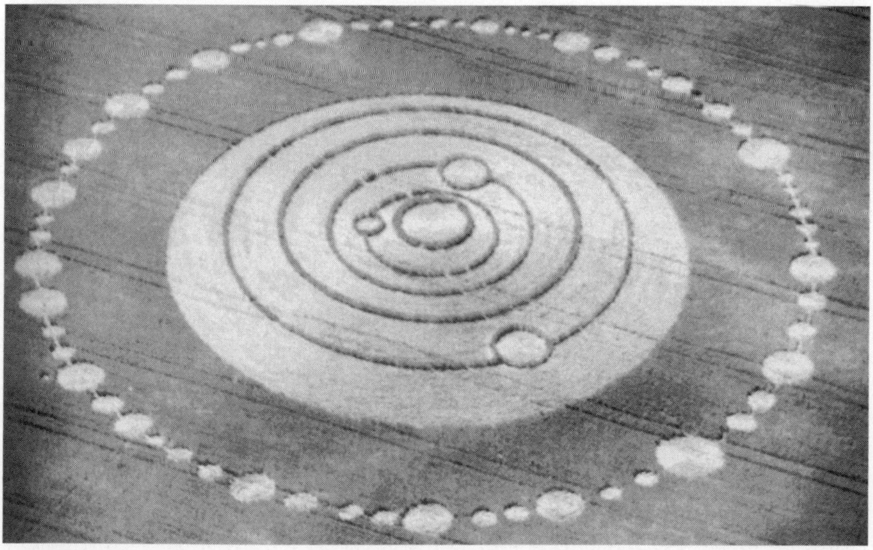

Fig. 16.9. Missing Earth formation at Longwood Warren, Hampshire, June 26, 1995. Earth is famously missing from the chart, but Venus appears to be the main subject. Photo: Werner Anderhub

There are 146 uinals between the Venus transit that occurred on June 8, 2004 (four days after the date encoded by the crop glyph), and the Venus transit due on June 6, 2012. The Venus transit in 2004 was the day 6 Ik in the tzolkin, and Ik, which means "wind," is the equivalent sign of the Aztec god Quetzalcoatl, who is associated with Venus. In 2012, the Venus transit on June 6 will also be an Ik day (1 Ik in the True Count).

MAYA CROP GLYPHS

On May 4, 1998, a crop formation consisting of thirty-three revolving lozenge shapes appeared near West Kennett Long Barrow, opposite Silbury Hill, in Avebury, Wiltshire. It became known as the Beltane Wheel. There was soon a rumor circulating that the crop glyph was the same as the Maya Sun glyph. Since the Maya Sun glyph is called Ahau, and doesn't look like this crop formation, I wondered where the information came from.

Fig. 16.10. The Beltane Wheel (in oilseed rape), Silbury Hill, Wiltshire, May 4, 1998. Photo: Andrew King

I eventually discovered that the source was Aluna Joy Yaxk'in, a New Age speaker and author who works with "the Mayan calendar" (the Argüelles version). Aluna (whose website is called Center of the Sun)[12] stumbled upon the formation on the day it appeared, while in the UK to attend the 1998 Mayan Dreamtime Festival in Glastonbury. She predicted finding a Sun crop circle shortly before seeing it, and has since said it connects to the Maya calendar, since there are "33 solar petals" on the formation, and the Maya tzolkin uses a combination of thirteen numbers and twenty day signs (thirty-three total), which she calls "Sun glyphs."[13]

On May 23, 1999, at Avebury Trusloe, in Wiltshire, the Fried Egg appeared, which contained eighteen "sperm." Beyond the sperm, around the edge, were twenty smaller circles. This could be interpreted as representing the three basic cycles of the Long Count: a tun, or 360-day year, which consists of eighteen uinals of twenty kin (days) each.

The following year, in 2000, the Waffle crop formation appeared at East Kennett, a 20 × 20 grid (see color plate 19) in which I personally

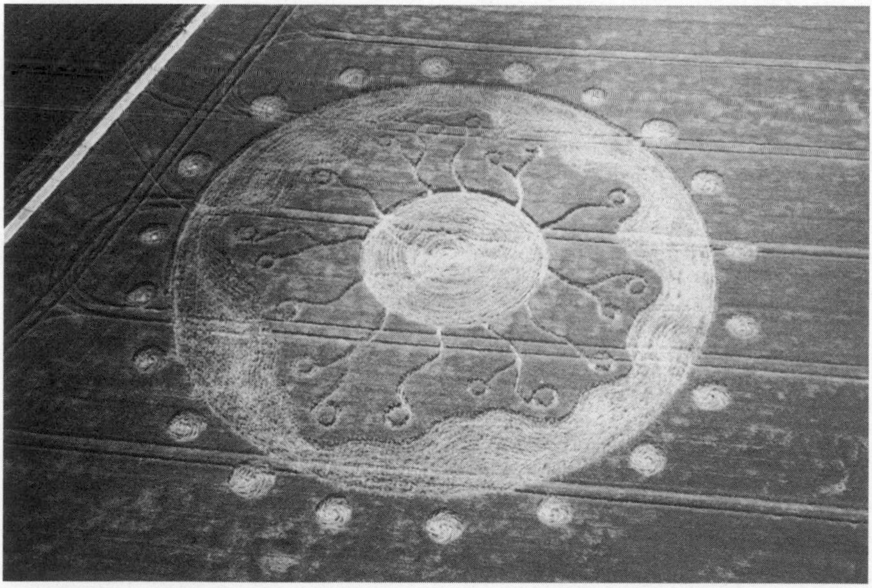

Fig. 16.11. The one tun, or Fried Egg, at Avebury Trusloe, Wiltshire, on May 23, 1999. Photo: Andrew King

found many examples of expulsion cavities. An expulsion cavity is a hole that can be found on cereal-crop stem nodes, which are formed when the stalk of a cereal plant is heated rapidly for a microsecond and all the moisture collected there turns to vapor and bursts out. The effect has been demonstrated by using an industrial-strength microwave oven, but apart from that has only ever been observed in crop formations. Many other anomalies have been found, such as a high density of melted meteoritic particles, microscopic pit holes in the stems, and growth abnormalities in seeds from crop formations (a thorough scientific investigation of these effects can be found at www.bltresearch.com). This second grid formation could again be interpreted as a calendrical diagram, but this time representing three cycles from the Long Count. Each square is a tun; each row of squares is twenty tuns, or one kaktun; twenty rows of one katun each are equal to one baktun (consisting of 400 tuns, or 7,200 uinals, or 144,000 kin).

On August 12, 2001, a huge, six-armed, eight-hundred-foot-in-diameter formation appeared on Milk Hill, near Alton Barnes, Wiltshire. It had thirteen major circles in each arm (six thirteen-baktun cycles?). On August 25, 2002, a five-pointed star (pentagram) appeared at Silbury Hill, near Avebury, Wiltshire. It had thirteen "scales" between each pair of arms, which makes sixty-five altogether. There are sixty-five Venus cycles (approximately 584 days each) in every Venus Round (Aztec and Maya), which is 104 haabs long (haabs are 365-day years). Venus makes a pentagram around the zodiac every eight haabs, so in the 104 haabs there are thirteen pentagrams. Also, there are sixty-five baktuns in five eras or thirteen-baktun cycles, making up a total of 26,000 tuns, or one cycle of precession.

Fig. 16.12. Diagram of crop formation at Silbury Hill, Wiltshire, on August 25, 2002. This remarkable design appears to embody both the cycles of Venus and precession. Diagram by Rob Seaman.

On July 20, 2003, the first-ever

thirteenfold star arrived at Huish, in Wiltshire (see color plate 18), looking like a clock face where each segment represents one baktun, combining to form a thirteen-baktun cycle (consisting of 7,200 tzolkins of 260 days each). The formation is uncannily similar to a diagram of the thirteen-baktun cycle that appears on page 47 of José Argüelles's *The Mayan Factor.* In April 2004, a crop formation was reported that turned out to be the "ghost" of the Huish formation, where the shape had been transferred to the new crop (oilseed rape) via bare patches, in which the stems had originally been laid down the year before, presumably through some unusual effect on the soil. In 2004, a huge wheel formation arrived opposite Silbury Hill utilizing Maya symbolism around its edge. The year 2005 saw similar examples. So every year now there are crop formations with connections to the Maya calendrical systems.

THE AZTEC PIZZA

I have already mentioned my animation of the Aztec Sunstone in chapter 1. The initial idea for the project happened around the turn of the millennium, and after three months of solid work on it, I completed the animation (apart from the proposed soundtrack) on Friday, June 1, 2001. On that very day, unbeknownst to me at the time, a crop formation appeared at Wakerley Woods, Barrowden, Northamptonshire, and was almost immediately nicknamed the Aztec Pizza by crop-circle researcher Karen Douglas.

The Aztec Pizza had a general resemblance to the Sunstone, in that there were concentric rings of glyphs; but the Sunstone has a ring of twenty day signs, surrounded by a ring of fifty-six quincunxes, surrounded by a ring of 104 "eagle feathers," within a ring formed by two "fire serpents" of twelve segments each (see color plates 1 and 3); the Aztec Pizza crop formation, on the other hand, had a ring of eighteen inward-pointing isosceles triangles (which actually do look like inverted versions of the solar rays on the Sunstone), surrounded by a ring of nine

Fig. 16.13. The Aztec Pizza, which appeared at Wakerley Woods, Northamptonshire, on June 1, 2001. Photo: Nick Nicholson

Fig. 16.14. The Secret Dial—found by cutting and pasting nine double-quincunx sections together on a photograph of the Aztec Sunstone.

wider triangle glyphs, surrounded by a ring of eighteen *G* glyphs, which is the Maya symbol for the Milky Way galaxy.

The thing that impressed me about the Aztec Pizza was that it arrived on the same day that my animation was completed, and it displayed the numbers nine and eighteen, which were the numbers involved with the discovery of the Secret Dial that made the whole animation possible. Nobody had ever found the Long Count in the Sunstone before, because the number eighteen could not be found on it, and since there are twenty days in a uinal and eighteen uinals in a tun, it was an essential requirement. I had spotted what looks like a visible segment of a hidden ring, showing two quincunxes. I then electronically cut and pasted eight copies of the suspected segment, each turned by multiples of a forty-degree angle, put them together, and found a perfect ring made from the nine segments of two quincunxes each, revealing the missing eighteen uinals. This does not work on some of the color paintings of the Sunstone, since the angle varies by up to two degrees on each side of the visible segment, and this can result in a twenty-quincunx ring. However, it works perfectly on a photo of the real thing.

As a result of this discovery, an animation showing the passing of the 5,125 years of the thirteen-baktun cycle can be shown on the Sunstone, starting in 3114 BC and ending on December 21, 2012, when the solstice Sun conjuncts the equator of the Milky Way. It also happens that there are eighteen years in a quarter degree of precessional movement, and since the Sun is half a degree wide, it will take eighteen years for the Sun to complete the second half of its crossing of the galactic equator, from 1998 to 2016. This is the event flagged by the Maya at 2012.

So in 1999 we had a tun represented in the Fried Egg crop formation at Avebury Trusloe; in 2000 we had a baktun represented in the Waffle formation at East Kennett; and in 2001 we had the secret knowledge of the thirteen-baktun cycle represented in the Aztec Pizza at Wakerley Woods. On the other hand, perhaps we just had an egg, a waffle, and a pizza.[14]

NIBIRU CROP CODE

In *Crop Circles, Gods and Their Secrets,* Robert Boerman, a Dutch crop-circle researcher, analyzes crop formations and discovers a kabbalistic code. The code gives hints about Nibiru, the planet on a huge elliptical orbit that, according to Zecharia Sitchin's translations of cuneiform tablets from Sumeria, returns to our solar system every 3,600 years, and houses the "gods" who crossbred our ancestors through genetic engineering. Boerman was constantly finding precessional numbers in his analysis of crop formations, by counting the number of circles and taking measurements, in yards, feet, and inches (which are very ancient units of measurement, as metrologist and archaeoastronomer John Michell has pointed out).[15] To understand how Boerman arrived at 2012 as the return of Nibiru, we must go to his primary source, Alan Alford's *Gods of the New Millennium,* which states that Nibiru itself would not return until circa 3400 AD,[16] but suggests 2012 as a date for the "return of the gods, "[17] who Alford suggests may no longer be living on Nibiru.

Alford's calculations are complex and his reasoning murky, and he actually made two errors in his calculation of 2012. Boerman then used Alford's "return of the gods" calculations for the return of Nibiru. This is simple but not very effective, since he has duplicated Alford's two errors. Using Alford's figures,* the correct calculation gives 2014, not 2012.[18] If we also take into account that if the current rate of precession has been slowing down from what it once was at the supposed time of the Great Flood, as implied by Alford, then there could be forty-two years or so to add on to 2014, bringing it to 2056.† We must therefore conclude that the evidence for a return of Nibiru in 2012, at least via Alford and Boerman, is questionable.

*See note 18 in chapter 16 (pages 442–44) for a complete description of Alford's calculations.

†The variation of the obliquity of the ecliptic, or axis-tilt cycle, causes the length of the precession of the equinoxes to vary over time: 2,160 + 2,158 + 2,156 + 2,154 + 2,152 + 2,150 represents the cumulative years of six zodiacal ages, while precession is changing from 2,160 years per zodiac age, to 2148.

17
SECRET GOVERNMENT

THE MONTAUK PROJECT

In 1979, Charles Berlitz and William Moore published a book titled *The Philadelphia Experiment,* describing a wartime attempt by the U.S. Navy to make a destroyer escort (the USS *Eldridge*) invisible to radar by using a very powerful magnetic field. The result was said to be the teleportation of the ship from its dock in Philadelphia to its other dock in Norfolk, Virginia, where it became briefly visible, then teleported back again to Philadelphia. The surviving crew members were reported to have been fading in and out of visibility in the aftermath of the experiment, or to have permanently disappeared, or gone insane. Some were said to have died.

The original source of the information, Carlos Allende, alias Carl Allen, said the experiment took place in October 1943, but investigation of naval records by Berlitz and Moore found that Allen was at the Norfolk, Virginia, dock in October, while the *Eldridge* was at sea. However, further investigation revealed that Allen stayed in Philadelphia on the weekend of August 13–15, 1943, just before joining his ship, the SS *Andrew Furuseth,* on August 16. It finally emerged that the *Eldridge* was in the Philadelphia docks at this time, and the experiment must have taken place between July 20 and August 20. The test with the most drastic side effects probably occurred just prior to the reported

Fig. 17.1. The USS *Eldridge.* From the National Archives.

public vanishing of two sailors, which happened on August 12 or 13.[1] Also mentioned by Berlitz and Moore is the claim that the experiment involved contact with "UFO occupants."[2]

An even more unbelievable sequel has arisen in more recent years. Two American seamen are alleged to have jumped off the *Eldridge* during the Philadelphia Experiment, on August 12, 1943. They found themselves not in the harbor, but on Long Island, New York, on August 12, 1983, having time-traveled to a point in the future where the government was conducting a similar experiment, the Montauk Project, which was connected by planetary magnetic fields that peak every twenty years on August 12. The scientists working on the project have allegedly been able to time-travel through some kind of hyperspace loop between the experiment dates, but for some reason can't go beyond 2012. The source of this story is Al Bielek, who claims to be Edward Cameron, one of the two seamen who jumped off the *Eldridge* (the other being his brother, Duncan Cameron).

Steven Gibbs is an inventor from Kansas who apparently makes, and sells, a time-travel machine called the Hyperdimensional Resonator, and a simpler version called the Sonic Resonator.[3] He says, "We can say that the machine seems to induce out-of-body experiences in most users almost immediately. When out of your body it is possible to travel in time and space with your astral body and witness or experience events in time. This technology is real and has been experienced by hundreds of people worldwide."[4]

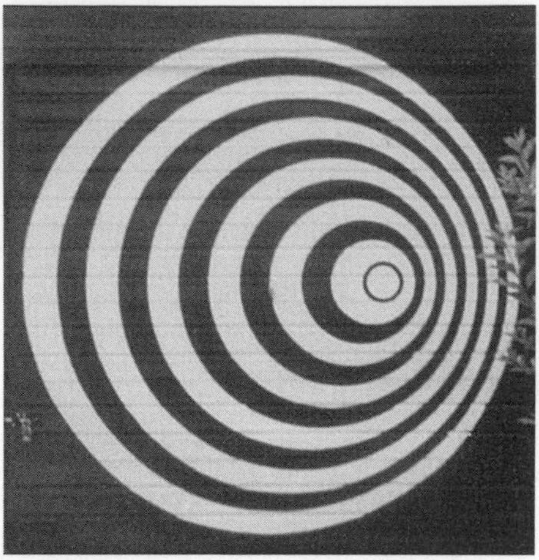

Fig. 17.2. The author's old garage door was a copy of the Time Tunnel crop formation that appeared at Cissbury Ring, West Sussex, on July 15, 1995. It is similar in appearance to the fictional time-travel device featured in the 1960s TV series *The Time Tunnel,* which in turn is similar to the description of the time-travel device supposedly used at Montauk.

Gibbs has been studying the time-travel concept since 1981 and has spoken to someone named Bruce Perrault, an inventor who gave him information on the equipment that was onboard the *Eldridge.* He says that when a Hyperdimensional Resonator is activated over a "natural grid point," or power spot, then actual, physical time travel is possible. He claims several people who have purchased his device have time-traveled, but they were unable to return to their own time (surely cause for a refund!).

When asked about time traveling into the future, Gibbs had this to say:

> When they sent some scientists from the Montauk Project into Earth's future, they hit a barrier in the year 2012. When they tried to go past the barrier from 2012 to 2013 they couldn't go through, they could only go around it. But after they went around it they found that all life on this planet had been wiped out . . . everything. Cities were all in ruins and there was no life found anywhere . . . at least they couldn't find any. They concluded that everything had been wiped-out.[5]

Bob Frissell, however, has a slightly less bleak interpretation of the same report. Frissell is the author of *Nothing in This Book Is True, but*

It's Exactly How Things Are, and he says that the reason the scientists can't go beyond 2012 is because the last time-travel experiment will be in 2012, and all the experiments are interconnected. Frissell says they could also travel into the past, but not beyond one million years ago, since this was the date that the alien race known as the Grays did their first time-travel experiment (New Age guru Drunvalo Melchizedek was Frissell's source).

An unnamed interviewee, who is presumably Al Bielek (a.k.a. Edward Cameron, one of the sailors who jumped off the *Eldridge*), gives us another version, in an online transcription of part of a nine-hour interview series. Bielek says[6] that beyond 2012, time travelers have found "a dreamlike reality" that continues until 10,000 AD, although in the same interview he says there is "a very abrupt wall there, with nothing on the other side." He says that humanity survives, but the future is just blocked from view. This is because of a change in vibratory rate: the end of third density and the start of fourth density. People will still have physical bodies, but be in a state of compassion, love, and understanding: "It also relates to the realization that one is not separate from the creator. It is that kind of spectrum which has been called by the Christians as [*sic*] the 'Second Coming.' The Second Coming is a state of being, not an individual arriving and establishing a power hierarchy."[7]

REMOTE VIEWING AND THE "DISCONTINUITY"

Remote viewing is a technique used by the military to gather information, and was originally known as psychic spying. It is a disciplined form of clairvoyance, and has had some amazing results. Ex-military remote viewers have set up their own training schools, and this is now becoming a fairly widespread technique. On the now offline but archived 2013.com website,[8] I found a statement that remote viewers have said "nothing can be 'seen' after the year 2012."[9] The explanation given on the site is that humanity's perception of time will no longer be so "linear and absolute." After the year 2012, "there is no lapse in time between a thoughtform and its corresponding 3D eventform. So, if a remote viewer in 1997 cannot see beyond 2012, it is to a large degree because the world(s) beyond 2012 have not been conceived of yet, or we are just perceiving/creating them now . . . THAT is the nature of the new time."[10]

Ed Dames claims he was in charge of a U.S. Defense Intelligence Agency team of remote viewers. He then became the head of his own remote-viewing company, called Psi Tech, and claims that his company's remote-viewing findings were 100 percent accurate. He is now executive director of the Matrix Intelligence Agency, a private consulting group (but has been criticized in some message groups for making predictions that don't come true). In an online transcript of a radio interview with Art Bell in 1997,[11] Dames answered a question about this 2012 "block," saying that the in-house term for it is the "discontinuity."

> Remote viewers run up against a point where things change dramatically and globally . . . and it's all at once. I can give you my best guess about what that is . . . We have explored it, its geophysical parameters, biophysical parameters, trying to discern what the heck is happening, and we have a working hypothesis now, and the working hypothesis (and it's that only, by the way . . . it may be something that is truly beyond our ken), is that time changes. That there is an event that is global, that something happens to the world that makes all the event timelines change.[12]

He goes on to say that it seems to be a spiritual event that happens at one point in time, and affects all subsequent experience: "You appear to be somewhere else, and, in fact, you've leapt onto a different event trajectory, a different time . . . parallel time, if you will."[13]

18

NOSTRADAMUS

KING OF TERROR

Michel de Nostredame was born in 1503 in Provence, France, and studied medicine and astrology. He had a natural talent for prophecy and used a bowl of water to aid his visions, similar to the way a crystal ball would be used (as a "screen" on which visions appear). He used astrology to determine the time sequence of his prophecies. Being Jewish, and living in a culture of enforced Christianity, his prophecies were influenced by kabbalistic thought and prophecies from the Old Testament and the New Testament. To avoid persecution as a witch or heretic, he deliberately obscured the prophecies and rearranged their sequence. He became quite renowned, and the queen of France, Catherine de Medici, hired him.

Fig. 18.1. Michel de Nostredame, 1503–1566

Nostradamus specialists[1] say that he predicted the Great Fire of London, the Second World War, and the assassinations of the Kennedy brothers, but the prophecies are easily misinterpreted (both the Allies and the Nazis used the prophecies as propaganda during World War II). There are a thousand prophecies, written in Old French, in a form of four-line poems called quatrains. They are grouped in ten lots of one hundred quatrains each, called Centuries (but Century Seven is incomplete, having only forty-two quatrains), and when a prophecy is quoted, the Century is given first, then the quatrain number.

Nostradamus made only one unambiguous prediction, in which he clearly identified the year in question. It is the oft-quoted 10:72, which is usually translated something like this (alternatives in parentheses):

> In the year 1999 and seven months ("*sept mois*")
> from the sky will come the great King of Terror
> He will bring back to life the great king of the Mongols.
> Before and after, Mars (war) reigns happily (by good fortune).[2]

If there is any substance to Nostradamus's prophecies, then surely this quatrain must be a major key. However, when nothing happened in July 1999, some said the prophecies were of no use. Others said that *sept mois* meant the original seventh month—September. Nothing happened then, either. One Nostradamus site, now unavailable, suggested that the seer was predicting the timing of a war, and that fortunately (since it would make it worse) Mars would reign before and after, but not during the war, which would occur sometime between 1996 and 2012. This was deduced by the fact that Saturn, the "kingmaker," left Aries in 1996 and will enter Scorpio in 2012, and both signs were thought to have been ruled by the planet Mars, also the god of war (though nowadays, Pluto is thought to rule Scorpio). However, this interpretation gets "war" and the planet "Mars" from the same word (Mars), which may or may not be valid. The site also explained another questionable technique of interpretation. In the actual number of the quatrain, it was claimed, two or three numbers may be included that actually comprise the date on which the event described will occur. So, in quatrain number 10:72, the 7 may refer to the seventh month, since it is mentioned in the verse, while the 2, 0, and 1 may refer to 2012 or 2001.

JOHN HOGUE

Prophecy scholar John Hogue, in his book *Nostradamus: The Complete Prophecies,* finds six quatrains that have a possible connection to the year 2012. Some of these connections are made as in the suggestion above, which links the number of the quatrain to the date of the prophecy. In quatrain 1:52, "the two evil influences in conjunction in Scorpio" are said to be a conjunction of Mars and Saturn in the constellation of Scorpio. Hogue links this to October 2012, when such a conjunction will occur. He links quatrain 2:12 to 2012 via the quatrain number. Quatrain 5:78 mentions a thirteen-year period that Hogue connects to the appearance either of Halley's Comet in 1986 or that of Comet Hale-Bopp in 1997, which, when thirteen years are added, renders the year 1999 or 2010, which is "closer to the end of time [2012] as we know it." In quatrain 6:24, a conjunction of Mars and "the sceptre" (Jupiter—in Cancer?) means April–May 2011, says Hogue ("near the end of the Mayan calendar"). Quatrain 9:12 links to 2012 via the quatrain number. In quatrain 9:55, a conjunction of Mercury, Mars, and Jupiter again refers to April–May 2011. I don't find these very convincing; if 2:12 and 9:12 link to 2012, what about prophecies found in 1:12, 3:12, 4:12, etc.? The quatrains Hogue connects to 2012 concern the end of the Catholic Church, the coronation of a new king who will bring peace to Earth, and a war followed by disease. However, the quatrains that are obviously based on the book of Revelation are not included.

TOUTATIS

The book of Revelation seemingly describes at least five meteorite showers,[3] four comet or asteroid strikes,[4] and four earthquakes.[5] Nostradamus also mentions a comet (a "bearded" or "burning" star, or meteor) in nine quatrains,[6] meteorite showers (or "stones from the sky," "red hail," or "fire from the sky") in ten quatrains,[7] and earthquakes in five quatrains.[8] Some of these are clearly based on St. John's book of Revelation.

Quatrain 1:69 describes a "great mountain" of size-seven stadia (about seven-eighths of a mile) that is "rolling along," and Michael McClellan's *New Prophecy Almanacs* says:

> Every four years an asteroid named Toutatis passes uncomfortably close to the Earth. This elongated rock—which measures nearly a mile across—also possesses another feature of interest: it rolls end over end in a tumbling motion. It is the only near-Earth planetoid that behaves in this manner, eerily similar to the "mountain" that "rolls along" in Quatrain 1.69 . . . If Toutatis is the asteroid predicted by Nostradamus, it will smash into the Aegean or Ionian sea part-way through the reign of Hieron (anagram of Chiren [*sic*]) . . . Evidence in the Epistle to Henry II and the quatrains place Chiren's rule as king and later as emperor in the 2002–2012 period.[9]

Although Toutatis was very close to Earth in 2004, it does pass again in 2012 (see color plate 11). McClellan also finds evidence in Nostradamus's quatrains that China will abandon Communism in December 2012, and Prince Edward will be made second emperor of Europe.

More and more Internet sites are interpreting various Nostradamus prophecies as describing events between 1999 and 2012, but they are not very convincing. Many sites offer such conclusions without giving any evidence. For example, on a site originally called Nostradamus 1999–2012 AD, a fundamentalist Christian, Pastor Harry of the Church of Philadelphia, says Nostradamus was "Satan's Prophet" (and the website has now been retitled Nostradamus: Satan's Prophet). The time window has recently been updated and another one added, since he now predicts two raptures: "Through Nostradamus, Satan sets a Window FOR THE COMING OF ANTICHRIST, soon after the Start of our

Millennium: seeming to focus between 2006 and 2012–2018 AD and not far into the Year 2050 or 2099 AD."[10] Another site, King of Terror, quotes the 10:72 quatrain (above), then says: "So warns Nostradamus in quatrain X.72, and thus heralds in [*sic*] the commencement of a window of prophecy originating from many cultures, religions and traditions, which runs 1999–2012."[11]

There is a book out in German by Manfred Dimde, the title of which translates as "Nostradamus: The Apocalyptic Decade—The Crucial Years Until 2012."[12] Unfortunately, my ability to read German is phrase-book level, so I don't know if he's actually found anything convincing. Is everyone desperately trying to find a way of decoding a 2012 prediction from Nostradamus's quatrains, or are we still missing something?

PHOBOS REVISITED

After the seventh month of 1999 came and went, some realized that although the prophecy mentioned the seventh month, the calendar has since been adjusted by eleven days (when it was converted from Julian to Gregorian), which brings the date forward, from July 31, 1999, to August 11, 1999, the date of the total solar eclipse (as seen over Cornwall in the UK). In fact, Nostradamus is said to have explained in a letter to Henri II of France that he was talking about an eclipse. Some said that since Phobos means "terror," one of the moons of Mars—the one called Phobos (see color plate 12)—would be released from its orbit by the supposed gravitational effects of the eclipse and the Grand Cross conjunction a week later, on August 18, 1999.

The Millennium Group (TMG—the online forum and repository for scientific research into the current transformation of Earth's systems and astrophysical environment) hosted an article by TMG team members Ray Ward and Gary Goodwin in 2001 titled "The Four Horsemen of the Apocalypse." It posed the question, "Where has Phobos gone?" The article suggested that during a CNN video clip shot in Turkey during the solar eclipse of August 11, 1999, "Shadows [were] also seen crossing the face of the Earth at a nearly right angle to the Moon's shadow."[13] They suggest, with supporting photographs, that this was Phobos, and they hint that perhaps Comet Lee had loosened it from its orbit in June 1999, so that it could be seen from Earth during the 1999 eclipse, as apparently predicted by Nostradamus.

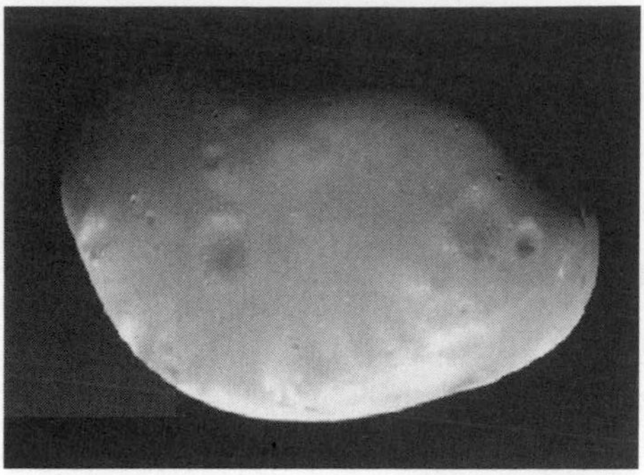

Fig. 18.2. Mars's moon Deimos, "panic." Image courtesy of NASA, JPL, and the Viking Project.

In July 2000, James van der Worp and Glen Deen suggested that the flyby of Comet 76P on June 5, 2000, shunted Phobos out of orbit and onto a collision course with Earth. In fact, TMG's Ward and Goodwin also pursued the 76P–Phobos connection in their 2001 article, before concluding that Phobos had been loosened from its orbit in 1999. Deen calculated September 18, 2012, as the projected time of impact, but then changed the theory to the "Planet X Cosmic Billiards Theory" (see chapter 7), which includes Phobos being loosened by Comet Kohoutek in 2001.

According to Ward and Goodwin, there is a connection between two very public failed Mars missions and the flyby of comet 76P, which they say triggered solar flares and dust storms on Mars. The last pictures of Phobos and Deimos (the Greek names of two moons of Mars, sometimes translated as "fear" and "panic") made public are from August 1998, and there is evidence that they are already orbiting Earth, say Ward and Goodwin.[14] The evidence includes photos taken from various satellites and from the ground showing anomalous dots. Phobos and Deimos are only ten miles and five miles in diameter, respectively, and are said to have been asteroids originally captured into a Martian orbit, now transferred to ours. Ward and Goodwin suggest there is a covert operation being mounted to blast these "Martians" out of the sky.[15]

You may recall the Bible-code prediction that there will be a collision between a comet and Earth in 2012, and that the comet will be "crumbled." At the Chapel Perilous website,[16] they have found that the code also conceals a previously undiscovered element, which they call the "Mars Code." Until recently, this Mars Code sounded fairly nonsensical, because the author concludes that the "comet" in question is actually the planet Mars. Thus, it will be Mars that is destroyed in 2012 when it passes near Earth. However, in light of the information above, "Mars" could just as easily refer to one of the small Martian moons, Phobos or Deimos (if you think it's possible that amateur astronomers could miss this for so long, and that NASA could keep a lid on it!).

NOSTRADAMIAN FUNDAMENTALISM

There is a website called An Ancient Prophecy Revealed for 1999 (or its duplicate site, The New World Order and Antichrist: The Years 2002–2012),[17] which makes the outrageous claim that the UK's heir to the throne, Prince William, is the "once-and-future king," that is, King Arthur, who is also the "Great Beast of Revelation" (who "was and is not and is to come") and the third Antichrist of Nostradamus's prophecies. The whole concept is supported by the prophecies of Nostradamus, and, preposterous as this sounds, if you spend some time looking at the material, you could end up feeling more than a little paranoid. When I discovered the site, it was just three weeks before the first of a series of prophecies was due to be fulfilled. It was predicted that Prince William would be crowned king of England on summer solstice 2002, and the royal family would be wiped out by a meteor strike just beforehand.

The timeline of prophecies went from the predicted 2002 royal wipeout up to the destruction of the Antichrist at Armageddon on December 21, 2012. Of course, when the events didn't happen, the entire timeline was changed and given much more leeway, so that most events on the timeline have between a three- and ten-year window, and now all events include 2012 as a possible date. The final destruction of the Antichrist is now scheduled for November 19, 2021[18] (don't ask me why). At least that should save on website updates! Obviously this is just another case of Nostradamus being used as propaganda.

NOSTRADAMUS MEETS MOTHER SHIPTON

So going back to the King of Terror quatrain, 10:72, if we go down the page to quatrain 10:74, we find:

> The year of the great seventh number accomplished
> it will appear at the time of the games of slaughter
> not far from the age of the great millennium
> when the dead will come out of their graves.[19]

Remembering that Nostradamus was Jewish, think back to the Judaism section of this book, and think about the Jewish calendar, which starts with the birth of Adam, year zero. Remember also the importance of one-thousand-year periods, which are the same as a day for God. So if year zero is the year of the "great first number," then year 6000 (2000 AD) is the year of the "great seventh number." As for the "games of slaughter," the Olympic Games started in Greece as a festival of athletics, but the Romans replaced the athletic games with violent gladiator contests. So not long after 2000 AD, "it" will appear, and this will be shortly before the start of the millennium (the thousand years of peace, the golden age, Day of the Lord, or seventh Day of Creation), when the dead rise (as described in Revelation 20). After 2000 AD, the Olympic Games were scheduled for 2004 (and they happened without mishap), and for 2008 (also passed without incident). They will take place next in 2012, so that gives us three possible years. This may sound tenuous, but most interpretations of Nostradamus are.

As to what "it" is that is to appear: when we glance up the page to the "Great King of Terror" that was loosened in 1999, we see that the quatrain between (10:73) is about judgment, which may connect the quatrains above and below it. In 1999, various people were saying that Comet Lee was the King of Terror (e.g., Howard Middleton-Jones reported that Comet Lee "fragmented on its trip round the Sun," and "three anomalous objects were observed during the solar eclipse on August 11th 1999").[20] But how could Comet Lee have been the King of Terror, since most people never heard of it? Perhaps there is another contender for that epithet that links 1999 to 2012, one that is more believable than Phobos or Comet Lee.

Now is a good time to introduce Mother Shipton, who was born in 1488 in Yorkshire, England, and made many prophecies that came true, including the Great Fire of London (also predicted by Nostradamus). Shipton's "Last Prophecies" have a dubious history, having been published by Nexus magazine in 1995. The woman who provided them said the originals were kept in a locked room with other prophetic writings at the Mitchell Library, Sydney (now the State Library of New South Wales). She claimed to have transcribed them and smuggled them out. One verse of them also appears in Charles Berlitz's 1981 book, *Doomsday 1999 AD* (p. 31). The Last Prophecies clearly refer to the book of Revelation, and verses three, four, and five of the eighteen are especially significant.

For those who live the century through
In fear and trembling this shall do
Flee to the mountains and the dens (Rev. 6:16)
To bog and forest and wild fens

For storms will rage and ocean roar
When Gabriel stands on sea and shore (Rev. 10:2)
And as he blows his wondrous horn
Old worlds die and new be born

A fiery dragon will cross the sky (Rev. 12:3)
Six times before the Earth shall die
Mankind will tremble and frightened be
For the six heralds in this prophecy

The inference here is that these three verses are describing the same events as those described in chapters six through twelve of Revelation. On a website called Myrddin's Warning,[21] these verses have been shown to act as a key to understanding Revelation and Nostradamus. The first line of the Last Prophecies mentions the year 1926, so the events happening to "those who live the century through" must therefore happen just after the end of the twentieth century. The third verse is interpreted as indicating six passes of a near-Earth object (NEO). It just so happens that Toutatis was discovered in 1988, and has a four-year orbit, which means that it will have passed six times and be on its seventh observed pass of Earth when it comes by on December 12, 2012. It passed in 2004, and came again in 2008, and is due again in 2012—the same years as the Olympic Games ("games of slaughter"?). But there is more . . .

Fig. 18.3. Mother Shipton: facsimile of her portrait from the cover of the 1686 edition of her prophecies.

The sixth chapter of the book of Revelation describes the opening of the seven seals, and when the sixth seal is opened, as the Myrddin's Warning website suggests, a solar eclipse is described ("the Sun became black as sackcloth"),[22] followed by a partial lunar eclipse ("the Full Moon became like blood"),[23] followed by a meteor shower ("and the stars of the sky fell to the Earth").[24] Incredibly, there is a solar eclipse due on November 13, 2012, followed by a "partial" eclipse, according to Myrddin's Warning (but it will actually be a penumbral eclipse—see below) on November 28, 2012, followed by the Geminids meteor shower on December 6, 2012.

Chapter 8 of the book of Revelation contains details of two more apparent meteorite showers and three meteor strikes; one of these is undoubtedly the one that Nostradamus referred to as the "great mountain" in quatrain 1:69, and another is the famous fallen star Wormwood. Myrrdin next brings our attention to quatrain 2:30:

> One that shall cause the infernal gods of Hannibal
> to live again, the terror of Mankind
> there was more horror not to say ill days
> did happen or shall to the Romans by Babel

Myrrdin explains that when Hannibal called on the Celts of Gaul to help him fight the Romans, any captured Romans would have been sacrificed to the Celtic god Toutatis, Lord of the World, after whom the asteroid is named. The quatrain also brings in a connection to the King of Terror quatrain (10:72), which implies that something happened in 1999 that will affect the orbit of Toutatis—and Toutatis is the Gallic equivalent of the Roman god Mars.

Myrddin differs by two days in his dating of the solar eclipse; he says it will be on November 15, 2012, and my two references say November 13.[25] He also says that the end of the thirteen-baktun cycle has been "recalculated by several experts" to be December 12, 2012. The real reason why some people are saying this is simply because they are confused.* A third error is that the partial lunar eclipse will actually be a penumbral eclipse, when the Moon only passes through a semishadow just beyond the main shadow of Earth, and this kind of eclipse is hardly discernible, since the brightness of the Moon is only slightly impaired, so it will not turn red. These three errors do not destroy the theory, but the fact that prophecy expert Peter Lemesurier says of Mother Shipton that "not a single one of the blood-curdling world predictions so often

---

*The American format for dates is month-day-year, while the British format is day-month-year. In this way, a quick glance at the end-date of 12/21/12 can be confused with 21/12/12, especially if you also confuse the start-date, when day number one (the day after 0.0.0.0.0) was August 12, 3114 BC.

attributed to her is actually to be found in the earliest edition of her"[26] might thankfully mean that this theory, though ingenious, is wrong. If that is the case, then we are left to wonder why Toutatis is also known as Lucifer's Hammer. I can actually answer that question.

In 1977, Larry Niven and Jerry Pournelle published a science-fiction novel called *Lucifer's Hammer,* about a dark-star companion to the Sun called Lucifer, which knocks a comet out of its orbit, sending it on an Earth-crossing trajectory, with catastrophic consequences. Website designer Stephen Miller made the connection when he saw an animation of Toutatis (since it is a very odd shape, like a tumbling dumbbell) and put up a Web page about it, entitled Lucifer's Hammer, on his www.mkzdk.org website.[27]

19

THE NEW AGE

THE NEXT PRECESSIONAL AGE

In *Gods of the New Millennium,* Alan Alford reports on a study he made of ancient Babylonian and Sumerian mythology and concludes that the gods of Babylon each ruled for one precessional age, which is a twelfth of an entire precession of the equinoxes. Combining "the latest scientific estimates" of the length of a precessional age (2,148 years, as quoted by archaeoastronomer Jane Sellers) with biblical and Babylonian information, Alford concludes that the precessional age of Marduk began in 2284 BC. The next one began 2,148 years later, in 135 BC. Counting forward another 2,148 years, says Alford, brings us to the next one, in 2012.

Alford does not think the planet Nibiru will return in 2012 as some have speculated, but predicts its appearance around 3400 AD. He does, however, see the start of the next precessional age as a likely time for the "return of the gods." This implies that they are lurking somewhere in the vicinity of the solar system, as predicted by Robert Temple in the latest edition of *The Sirius Mystery*. As already mentioned, there is an error of between two and forty-four years in Alford's calculation, but that could just mean that his date for the Great Flood is off by that much.

AQUARIUS NOW

Most astrologers place the juncture between the Age of Pisces and the Age of Aquarius between 2060 AD and 2160 AD, but with the popularity of the New Age movement, it has been brought back by many to the year 2000 or thereabouts. Many have simply stated that the Age of Aquarius has recently arrived, but don't give any credible proof.

The problem is that there are two ways of measuring the boundaries between the signs of the zodiac: one is by agreeing that each of the twelve signs of the zodiac is an equally sized "slice," in which the zodiac is split into twelve slices of thirty degrees each; the other is by agreeing that the boundaries of the signs vary according to the size of the constellations. Since the Age of Aquarius starts when the spring equinox Sun enters the sign of Aquarius, we find that another problem (even when using the equal-slices version) arises when trying to agree on exactly when and where one of the boundaries between the constellations (i.e., the cusp) is. Since there is so much disagreement as to exactly where the Pisces-Aquarius boundary is, a logical approach is to use astronomical software to reverse the spring equinox back through time, until it coincides with a more obvious

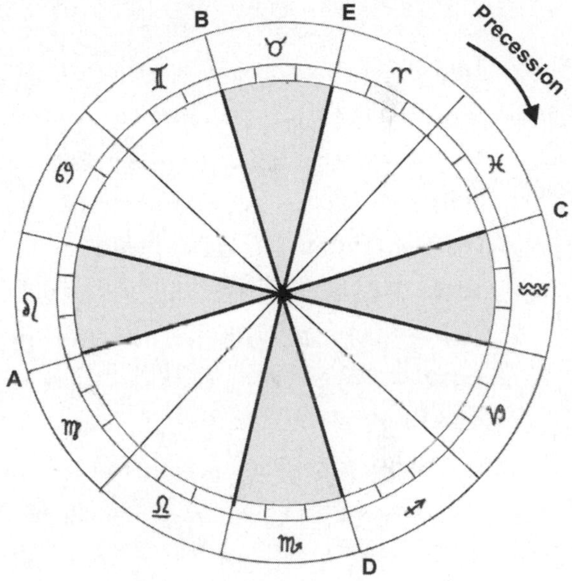

Fig. 19.1. The precessional cross displayed on the astrological zodiac (signs arranged clockwise). **A:** Virgo–Leo cusp—spring equinox circa 10,958 BC and autumn equinox today. **B:** Gemini–Taurus cusp—spring equinox 4480–4468 BC and summer solstice today. **C:** Pisces–Aquarius cusp—spring equinox today. **D:** Scorpio–Sagittarius cusp—winter solstice today, 2000–2012. **E:** Taurus–Aries cusp—start of the age of Marduk, spring equinox 2284 BC.

constellation boundary, and then wind it forward in time, counting 2,160 years per sign.

However, we also have the problem of agreeing on the length of the cycle of precession and, therefore, how long one-twelfth of it would last. As we have seen with Boerman and Alford, it is usually taken as 2,160 years per sign, though it is probably now nearer to 2,148 years per sign. Weidner and Bridges use the spring equinox on the cusp of Leo–Virgo in 10,958 BC to work forward half a precessional cycle, to 2002. This assumes six equal-sized ages of 2,160 years each.

The Aquarian Age Network[1] has issued a statement explaining why it is now declaring that the Age of Aquarius started on the spring equinox of March 20, 2000.* In layman's terms, this translates as: "The point of the Spring Equinox (once being between the constellations Gemini and Taurus, on the Galactic Equator) already has moved more than 90° from this point on the Galactic Equator, which is opposite the center of our galaxy . . . The 90° relates to three constellations, each 30°, and therefore the constellation Aquarius enters the Spring Equinox dawn and the New Age arrives."[2]

A Christian magazine, *Testimony,* has actually modified this by twelve years, so that the Age of Aquarius is due on spring equinox 2012:

> We now need to consider the concept of the meridian. This is an imaginary line running from the horizon due north of an observer

---

The statement: "While acknowledging the zero point of the zodiac to be defined as the point where the true galactic equator crosses the ecliptic between Taurus and Gemini, and recognizing the J2000 galactic equator defined by the galactic north pole of 12h 51.4m, 27° 8' to be representative of the B1950 definition of the galactic coordinate system, by virtue of Sagittarius A being recognized as the dynamical center of the galaxy with the true galactic equator passing through Sagittarius A*, and by virtue of the Sun being at 90° 00' 32" past the point of the crossing of the J2000 galactic equator and the ecliptic at the Spring Equinox of 20 March 2000, 0735 GMT, we declare the Age of Aquarius to have unequivocally arrived. Pax Aetatis Aquarii vobiscum!"

through the heavens to the horizon due south of an observer. The meridian at the vernal equinox is called the Zero Meridian, and it runs through the constellation of Pisces in which the Sun is situated. At the time of the birth of Jesus Christ, the nearest visible star in Pisces to the Zero Meridian would have been that designated Alpha. However, due to the slow movement of the Sun through the zodiac, mentioned above (the precession of the equinoxes), the nearest star in Pisces to the Zero Meridian, in fact the nearest of all the 1,800 brightest stars as listed in a star catalogue, is now that designated Omega. It is calculated that it will be exactly on the Zero Meridian early in 2012.[3]

THE GREAT CONJUNCTION

On a website called Audrey's Ancient Egypt,[4] amateur Egyptologist Audrey Fletcher explains how she used the Narmer Palette from First Dynasty Egypt, plus SkyGlobe software, to figure out that the Age of Aquarius starts on December 21, 2012 AD.

She first ascertained that the transition between the Age of Gemini and the Age of Taurus was in 4468 BC. Then, by working forward in 2,160-year jumps (since in a Platonic year of 25,920 years each zodiacal age lasts 2,160 years), she found that the Age of Aquarius would start in 2012. Since the Platonic year is made up of four "great seasons," and the Gemini–Taurus juncture was exactly a quarter of a Platonic year ago and represented a "great autumn equinox," it follows that the "great winter solstice" would be in 2012.

Fletcher calls winter solstice 2012 the "Great Conjunction," since SkyGlobe reveals that six of the planets plus the Sun and Moon will all be in the same 180-degree section of the sky. However, the May 2000 conjunction, when four planets plus the Sun and Moon were aligned, was widely expected to bring massive changes, but no such changes occurred. Astrologer Helen Sewell acknowledges that the "Yod," or "finger of God," configuration in December 2012 of

Saturn and Pluto, with Jupiter (spiritual enlightenment) as its central focus, could be a significant pointer to any events that occur at this time, especially as the Sun also enters Capricorn (the sign most associated with manifestation) at precisely 11:11 Universal Time, on December 21, 2012. However, she says this does not constitute a Great Conjunction.

I asked Fletcher, "You don't mention the Maya—is it the Mayan Long Count end-date that led you to 2012 as the dawning of Aquarius, or did you deduce it independently?" She replied, "Yes, the dating for the beginning of the Age of Aquarius is the same as the Mayan forecast, but I did arrive at that date independently."[5]

Barbara Hand Clow, in her book *Catastrophobia,* has also used the Narmer Palette to deduce the date of the Gemini–Taurus transition, but she dates it 4480 BC, which is twelve years before Fletcher's date. (Clow and Fletcher's calculations omit the lack of a year zero in our calendar, so they are both a year off.) This results in a dawning of Aquarius in the year 2000, but on the spring equinox. The Pisces–Aquarius boundary is very vague, so to get a reliable calculation one must go back to a time when the boundary is less disputable. Weidner and Bridges found the Leo–Virgo boundary to fit the bill:

> We discovered that the ancients had a very precise delineation between the two, one that has been maintained, pretty much at the same point, into modern usage. With the help of Voyager II and our local planetariums, Chapel Hill and Boulder, we determined the point when the ancient Leo/Virgo cusp point fell on the Spring Equinox, March 22, 10,958 BC. We used that as our marker and looked for the moment when it fell on the Fall Equinox, which turned out to be September 22, 2002.[6]

By this calculation, this also means that the spring equinox of the year 2002 would fall six zodiac boundaries away from the autumn equinox (since the Sun moves through six constellations during half its annual journey, over six months), which means that the Pisces–Aquarius boundary coincided with spring equinox 2002.

CROSSING THE THRESHOLD

In the summer of 2002, I heard philosopher Stanley Messenger give a talk about Rudolf Steiner's prediction of "mankind crossing the threshold at the end of the century." Messenger said that all the people who have completed a struggle toward enlightenment throughout the ages have formed a critical mass that will allow the majority of humankind to cross this threshold together, that Steiner foresaw this, and that the time is now. Messenger sees the event as reaching its climax by 2012, and sees a connection to the crop-circle phenomenon.[7] Although some might feel it is unfair that the masses should reap the benefits of the sacrifices of a few, it would seem to be a natural process, like Rupert Sheldrake's Morphic Resonance theory, or the so-called (now discredited) hundredth-monkey effect, in which an entire species was claimed to have somehow miraculously gained a behavioral attribute when a certain critical percentage of individuals pioneered it.[8]

PERIOD OF JUSTIFICATION

Visionary hugmaster Robert Gilson described an experience he had in January 1992 in which, after attending a weeklong intensive course for spiritual growth in Phoenix, Arizona, he was "zapped by the Michael energy" (in a similar manner to the Diagnosis 2012 image on the index page of my website—see color plate 15). In a state of heightened awareness, Gilson was informed intuitively that there was to be a "period of justification" lasting twenty years before the advent of a new consciousness in humanity. During this time, people should start to sort out all their personal issues and emotional problems (or "get our hearts in order") to prepare to receive the new energy, "in order to make the shift gently, rather than in a mad panic."[9] Twenty years from 1992 is, of course, 2012, but Gilson knew nothing of the significance of 2012 at the time.*

Coincidentally, this happened at the same time as the 11:11 events, when worldwide meditations were performed to open a "doorway." The doorway is an opportunity to evolve along with Earth, and will stay open for twenty years, from January 11, 1992, to November 11, 2011, according to New Age teacher Solara,[10] who started the 11:11 "movement." She claims that she is not a channel, but "simply embodies her vastness," or, in other words, she just knows, and we should take her word for it. She says that the sign of 11:11 is "a pre-encoded trigger placed into our cellular memory banks prior to our descent into matter," and that this is why so many people responded to the 11:11 concept. It aroused a lot of interest around 1992, when various people started noticing that number combination (for example, in crop formations, or 11:11 being the exact clock time of synchronicities occurring).

The doorway consists of eleven gates, and the first is the healing of the heart, after which it "seems to expand," according to accounts on the official 11:11 website, www.nvisible.com. One thing that many of the 11:11 enthusiasts have missed is that not only (in the tropical zodiac) does the Sun enter Capricorn at 11:11 on December 21, 2012 (see above), but winter solstice, the same day, will also occur at exactly 11:11 Universal Time, according to the U.S. Naval Observatory.[11]

*José Argüelles gives 1992 as the start of the final katun in the last baktun of the thirteen-baktun cycle, and this date is quoted by Weidner and Bridges in their book *A Monument to the End of Time.* However, they have confirmed to me that this page is flawed, since they agree that the last katun started in April 1993 (e-mail from Vincent Bridges, May 12, 2001). Unfortunately, the error is also repeated on page 335 of the revised and expanded edition of *The Mysteries of the Great Cross of Hendaye.* However, in the Long Count correlation invented by Carl Calleman—the Calleman correlation—the final katun begins on February 10, 1992. In Calleman's system, the thirteen-baktun cycle ends on October 28, 2011, not December 21, 2012. There is more on this later in this chapter.

GALACTIC SYNCHRONIZATION

In *The Mayan Factor,* back in 1987, José Argüelles said that the 5,125-year "Great Cycle" (i.e., thirteen-baktun cycle) marks the end of Earth's passage through a "resonant frequency synchronization beam," which will prepare our planet and humanity for an ascension, or evolutionary leap, into the next dimension. Argüelles's book has been an inspiration to many people and was responsible for bringing the Maya calendar to the attention of the world at large. In it, Argüelles synthesized Benjamin Franklin's magic square (see chapter 4), the I Ching, sunspot cycles, the tzolkin, the Long Count, and the law of Complexity/Consciousness of Teilhard de Chardin.

The galaxy is seen as an immense living being that communicates with its "organs"—the solar systems—through pulses of energy sent from Hunab Ku (Galactic Center—see chapter 16, note 10). The Sun receives these pulses and transmits them, via the twenty-two- to twenty-three-year binary sunspot cycle, to the planets and life-forms, in order to cause a harmonic resonance that will allow a progressive evolution toward global, then solar, then galactic consciousness. This is achieved via the evolving noosphere, which Argüelles calls the planetary memory program, or psi bank, which is situated in the Van Allen radiation belt surrounding Earth.

The thirteen-baktun passage through this beam is the gestation time for the creation of a planetary light body that takes the form of a dodecahedron-shaped etheric geomantic grid[12] on the surface of Earth. UFOs are defined as "Unified Field Organizers, an intelligent release of galactically programmed, psychically active, radiant energy simultaneously attracted to and emanated by Earth's resonant etheric body."[13]

The period from 2007 to 2012 sounds very similar to Jon King's 1998 to 2004, the 5D World Wide Web positioning and synchronization period. Having compared King's fictional TUMI column in *UFO Reality* to Argüelles's *The Mayan Factor,* the influence seems obvious (but it might just be that they have both been influenced by Hurtak's *The Keys of Enoch,* cited by Argüelles in his bibliography. See note 11 of chapter 15). However, King[14] makes it clear that he was given the dates of the timegates during his contact experiences, which predated the publication of Argüelles's book. *The Mayan Factor* version goes like this: "The final five-year period, 2007–2012 AD, will be singularly directed to the emplacement of galactic synchronization crews at all the planetary light-body grid-nodes."[15] Unfortunately, there are some errors and inconsistencies in *The Mayan Factor* concerning the Maya calendars (see note 16 of chapter 19). Due to the book's popularity, these have caused subsequent errors and confusion in the theories of other people who used it as their main source on Maya calendrics[16] (see also chapter 3).

This is how Argüelles visualizes the final moments of the thirteen-baktun cycle:

> Then, as if a switch were being thrown, a great voltage will race through this finally synchronized and integrated circuit called humanity. The Earth itself will be illumined. A current charging both poles will race across the skies, connecting the polar auroras in a single brilliant flash. Like an iridescent rainbow, this circumpolar energy uniting the planetary antipodes will be instantaneously understood as the external projection of the unification of the collective mind of humanity. In that moment of understanding, we shall be collectively projected into an evolutionary domain that is presently inconceivable.[17]

PHOTON BELT

The photon belt, or photon band, is a persistent myth that is always being associated with 2012. By tracing it back to its origins and seeing how it has developed, we should be able to understand how much of this story is based on fact and how much is pure fiction.

In the February–March 1991 issue of *Nexus Magazine* there was a reprint of "an obscure but popular 1981 article"[18] that said that in 1961, "satellite-borne instruments" had discovered a "photon band" around

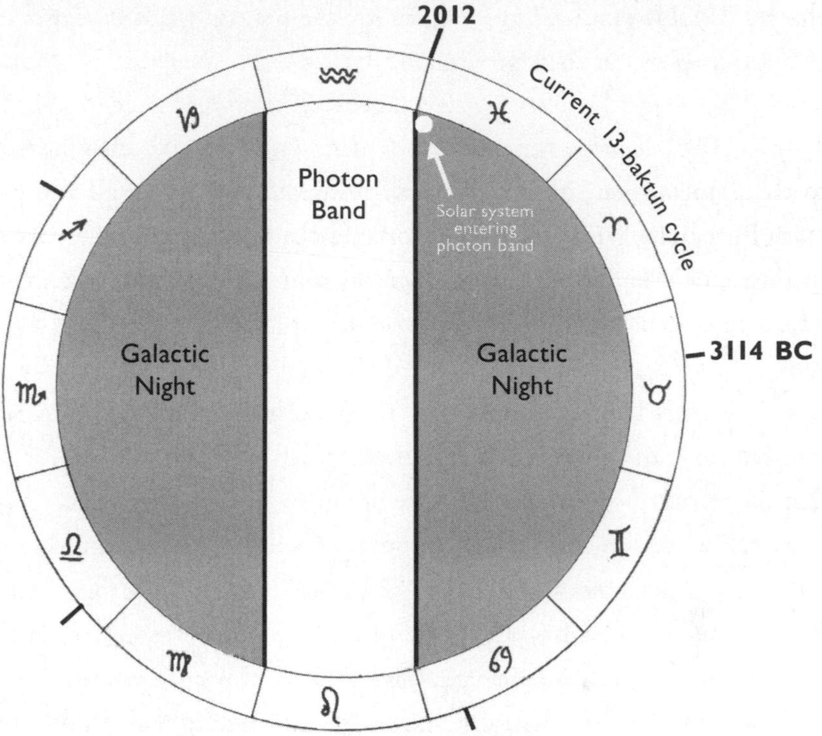

Fig. 19.2. The photon band/belt concept displayed on the astronomical zodiac (signs arranged counterclockwise), showing the current thirteen-baktun cycle and the positions of the previous four. Adapted from Clow (1995) and other sources.

the Pleiades star cluster. The article claimed that someone named José Comas Sola (a Catalan astronomer who is more famous for his observations of Saturn's largest moon, Titan, circa 1908) had discovered that our solar system is part of the Pleiades constellation, and that Paul Otto Hesse was the discoverer of the photon belt: "Paul Otto Hesse also made a special study of this system, of which our Sun is a part, and discovered, at absolute right angles to the movements of the suns, a PHOTON BELT or MANASIC RING, a phenomena [*sic*] which scientists have not yet been able to reproduce with laboratory experiments."[19] One revolution of our solar system around the Pleiades was described as 24,000 years, which is made up of a 10,000-year period of darkness, followed by a 2,000-year period of light (passing through the band), followed by another 10,000-year period of darkness, and a second 2,000-year period of light. The article then proclaimed:

> WE ARE NOW POSED [*sic*] TO ENTER THAT PHOTON BELT. It is inevitable . . . between now and the end of this century—but it is inevitable! We have completed the full circle and are back at the beginning. It is described in detail in your Bible, by all books on mythology, by Nostradamus, and by modern day scientists.[20]

The article predicts what will seem to be falling stars, as solar radiation interacts with the photon belt, and the sky will appear to be on fire. This may be preceded by 110 hours of darkness if the Sun enters first. People will experience something akin to putting their finger into an electrical socket, and then be converted from "Corporeal" people to "Atmospherean" people, but not all will survive the shift. There is also the following total misquote of the Nostradamus quatrain 10:72: "Nostradamus, in his quatrain about the end of the world as we know it in 1999 [says] 'and it will rain no more, but in 40 years, all will be normal.'"[21]

The modern form of mediumship called channeling lent support to the photon band theory. As time went on, channeled information was published supporting the story, enlarging on it but changing some of the details to match up with other prophecies. In 1991, a book, *God I Am—From Tragic to Magic,* by Peter O. Erbe was published in Australia (home of *Nexus Magazine* and the original *Flying Saucer Research Magazine,* where the first article appeared), in which "the Great White Brotherhood" channeled through a message saying that Earth completes its orbit around the Pleiades not every 24,000 years, but every 25,860 years: "This particular cycle is nestled within a number of greater cycles, of which it is but a part—including the overall cycle of 206 million years—in one single point of convergence. This is also called the Harmonic Convergence."[22] Two hundred six million years equates approximately to 8,000 of the 25,860-year cycles, which are each close to the accepted current length of precession.

We were further informed that we would enter the photon belt in 2011, which would trigger an ascension to "fourth-level density." Erbe's book also noted that St. Germain (the famous immortal, who is one of the Great White Brotherhood, as related by the Theosophists) "stipulates the month of December 2012 as the approximate date of the actual transition."[23] The alien race known as the Grays (usually thought to be from Zeta Reticuli—since the Betty and Barney Hill case, the first-ever-recorded alien abduction with subsequent hypnotic regression, happened in September 1961) was said to be in league with governments trying to abort the process. In some New Age circles, the Grays are seen as being the "opposition," who made a deal with the American government after the Roswell UFO crash in 1947, or the Philadelphia Experiment in 1943, trading alien technology for permission to abduct humans for genetic experiments. According to Erbe, the Grays feed on human fear and do not want to lose their food source.

In Azena Ramanda's *Earth's Birth Changes,* published around the same time, the author channels St. Germain, who says we will enter the photon belt in December 2012, attaining a "parallel universe." This will represent "the culmination—The Harmonic Convergence—of a 200-million-year evolutionary cycle."

In 1994, Sheldon Nidle, who channeled "the Sirians," published the results in *You Are Becoming a Galactic Human.* The Sirian version was that we were scheduled to enter the photon belt by the end of 1996.[24] After five or six days in the "Null Zone" on the edge of the belt, while the geomagnetic field would be collapsed, the atoms of our bodies should have been modified, changing our physical bodies to semietheric bodies, at which point there would have been a mass landing of the Galactic Federation. In 2012–2013, our whole solar system was scheduled to travel down a wormhole in the direction of the Sirius system (however, all the other predictions have failed to happen, so this one also looks highly unlikely). The photon belt will alter DNA and give us six extra chakras for a total of thirteen, allowing us to use our interdimensional soul bodies for existence in a timeless higher dimension—the fifth.

Barbara Hand Clow, whose grandfather was Cherokee, is an astrologer and was the editor of Bear & Company, which has published a wide selection of alternative books. In her 1996 book *The Pleiadian Agenda,* she channels Satya, a Pleiadian goddess who gives a version of the forthcoming shift that seems to unify all the other conflicting versions.

In Clow's view, Nibiru was originally a planet from the Sirius star system,[25] and travels from there to here and back every 3,600 years. It was last here two thousand years ago and appeared as the Star of Bethlehem. In 1987 (year of the Harmonic Convergence), the galactic synchronization beam (Argüelles's "resonant frequency synchronization" beam) reacted with the photon belt as we entered it. On winter solstice 2012, Earth's entire orbital path will be engulfed by the belt: this is to be the ascension, or "galactic orgasm." We must prepare, by meditating at sacred sites, to modify our DNA and emotional bodies. In 2013, "the third dimension will dissipate on Earth."

Now it is time to burst the photon belt bubble with a few quick words from Cecil Adams. Adams, who claims to be the "World's Smartest Human Being" (he also has a sense of humor), has a syndicated newspaper column in over thirty U.S. newspapers, and a website, www.straightdope.com, that aims to clear up misconceptions. So here is the "straight dope":

1. No Photon Belt or other such region of increased energy has been discovered. Photons in any case are merely particles of electromagnetic energy, which we commonly experience as light. Upon exposure to excess photons, the most common transformation of your being is sunburn.
2. There's no "anomaly" near the Pleiades star cluster. The Pleiades are surrounded by a nebula, or gas cloud. This cloud is composed not of photons, but of dust and hydrogen gas.
3. The Earth isn't heading toward the Pleiades, but away from them [*sic*]. In the 1850s it was conjectured that the Earth orbited the Pleiades, but this has long since been discredited.

4. Paul Otto Hesse is a German author of a book on Judgment Day and is unknown to astronomers.[26]

In 1973, J. J. Hurtak wrote *The Keys of Enoch,* a book purported to have been projected into its author's third eye while he was "before the throne" (not channeled, therefore, but revealed, like the book of Enoch and St. John's book of Revelation). Hurtak says,[27] "Thus, this precession of the equinoxes is the length of time required for our solar system to make one complete revolution around the Pleiades, our greater parent sun, which calculates to a period of 25,827.5 years."[28] So this would seem to be the earliest provable reference to the solar system revolving around the Pleiades. The book also refers to an "electromagnetic Null Zone,"[29] which will propel the solar system into a higher frequency, where fifth-dimensional bodies will be needed.

The appeal of the photon belt concept could perhaps be explained by its simplicity and its similarity to the scientifically feasible theories of galactic superwaves and magnetized plasma bands, and its common points with the popular synchronization-beam theory of Argüelles. It also retains some validity as a modern myth encoding a cyclical encounter between Earth and some kind of radiation connected with Galactic Center, which is governed by, or coincidental to, precession, and which in turn may cause an evolutionary leap in consciousness, along with changes in the atmosphere and the planet itself. However, considering its misleading association with photons, its association with Hurtak's "Null Zone," the Pleiades, and failed prophecies, it is, on the whole, best forgotten.

REALM BORDER WAVE

The Cassiopaea website[30] holds a colossal repository of channeled messages from "the Cassiopaeans," who claim to be "sixth-density beings." In the section of the site called "The Wave," it is explained that there is a cluster of comets on a 3,600-year orbit that appears as a single body. Every time the comets arrive, the Annunaki arrive just beforehand, to feed on the soul energy created by fear. The Annunaki are from Zeta Reticuli (which effectively makes them Grays to some, as they allegedly originate from the same star system). Right behind the comet cluster is a "Realm Border Wave," which returns every 309,000 years (twelve precessional cycles) and will merge this third density with the next one up, the fourth. This is called a "Realm Border Crossing."

It is said that riding on this wave is a colossal fleet of spaceships. Three immense ships are also coming from the direction of Orion, each carrying twelve million Nephilim. The Nephilim (literally, "those who fell down" in Hebrew), are mentioned in Genesis 6, and are usually thought to be the offspring of the union between the "sons of God" (or Annunaki—see chapter 7) and the "daughters of men." However, a close reading of Genesis 6:1–4 reveals three nonhuman species, since the Nephilim were there before the sons of God arrived, and the offspring were the "mighty men of old," who were giants, according to the book of Enoch.

The Cassiopaeans and Pleiadians are helping us to prepare for this shift, according to the channel Laura Knight Jadczyk. Those who successfully make the transition will merge with all other densities of their being for a short moment, which may seem to last forever. This sounds very similar to Ken Carey's *Starseed* scenario. Those who fail to make the transition will have to repeat the whole 309,000-year third Earth cycle. There will be physical side effects of this wave, such as a shift of the "magnetic axis" and ozone depletion, which are already occurring. The wave will arrive between now and 2012 (although in some versions it could be as late as 2018); timing depends on whether it is speeded up by certain scientific projects (which sound similar to the Philadelphia Experiment) that are currently under way.

Most of the Cassiopaean information was obtained using a planchette (a small board with wheels and a pencil that supposedly writes spirit messages when a person's fingers rest lightly on it) or by using a Ouija board. It is said in spiritualist circles that only "lower astral shells" can be contacted through this method, and that "higher spirit guides" never make contact through it. Whether you agree with that statement or not, there is now a mass of evidence that the Cassiopaean information is unreliable, to say the least. Unfortunately, anybody who points this out is liable to be labeled a "psychopath."[31]

THE CALLEMAN "SOLUTION"

Swedish researcher Carl Johan Calleman noticed that since the tzolkin can be used as a map of the thirteen-baktun cycle, (they both consist of 260 units that can be mapped on a 13 × 20 grid), it could theoretically also be applied to the larger cycles that have been found on a few stelae (a stela, or stele, is an upright slab or pillar with an inscription or sculpture; the plural form is *stelae*), such as stela 1 at Coba, a ruined Maya city in the forest sixty miles east of Chich'en Itzá, in the Yucatán. Then he noticed that if the larger cycles above baktuns, which consist of four hundred units each, are also restricted to 260 units, then four levels above the baktun, a cycle is discovered that is close to the scientifically accepted age of the universe. To understand this, we must first look at the extended version of the Long Count calendar:

Fig. 19.3. A stela at Tonina, showing 4 Ahau 8 Cumku at the top—the origin point of the thirteen-baktun cycle

20 kin or days is one uinal—20 days
18 uinals is one tun—360 days
20 tuns is one katun—7,200 days (about 20 years)
20 katuns is one baktun—144,000 days (about 400 years)
20 baktuns is one pictun—2,880,000 days (about 8,000 years)
20 pictuns is one kalabtun—57,600,000 days (about 160,000 years)

20 kalabtuns is one kinchiltun—1,152,000,000 days (about 3.154 million years)
20 kinchiltuns is one alautun (about 63.08 million years)
20 alautuns is one hablatun (about 1.26 billion years)
20 hablatuns is about 25.2 billion years

What Calleman realized is that thirteen hablatuns would be equal to about 16.4 billion years, which is close to the usual estimate given for the age of the universe, about fifteen billion years. If the pattern of the 13 × 20 Sacred Calendar, or tzolkin, encodes larger sacred cycles, such as the thirteen-baktun cycle, he reasoned that a whole cosmology of thirteen-unit cycles, rather than twenty-unit ones, could be represented in a nine-step pyramid form, such as the Temple of the Giant Jaguar, at Tikal, Guatemala; the Pyramid of Kukulcan, at Chichén Itzá, Mexico; or the Temple of the Inscriptions, at Palenque, Mexico (the one that

Fig. 19.4. The Pyramid of Kukulcan, Chichén Itzá, Mexico. Nine-step pyramids are thought by Carl Johan Calleman to represent nine evolving states of consciousness. Photo: Lee Anderson

conceals the tomb of the ancient Maya ruler Pacal). In this way, the top, or ninth, level would represent a thirteen-uinal period, while the bottom, or first, level, would represent a thirteen-hablatun period.

Thus the topmost layer represented the 260-day tzolkin of thirteen uinals. The next level down would be a 260-uinal cycle of thirteen tuns. The next level down after that would be a 260-tun cycle of thirteen katuns (this is a cycle normally known as the Short Count). Under this would be a 260-katun cycle of thirteen baktuns (this is the familiar thirteen-baktun cycle). Below this would be a 260-baktun cycle of thirteen pictuns, and so on. In this way we would have a nine-level pyramid, where each layer is structured like the tzolkin, but twenty times larger than the layer above.

This is a very intriguing idea, showing how the development and evolution of the universe could be encoded in the structure of the Maya calendrical systems, from the creation of the universe until its final consummation at the end point of 13.0.0.0.0. At the juncture where one level of the layered pyramid meets the next, humanity apparently gains another "frame of consciousness," and although Calleman says that speculation on the consciousness that will be accessed at the "summit of the pyramid" would seem to be "almost pointless," he does indulge the reader a little. It will be an end of time, in which no more calendars will be needed, thus leading to:

> the attainment of a time-less cosmic consciousness, and an experience of unity with All-That-Is in its fullest sense, a disappearance of the line dividing life from death. Although many that [*sic*] live today have had temporary experiences of such a cosmic consciousness, these have only been temporary, due to our limiting frame of consciousness. At the end of Creation this may no longer be so and the eternal life may in fact manifest as a time-less cosmic consciousness.[32]

If this sounds quite similar to other interpretations we have covered, maybe that is because it is based on "several ancient scriptures" that turn out to be mainly the biblical books of Genesis and Revelation.

Anyway, you might wonder what it would imply for the Long Count notation if there were cycles larger than the thirteen-baktun cycle. The Maya usually just used the last five positions; in fact, stela 1 at Coba implies that all the larger cycles have already reached thirteen, so you might think the 13.0.0.0.0 date should actually be written as 13.13.13.13.13.13.0.0.0.0 in Calleman's system. However, when you look closer, the logic breaks down a little. In fact, stela 1 at Coba shows cycles that are longer than the thirteen hablatuns that have passed since the start of the universe at the time of the theoretical Big Bang. The full Long Count date on the stela is written as:

13.13.13.13.13.13.13.13.13.13.13.13.13.13.13.13.13.13.13.0.0.0.0
4 Ahau 8 Cumku

Calleman says this stela implies a hierarchy of creations, i.e., his multilayered pyramid.[33] But this Long Count date actually signifies a time span equivalent to about 6.5×10^{22} years. This is trillions of years longer than the age of the universe, or, in other words, the current universe is about the four billion-billionth universe of 16.4 billion years each! Calleman's pyramid would need twenty-four levels to show this.

Although the Creation date of 3114 BC is theoretically written as 0.0.0.0.0, wherever it is referred to on a stela it is written as 13.0.0.0.0. (From the accompanying Calendar Round date of 4 Ahau 8 Cumku, Mayanists know this refers to the start of the current thirteen-baktun cycle in 3114 BC.) Most Mayanists have thus concluded that every time we get to the end of the thirteenth baktun (which is baktun number twelve, since the first one was baktun zero), we reach a new Creation date, and the count of baktuns, katuns, tuns, uinals, and kin go back to zero (the day after 13.0.0.0.0 is 0.0.0.0.1). Since there are no known inscriptions referring to the first baktun in the current era, some Mayanists propose that the first baktun would have been recorded with the 13 throughout the whole baktun, not just the first day. In this case, the day after 13.0.0.0.0 would have been 13.0.0.0.1 and so on, until the following baktun—1.0.0.0.0—commenced. A record on a temple stairway at Yaxchilan[34] is said to show eight levels above baktun, and they are all still at 13:

13.13.13.13.13.13.13.13.9.15.13.6.9
3 Muluc 17 Mac

So, the implication of the Coba and Yaxchilan dates, above, means that on the 4 Ahau 8 Cumku Creation date, the full Long Count date turned over from the previous day's date of 13.13 . . . . 13.12.19.19.17.19 . . . or that all the higher cycles clicked up to 13 on that date. However, Eric Thompson's work with many inscriptions showing dates in these higher cycles shows that the full Long Count date for the 4 Ahau 8 Cumku Creation date would be expressed as 0.1.13.0.0.0.0.0.0 . . . (or with the first baktun expressed as 13 rather than zero: 0.1.13.0.13.0.0.0.0); thus the previous day would have been 0.1.13.0.12.19.19.17.19. So the Coba and Yaxchilan inscriptions do not tally with Thompson's findings and must then be simply depicting mythic time and underscoring the importance of the Creation dates.

Calleman supports the True Count, as opposed to the Lounsbury correlation, but he realized that his theory would be much neater if only the Long Count had ended on day 13 Ahau in the tzolkin instead of 4 Ahau. Incredibly, he decided that the Maya had it all wrong, and said that although he still agreed with the True Count of the tzolkin, the correlation between the Long Count and the tzolkin was wrong.

This is something nobody had questioned before, since there has never been any doubt of the tzolkin-to-Long Count correlation; it was set by the Maya and wherever a stela shows the Long Count, it shows the relevant tzolkin date. Previous arguments had concerned how the calendars relate to the Gregorian calendar, not how they relate to one another. Calleman criticized Argüelles for creating his own New Age Dreamspell count,[35] yet has not seen the parallels to his own creation. In fact, he says in reference to later "invented" tzolkin correlations,

"Such disregard of the calendrical knowledge of the Classical Maya is not likely to lead to truth,"[36] but this applies equally to his own invented Long Count correlation. The result of the change in correlation means that Calleman sees the thirteen-baktun cycle as starting on June 17, 3115 BC, and ending on October 28, 2011.[37]

There is a problem, however, because dividing by twenty works fine until we get to the level beneath the summit of the pyramid, when we find that thirteen tuns is 4,680 days, and a twentieth of this is 234 days, not 260 days. Calleman admits this where he refers to "the short universal cycle of 2011, probably totalling only 13 × 18 = 234 days, but possibly 260 days."[38] Two diagrams on page 77 of Calleman's book differ on the length of this last cycle—one says 13 × 20 days (260) and the other says 13 × 18 days (234). If the tzolkin level (the Universal Underworld) is shortened to 234 days, so that the pattern (of each level being a twentieth the size of the one above) is retained, then the tzolkin as the key 260-unit "filtration pattern of divine light" is unable to form the top level of the pyramid. Alternatively, if the 260-unit tzolkin is retained as the top level of the pyramid, so that each level consists

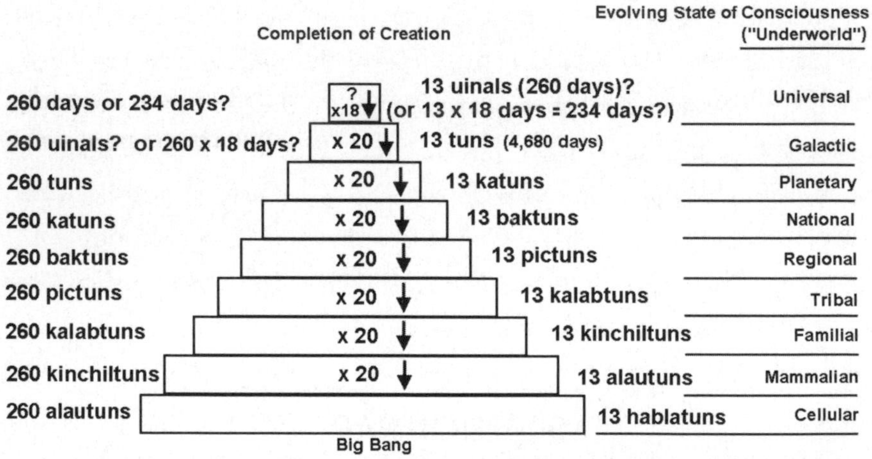

Fig. 19.5. Calleman's nine-level pyramid model of the nine evolving states of consciousness illustrates that if we retain the twentyfold pattern up to the top, we lose the pattern of 260 components, but if we retain the 260 components, we lose the twentyfold pattern.

of 260 subunits, then the top level will be only an eighteenth the size of the level below, whereas all other levels are a twentieth the size of the level below. Since both solutions lose the symmetry of the pyramid, Calleman is clearly undecided as to which is correct. This brings the validity of the whole pyramid construct into question.

If Calleman had stuck with the Maya Long Count correlation, the idea would have remained an interesting concept. But the idea that the tzolkin ends on 13 Ahau and begins on 1 Imix is, in fact, "based upon a misconception of the tzolkin calendar"[39] due to modern conventions in representing it, according to John Major Jenkins. Therefore the alteration to the Long Count was not necessary, and Calleman could have presented the idea without resorting to inventing his own correlation. Calleman now has quite a following, and supporters are presenting his theory as "The" Maya calendar, or stating "The Maya say . . ." when they should be saying "Carl Calleman says . . ." His book is titled *Solving the Greatest Mystery of Our Time: The Mayan Calendar,* but his "solution" is being taken as the genuine Maya calendar, when it is an attempted explanation that got out of hand and ended up "improving" the facts to fit the explanation. Amazingly, José Argüelles has written the introduction to Calleman's sequel, just as several websites are attempting to combine the theories of Argüelles and Calleman . . . and even Jenkins. It seems that people are so wary of throwing out the baby with the bathwater that they have poured all the water into one bath, inadvertently drowning the baby.

For further analysis of Calleman's theory, see the Jenkins–Calleman debate on the Diagnosis 2012 website, in which each author simultaneously submitted a series of three articles.[40]

SCALLION RAP

Modern prophet and self-styled futurist Gordon Michael Scallion publishes his *Earth Changes Report* nine times a year. It is an eight-page newsletter focusing on his most recent visions. In his 1997 book *Notes from the Cosmos,* he recorded an impressive list of uncannily accurate predictions that he made for the years 1991–1997 (which were published in ECR), plus the facts of what actually occurred. He is most famous for his prediction, in 1993, of the epicenter of the 1994 earthquake in Northridge, California.

Scallion predicts that solar flares and sunspots will precede a reversal of the Sun's magnetic field, in preparation for a new age. This is the time of the Tribulation, and like Native Americans, Scallion says that human consciousness is triggering Earth changes. The Blue Star Kachina of the Hopi Indians will appear from behind the Sun, traveling toward Arcturus (a star in the Bootes constellation, near Virgo). This blue star is a "companion to Sirius B" and will cause Earth's frequency to double to over fifteen cycles per second, or 15 hertz (this is the Schumann resonance, consisting of electromagnetic waves in the cavity between Earth's crust and the ionosphere, which, as we saw in chapter 9, Gregg Braden predicted would be at 13 hertz by 2012). Our subtle bodies will change, and as we adjust we may experience palpitations, electrical sensations in the limbs and spinal column, migraines, flu symptoms, and intense dreams.

Scallion predicted that between 1998 and 2002, Earth's magnetic field would begin to shift two or three times, by six or seven degrees each time. This would be the result of the interaction of an 11,600-year "critical-axis-tilt-point" cycle and a 3,750-year "solar-magnetic reversal cycle." The "magnetic polar shift cycle" is caused by "the interaction between Earth, the Sun and other planets," possibly implicating the 1999 total solar eclipse and the May 2000 planetary alignment. However, Earth's magnetic field has not shifted by twelve to twenty-one degrees between 1998 and 2002, so this is a failed prediction.

Scallion says that in a vision he saw the planet Mars wobble on its axis, causing its smallest moon, Phobos, to move in a "snake-like motion," pushed out of its orbit by the approach of the blue star, now like a second sun. Phobos spirals toward Earth's orbit and ricochets off our atmosphere, causing hurricanes and tidal waves, then earthquakes and volcanoes, then a melting of the ice caps and global flooding. He seems to have covered it all here! (See chapter 18 for other Phobos predictions.)

While in Wiltshire, UK, in 1993, looking at crop formations, Scallion received a "direct communication" from "the Etherians," who told him that the culmination of these Earth changes was to be 2012, but that the cycle has been affected by humanity:

> Solar-magnetic cycles—pole shifts—occur at the 11,600-year interval, with lesser cycles occurring at harmonic sub-junctures in the greater cycle. A full magnetic reversal would regularly occur on the Sun in the year 2012. However, due to the expansion of thought-forms and nuclear activity in your world, the cycle has been modified, and now occurs earlier, before preparations are completed. But knowing these cycles can enable shifts to occur in your world, so that the cycle can be corrected. In doing so, preparation will be facilitated in our combined worlds. The effect of such a solar reversal brings great changes to our combined worlds, making for new seasons, and, indeed, new species.[41]

On the other hand, if you can find early copies of *Earth Changes Report,* some researchers[42] claim that Scallion made many dire predictions pre-1988 that have failed to happen, such as Mount Rainier exploding, huge fractures causing the west coast of America to split into islands, Japan sinking into the sea, and finally a "shift in the Earth's poles" that was predicted to follow all these events by the year 2001 at the latest. Shifting this forward to 2002 didn't work either, so it seems he shifted it even further forward, to 2012.

CRYSTAL SKULLS

In 1924, British explorer Frederick Mitchell-Hedges was excavating a Maya city called Labaantun, in the jungle of what is now Belize. His seventeen-year-old daughter, Anna, climbed to the top of the tallest

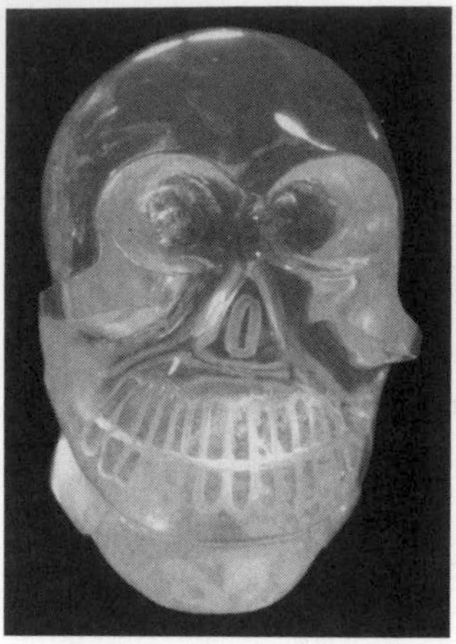

Fig. 19.6. Kathleen Murray's crystal skull, Mahasamatman

pyramid to admire the view and noticed something reflecting the sunlight between the stones at the pyramid apex. When the top stones were removed, she was lowered on a rope through the hole and came out clutching a life-size replica of a human skull, carved out of a solid piece of quartz. Three months later, the lower jaw of the crystal skull was found in the main chamber of the pyramid.

Mitchell-Hedges was told by a Maya priest that the skull had magical powers. It remained in the possession of Anna Mitchell-Hedges for the rest of her life, and various psychics who spent time with it were said to have gone into trances, channeling volumes of bizarre information.

Over the years, more of the skulls have come to light, including one in the British Museum in London, which came from Mexico and was purchased by the museum in 1898. However, recent tests on this particular skull suggest the visible grinding techniques it exhibits are more in keeping with it being less than two hundred years old, and it has been dismissed as a fake.

In *The Mystery of the Crystal Skulls,* Chris Morton and Ceri Louise Thomas interviewed North and South American native peoples who claim that there were originally fifty-two crystal skulls, of which the Maya had thirteen. They say the skulls were brought to Earth in the distant past by extraterrestrials, having been encoded with information from twelve different planetary civilizations: "The whole layout of Teotihuacán was like a huge clock-face, centred around the Pyramid of the Sun. The Avenue of the Dead, like one hand of a clock, pointed at where the Pleiades would have set on the southern horizon on 12th August 3114 BC, whilst the skull beneath the Pyramid of the Sun, like the other hand of the clock, pointed at the place on the horizon where the Pleiades set today."[43]

Morton and Thomas found that the three pyramids of Teotihuacán were a reflection of Orion's Belt, similar to Robert Bauval's saying the Pyramids of Giza were a reflection of the constellation (there is disagreement on this from Egyptologists, but Dr. Jan Harding announced in 2003 that Thornborough Henge, in Yorkshire, England, which predates the Giza pyramids, also symbolized the stars of Orion's Belt (see "Ecstatic Fuse Box" in chapter 21). Morton and Thomas say they were told by Hopi elders that the extraterrestrials who brought the crystal skulls were from Sirius, Orion, and the Pleiades (where all that channeling comes from!). The information is held in the molecular crystal lattice, as in the silicon chips we use today, but can be accessed by consciousness. The Mitchell-Hedges skull was tested by Hewlett-Packard (experts in crystallography, since they use them in their computers and have a specially equipped laboratory for testing crystals), but they could find no tooling marks on the skull, and so they concluded that it would

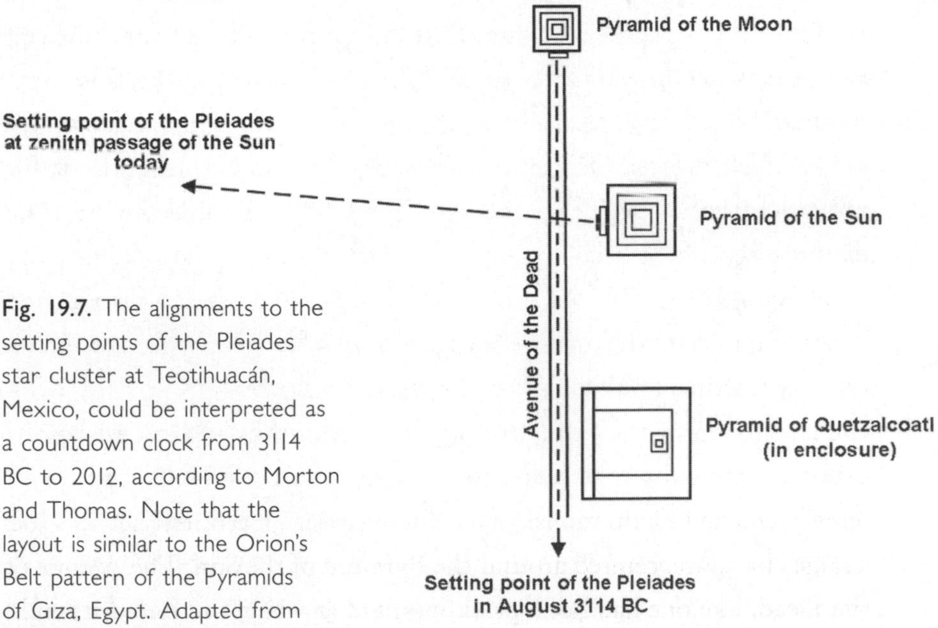

Fig. 19.7. The alignments to the setting points of the Pleiades star cluster at Teotihuacán, Mexico, could be interpreted as a countdown clock from 3114 BC to 2012, according to Morton and Thomas. Note that the layout is similar to the Orion's Belt pattern of the Pyramids of Giza, Egypt. Adapted from *The Mystery of the Crystal Skulls: Unlocking the Secrets of the Past, Present, and Future,* by Chris Morton and Ceri Louise Thomas.

have taken three hundred man-years to make it. Harley Swiftdeer, who claims to be a "Cherokee medicine man" and "war-chief of the Twisted Hairs Society's council of elders," told Morton and Thomas that originally the Olmecs had brought the skulls: "The skulls were kept inside a pyramid in a formation of tremendous power, known as the Ark. The Ark was composed of the twelve skulls from each of the sacred planets kept in a circle, with a thirteenth skull, the largest, placed in the center of this formation. This thirteenth skull represents the collective consciousness of all the worlds. It connects up the knowledge of all the sacred planets."[44]

Don Alejandro Cirilio Pérez Oxlaj (the Maya elder mentioned in part 1 of this book), head of the National Council of Elders of Guatemala, was the source for most of the material in the penultimate chapter of Morton and Thomas's book. Don Alejandro says that now is the "Time of the Warning," also called the Time of the Awakening, or the Time of the Quickening—the transitional period between the end of the Nine Hell cycles, which ended August 17, 1987, and the time of the Thirteen Heavens, which begins "after sunset" on December 21, 2012. The crystal skulls were left here to help us in this time of transition, he says. We must change before 2012, showing love and respect for one another, for life, and for the planet, or we shall be cleansed from it by cataclysms. The modern Maya have been arranging a series of meetings for elders from indigenous tribes of North and South America and other continents, to culminate in a reunion of all the crystal skulls, and this will happen at a predestined time known only to the Maya daykeepers.

On page 349 of *The Mystery of The Crystal Skulls,* Don Alejandro is quoted as saying that "the prophets said the Nine Hells would end between Long Count kin 1863022 and 1863023, between the days you call 16 and 17 August 1987, that the Time of the Warning would come to pass and that now is the time to prepare for the Thirteen Heavens. The Thirteen Heavens will begin after sunset on 21 December in the year 2012. The next day after that is written in our Long Count calendar as 13.0.0.0.0."[45] An analysis of this statement is very revealing.

The first thing we notice is that the end of the Nine Hells is said to be exactly the same days that Tony Shearer calculated, and which were also repeated by José Argüelles in *The Mayan Factor.* If you look at note 21 for this chapter of the book you hold now (part 11 of note 21, dealing with the Harmonic Convergence), you will see that there was an error of 221 days in this calculation.* The next thing to note are the kin numbers (days since the beginning of the count), 1863022 and 1863023. Again, these are exactly the same as the kin numbers given in *The Mayan Factor* for the days of the Harmonic Convergence. The trouble is, Argüelles made an error of 281 days (see note 16 of this chapter) in calculating these numbers, and Don Alejandro has the same numbers. Since there are 1,872,000 days in the thirteen-baktun cycle, if we calculated forward to find the end of the cycle, we would arrive at a day 281 days short of December 21, 2012 (i.e., March 15, 2012). Since Don Alejandro goes on to give the correct end-date, we can only assume that the "prophet" who supplied him with the information is none other than José Argüelles (or Valum Votan, a.k.a. Lord Pacal, who Argüelles claims is inhabiting his body nowadays).[46]

AEON OF MAAT

Magick is defined as "the art of causing change to occur in conformity with the will" (the spelling with a *k* is a modern convention among occultists to avoid confusion with sleight-of-hand entertainers). Thelemites are magic(k)ians who follow the teachings of Aleister Crowley's Law of Thelema (Greek for "will"), which is centered on the "Do what thou wilt shall be the whole of the law" doctrine. Maat Magick is an extension of Thelemite ideas and consists of techniques for accessing the power of the future Aeon of Maat, the Egyptian goddess of truth and justice, who is

*Fifty-two haabs of 365 days makes a Calendar Round, and nine of these add up to 170,820 days. If we count that number of days forward from the day that Cortés landed, we come to January 7, 1987, in the Gregorian calendar—an error of 221 days. See note 16 for more.

Fig. 19.8. The magick practices and beliefs of Aleister Crowley have entered popular culture in more ways than one: here is Crowley's face appearing on a boot.

the sequel to the current Aeon of Horus. Maat Magick lists six aeons: the Nameless Aeon (prehistory, hunter-gatherers, shamanism); the Aeon of Isis (beginnings of agriculture, goddess worship, paganism); the Aeon of Osiris (cities, war, Abrahamic religions); the Aeon of Horus (atomic energy, radio, TV, atheism, existentialism, Thelema); the Aeon of Maat (quantum theory, genetic engineering, Internet, various Magicks); and the Wordless Aeon (when a new species emerges from *Homo sapiens*).

The Aeon of Horus, according to Thelemites, began in 1904 when Crowley received the book of the Law (his wife, Rose, went into a trance in Cairo and channeled it). In 1974, when Priestess of Maat, Andahadna, received the "Liber Pennae Praenumbra" (see the discussion that follows), the Aeon of Maat began to run concurrently with the Aeon of Horus. Since Maat is "the Daughter, twin to Horus," the influences will work together to change humanity by "the awakening of the Racial [human race] Unconscious to Racial Consciousness."[47] When humanity is thus transformed into a "doubly conscious species," the Wordless Aeon will start, and this will be an aeon in which "physical speech is unnecessary with a shared consciousness."

An even more recent development, Chaos Magick has developed as an unstructured approach to magick using any practices that get results, but without adopting the associated belief systems. Even here, there is an accepted aeon structure, according to Peter Carroll's book *Liber Kaos*.[48] In this system, there are four aeons: the Shamanic Aeon, Religious Aeon, Rationalist Aeon, and Pandemon Aeon (which is equivalent to the Wordless Aeon of the Maat Magick system).

On the Mutation Parlour[49] website, Orryelle Bascule connects

Chaos Magick and Maat Magick to the work of Jenkins, McKenna, and Argüelles, plus the Hopi prophecy of the Thirteenth Tribe, or Rainbow Warriors, to conclude that the return of Quetzalcoatl will be the activation and rising of kundalini and an awakening of collective consciousness for humanity. This will also be an activation of Gaia's kundalini, the serpentine energy lines that connect Earth's sacred sites, the chakras of our planet. This is explained in his "Zuvuya" essay.[50]

Bascule has formed a microcosmic thirteenth tribe of Rainbow Warriors from a variety of nationalities to perform a series of rites—"the 13th Tribe Weaving"—involving "chakra-piercing," which is explained in his "Liber XIII."[51] Magickal works are traditionally titled in Latin, hence "Liber XIII" means "Book 13," since the concept of twelve tribes is not just concerned with Jewish ancestry, but, says Bascule, also exists in Australian aboriginal mythology, as well as that of the Hopis, and represents twelve races and a thirteenth that unites them. Thirteen is also a most significant number for the Maya. The anahatas, or heart chakras, of the twelve initiates are woven together "both etherically and with physical threads through piercings ritually executed in the Anahata chakra of each, by a Thirteenth." These rites were to be performed at "Gaia's seven chakra points" in the years approaching 2012: "Activating group Kundalini at each of Gaia's chakras via these weavings, progressively towards 2012, will help to resonate humanity's awakening Groupmind with the intense awakening of these Great Serpents, the Ida and Pingala of Gaia's Kundalini."[52] Bascule thinks that in 2012, the Aeon of Horus will fuse with the Aeon of Maat. In light of the aeon information above, this could also be interpreted as the arrival of the Wordless Aeon, or at least a possible access point to it, and the start of a new species of telepathic human.

HIGH PRIESTESS OF MAAT

There are currently several organizations calling themselves the Ordo Templi Orientis, or the Order of the Temple of the East, but the branch run by Kenneth Grant, who is one of Crowley's executors, is called the Typhonian OTO (Typhon was a Greek goddess, according to Grant, though in standard Greek mythology Typhon was the son of Gaia and a snake-legged monster who attacked the gods), to distinguish it from the other, "spurious organizations." It is also called the Temple of the Rising Star, Sirius.

Soror Andahadna, priestess of Maat in the Typhonian OTO, who is now known as Nema, published the *Liber Pennae Praenumbra*[53] in 1976. The information was received while out of body during magickal workings, and concerns the Aeon of Maat, an age of truth and justice that most occultists do not expect for another two thousand years.[54] However, since the Manhattan Project, during World War II, when nuclear explosions caused rips in space-time, a "seal" was dissolved on a gateway to another universe. This is where some very nasty beings live, known as "the Elder Gods," and if you have ever read any H. P. Lovecraft, that may sound familiar. For this reason, our evolutionary progression must be speeded up so the gateway can be resealed before the Elder Gods, who are "alien beyond human imagining," enter and devour us.[55]

The magickians have, since 1975, been working with Maatians (i.e., our future selves) to try and birth an evolving "cosmic net": "Once established, the Net will radiate a field akin to magnetism that will induce mass mutation in the genes of the Race of Man. By subtle atomic and molecular adjustments, it will unlock the [as] yet unused potential of the human brain, stimulate the human quality called 'genius,' and vastly enlarge the capacity of Man on all planes."[56]

The next phase would be the development of a "world mind": "The Mutants will have the additional sense-faculty of the pineal eye reawakened, a material manifestation, as it were, of the Ajna Chakra . . . This Eye will perceive the Net; perception leads inevitably to union. In uniting with the Net, they shall unite with each other, and Man shall achieve godhood as a race."[57]

PART 4

Diagnosis

The rise in levels of beta-carbolines seen in the pineal glands as one ascends primate phylogeny—with the highest levels occurring in Homo sapiens*—lends credence to the idea that the adaptation called consciousness may involve mutation of the metabolic pathways associated with serotonin, other tryptamines, and harmine. The shift of emphasis from serotonin pathways to beta-carboline and methylated tryptamine pathways is, we speculate, the molecular evolutionary event that is responsible for the intimations of transfiguration that have recently characterized mass consciousness.*

Terence and Dennis McKenna,
The Invisible Landscape

OCRUXAVES
PESUNICA
MMXII
MCMXCIII
MCMLXXIII
MCMLIII
MCMXXXIV
MCMXIV
MDCCCXCIV
MDCCCLXXIV
MDCCCLV
MDCCCXXXV
MDCCCXV
MDCCXCVI
MDCCLXXVI

20
ANALYSIS

THREE QUESTIONS

In part 1, we saw that in addition to the thirteen-baktun cycle of the Maya, other calendars, such as the Tibetan, Vedic, and Jewish, indicate that 2012 is the end of a long-term cycle. We also saw how many indigenous people from the North, Meso, and South Americas, and even Africa and New Zealand, have recently been saying that now is the time to reveal their ancient prophecies. They invariably mention dates that correspond to the Western year 2012 as the time of either catastrophic Earth changes and/or a spiritual leap for humankind. These prophecies seem to validate the conclusions that arise from study of the various calendars, but how reliable are the prophecies?

In part 2, we looked at theories in which researchers have tried to find a possible cause for the predicted changes. A lot of the interest shown in the year 2012 has originated from two best-selling books. These are *The Mayan Factor,* by José Argüelles, and *The Mayan Prophecies,* by Maurice Cotterell and Adrian Gilbert. As we have seen, despite some good points, both books have errors that have led to confusion about the Maya calendars. This confusion has infected the theories of several researchers who have been influenced by the books. A third best seller, Graham Hancock's *Fingerprints of the Gods,* has further confused matters by quoting a defunct 2011 end-date for the thirteen-baktun cycle.* The problem here is to sift out the infected parts of the theories, along with the other errors. We don't always have to throw the whole theory out, but how much is left at the end of this process?

In part 3, we looked at modern sources of confirmation about 2012, including sacred plants, near-death and out-of-body experiences, UFO contact experiences, remote viewing, prophets, visionaries, and magic(k)-ians. The question now is how many of these sources have also been infected, whether consciously or subconsciously, by existing prophecies and theories.

ARE THE PROPHECIES RELIABLE?

In New Age circles, if a tribal elder or shaman gives a talk, she or he is usually surrounded by an aura of wonder. The assumption is that such an elder is a repository of knowledge that has been carefully preserved by word of mouth, and that the elders have met and agreed that now is the time to release the wisdom and prophecies for the benefit of the world.

In 2003 I spent a day with Tlakaelel, a Nahuatl "man of knowledge" from Mexico. Speaking through an interpreter, he explained that he did not want to be worshipped. A short while later, someone arrived who had spent time with Tlakaelel at his educational center near Teotihuacán, and greeted him by kneeling on the floor and kissing his hand. I cringed. Tlakaelel said that he "holds forty-eight thousand years

*The 2011 end-date was inherited from Frank Waters's 1975 book, *Mexico Mystique,* which in turn got the start-date from *The Maya,* by Michael Coe (1966). The start-date of 3113 BC should have said year –3113, or should have clarified that it referred to 3113 BC astronomical, as opposed to 3113 BC historical. To get an end-date of 2011, you would think the calculation must have been made from a 3115 BC start point, which is equivalent to year –3114 (3114 BC astronomical), but it is probably just a case of error due to the missing year zero: 3113 + 2011 + a missing year = 5,125 years, whereas the correct calculation is either the historical one (3114 + 2012 = 5,126 minus year zero = 5,125) or the astronomical one (3113 + 2012 = 5,125).

of history" in his memory. He said this relates to four 12,000-year periods between Earth's passing through a photon band that goes around Alcyone, the central star of the Pleiades. My heart sank at mention of the very dubious photon band/belt concept (see chapter 19). He said that this is a time of purification and cataclysms, and that the last time we crossed the band, the year changed from 360 days to 365 days, and that this time it will change back again. This provides a convenient explanation for the lack of a Long Count calendar among the Aztecs—in fact, Tlakaelel said that the 360-day, tun-based Long Count calendar was "superseded" by the 365-day haab calendar, whereas Mayanists tell us that the haab and tzolkin both preceded the Long Count.[1]

Tlakaelel said that Tenochtitlán, the Aztec capital that once stood where Mexico City now stands, was founded on July 26, 1325, and finally fell to the Spanish on August 13, 1521, and that it lasted for a period of four Suns of fifty-two years each. Unfortunately, the period is not in actual fact 4 × 52 (208) years, but around 196 years. The year of 1325 as the founding of Tenochtitlán agrees with what the historians say. July 26 is the date of the zenith passage of the Sun at Teotihuacán and it is familiar as the "synchronization date," or New Year's Day, adopted by José Argüelles for his Dreamspell count, after the fashion of the postconquest calendar reform of the Yucatec (from the Yucatán Peninsula) Maya. The date of the fall was correct at August 13, 1521.

This was as much as Tlakaelel would say about the calendar. In fact, when I showed Bert Gunn, his interpreter, my booklet on the Aztec Sunstone (which accompanies the Aztec Sunstone animation I put together—see chapter 16, Aztec Pizza), he asked me if I could give them any information about fifty-two-year and 12,000-year astronomical cycles. In other words, it seems that these prophecies are already tainted by New Age sources, and are still incomplete.

An Internet search provided some additional information from an Aztec "carrier of the Word" named Ehekateotl Kuauhtlinxan,[2] who says that from the date of the fall of Tenochtitlán on August 13, 1521, there were nine periods of fifty-two years that add up to 468 years (together

Fig. 20.1. An Aztec priest apparently adjusting a ratchet on a calendrical device. Museum of Anthropology, Mexico City.

making up a period of darkness), until August 13, 1989. He then explained that because the 1521 date was recorded in the Julian calendar, the equivalent date today would be August 28, 1989, and that this was the day on which the Aztec people started to "come out," to claim recognition as an ethnic group.

The period described by Ehekateotl is the same as the Nine Hells that we have already discussed, except that these Aztecs are measuring them from the fall of Tenochtitlán rather than the landing of the Spanish, which means that their equivalent of Harmonic Convergence was on August 28,

1989. However, the first thing we notice is that there was a thirteen-day discrepancy between the Julian and Gregorian dates by 1989, not a fifteen-day one. The next thing is that the calculation would be based on 365-day years—the Aztec *xiuhpohualli* (which the Maya called haabs), rather than solar years. When we then calculate 9 × 52 365-day years from August 13, 1521 (Julian date), we arrive at May 1, 1989 (Gregorian date).[3] John Mini, in *Day of Destiny,* also made an error with this calculation, somehow coming up with August 13, 1999—over ten years in error—but then he never did explain how he came to this conclusion.*

So having found that the New Age concept of a photon band had tainted the "knowledge" of the man of knowledge and that the carrier of the Word had tripped up on his numbers, I realized that Britt and Lee Elders, who made the Mexican UFO videos that incorrectly gave 1991 as the advent of the Sixth Sun (see chapter 15), were not the only "elders" who had shot themselves in the foot.

In the previous discussion of the crystal skulls, in chapter 19, we saw how Don Alejandro was quoting *The Mayan Factor,* implying that this was ancient prophecy. It looks as if some of the elders have been looking to nonnative researchers to pep up their knowledge base, while some of those same researchers have been changing their names to Maya names, Native American names, ancient Egyptian names, or even ancient Hebrew priest names (see note 5 of chapter 9) to give their theories and prophecies some credibility. Another of Morton and Thomas's main sources of crystal-skull information was the Cherokee

*John Mini hints that the day 13 Reed represents the end of the Fifth Sun, and that it was on August 13, 1999 (hence its position among the snake rattles, which are a warning). He probably arrived at this date by counting the twelve years, from 87 to 99 (if 87 = 1 Reed, 88 = 2 Knife, 89 = 3 House, 90 = 4 Rabbit, 91 = 5 Reed . . . 99 = 13 Reed), then subtracting three days from August 16, to compensate for the three intervening leap days. Day 13 Reed in Year 13 Reed does work out as 1999, but either January 17, 1999, or October 4, 1999, not August 13. Mini also said on page 173 that there are only eight dots on each of the fire-snake segments surrounding the Aztec Sunstone, and included a fold-out diagram at the back of the book showing only eight dots per segment. It is plain, even by looking at a photo of the real Sunstone, that there are ten dots per segment.

medicine man Harley Swiftdeer—someone various Native Americans say is not recognized by the Cherokee Nation, and whose practices are not authentically Cherokee.[4]

There are now websites with information on what are called "plastic medicine men," although in my opinion some of this material is over the top, even listing anthropologist Michael Harner as an offender, for no particular reason other than that he has published a study on shamanic practices that is used as source material for workshops by self-taught technoshamans. All I am saying is that we should proceed with caution concerning indigenous prophecies, and cross-check them rather than giving our power away.

Having said that, if we look back through the prophecies we've examined, we notice that some predict a catastrophe of huge proportions, with just a handful of survivors, while others play down this element, emphasizing only the changes to humankind. Many versions include the concept of a period of purification or purging prior to 2012, but the start-dates vary: 1985 (Patricio Dominguez—see note 3); 1987 (Senecas, Cherokees, and Carlos Barrios); 1989 (modern Aztecs such as Ehekateotl Kuauhtlinxan, the "carrier of the Word," mentioned previously); 1990–1993 (Q'ero); 1991 (Quetza-Sha); 1991–2001 (Gerardo Barrios); 1998–2000 (the Pueblo); 2000 (Q'ero, interviewed by Alberto Villoldo).[5] Others have said that the last katun of the thirteen-baktun cycle is the "time of purification," but have then given 1992 as the year in question, repeating an error from *The Mayan Factor*.[6] A katun is twenty tuns of 360 days, totaling 7,200 days, or approximately 19.72 years. The final katun started on 12.19.0.0.0, which was April 5, 1993. So the prophecies agree that there will be a period of purification lasting anywhere from twenty-seven years to eleven years before the end of the final katun, in December 2012.

If we filter out the prophecies that are infected with New Age dogma, we are still left with some impressive predictions. The Seneca medicine man Moses Shongo, for example, made his prediction around eighty years ago, and told of a twenty-five-year period of purification before 2012. I recently received an e-mail from my friend Len Stevens, who has been living at a meditation retreat at a temple in Thailand, where he told me of the predictions of his teacher, who lives in northern Thailand. He has three times received detailed information while meditating concerning forthcoming acts of war:

"[These events culminate] in a pole reversal in 2012—the destruction of everyone who isn't at least partially enlightened or in the presence of an enlightened one. I really don't think these people are aware of the other prophecies regarding 2012."[7]

The Maori prophecy (see chapter 2) is instructive here, showing how a misinterpretation can arise. The "fall of the curtain" was interpreted as the end of the world following the final scene of the play, but on further investigation it actually means the dissolving of the veil. In fact, many people are worried that in 2012 the world will end. They are right. The world *will* end: it will be the end of the Fourth World of the Hopis and the Maya, and the start of the Fifth World. It will be a world transition, or the end of one world era, and perhaps the end of the world as we have known it. This is the same as the end of the Fifth Sun of the Aztecs and the start of the Sixth Sun.

The Earth changes have begun and there is a general agreement in the prophecies that the purification period will be a time in which humankind will have the opportunity to evolve, and that we can consciously help the process by visiting sacred sites to strengthen our energy bodies, or do exercises that engage our energy bodies. Some say that the progress we make on ourselves and on lessening our ecological impact on the planet could lessen the Earth changes. Phrases like "destruction of unenlightened ones" may be due to the insidious percolation of the Christian concept of the Judgment Day, but they may also be a warning to clean out our cerebral closets before the locks are broken open and we are confronted with our own suppressed complexes. Judging from the results of various self-help therapies in exorcising personal demons, this could be like all hell breaking loose in the mind before equilibrium is achieved.

Fig. 20.2. Mural of the Four Suns at Palenque. The tassles around the neck of the central god are Suns, or gods, or eras. The central god may be the equivalent of the central controlling deity on the Aztec Sunstone, who controls the fifth and final era in the set of five.

THEORY FILTRATION

We have seen how Cotterell and Gilbert's interpretation of the "birth and death of Venus" is flawed, and their supposed proof of Maya knowledge of sunspot cycles is a mess of graphs and calculations that lead nowhere. These misconceptions have infected all of Cotterell's later books—the whole series is based on *The Mayan Prophecies.* Also infected are later books by Gilbert, including *Signs in the Sky;* Geryl and Ratinckx's *The Orion Prophecy;* Michael Poynder's *Pi in the Sky;* Morton and Thomas's *Mystery of the Crystal Skulls;* Nigel Appleby's *Hall of the Gods;* Elliott Rudishill's *E.Din: Land of Righteousness;* William Gaspar's *The Celestial Clock;* and even *The Unity Keys of Ananda* and David Wilcock's *Shift of the Ages.* However, Cotterell and Gilbert were right about the increase in solar activity, and also in general about a solar megacycle, as we have seen. Also relevant is their mention of the effects of fluctuations of

Fig. 20.3. There were originally four solar deity heads in the mural of the Four Suns at Tonina (see the illustration facing chapter 6, which includes a foreshortened partial reconstruction of the mural). Note the similar style to the tassles in the previous picture. Notice the skull-like central controller at the top, perhaps governing the fifth and final period of the precessional era (26,000 tuns).

the geomagnetic field on the human endocrine system. They mention this as part of an attempt to prove that changes in the solar wind that appear on their graph in the year 627 AD were responsible for a drop in the fertility of the Maya, causing them to abandon their cities. Jenkins points out[8] that this actually corresponds to "the Classic Period florescence of Mayan culture," rather than the decline, which started around 800 AD. Cotterell indulges in some dubious graph work to make this point,[9] involving a forty-degree swivel to a graphed line of the solar cycle to align it with another line showing the production of follicle-stimulating hormones (FSH).

The errors in *The Mayan Factor* (in this case, the 3113 BC start-date not being clarified as an astronomical rather than a historical date, but other errors are fully listed in note 21, chapter 19) have caused a problem for Sergey Smelyakov and his Auric Time Scale theory, as we have seen. They also caused a problem on page 173 of Weidner and Bridges's *A Monument to the End of Time,* where the Argüelles date for the final katun was used—1992 instead of 1993 (and unfortunately this has been repeated on page 335 of the latest, expanded edition).[10] However, Argüelles's simple discovery that the tzolkin is a microcosmic version of the thirteen-baktun cycle (of 260 kins and 260 katuns, respectively) is invaluable. His speculations concerning the I Ching and the Franklin magic square (first discussed in his *Earth Ascending*) were prophetic, and his application of de Chardin's noosphere concept is inspirational.

Even that dreadful book *Giza-Genesis,* by Wilkie and Middleton-Jones, managed to get something right: Rush Allen, the astromythologist who runs the SiLoaM.net website mentioned in chapter 8, agrees that the theory was misinformation, but concedes that somehow James Wilkie and Howard Middleton-Jones came to the same conclusion he arrived at—that the Hall of Records will be opened on December 21, 2012, at 10:18:13 local Cairo time—though his interpretation of the event is a mass illumination of humankind.

COSMIC DUST INCREASE

Paul LaViolette's galactic core explosion theory is a science-based concept that sounds feasible and connects several ideas we have discussed, although only peripherally in some cases. A long-term solar cycle, geomagnetic reversals, legends of the Sun going dark and other Egyptian mythology, Maya and Aztec mythology, the precession of the equinoxes—all these are explained by LaViolette's theory. He even points out[11] that José Argüelles's notion of a galactic synchronization beam emitted from Galactic Center has certain points in common with the galactic-core explosion theory. He also has an essay on his website[12] in which he makes a case that the photon belt is a metaphor for a galactic superwave.

LaViolette's theory also fits precisely into the predictions of the alchemist Fulcanelli, and the Cross of Hendaye, as recounted by Weidner and Bridges. However, despite the fact that LaViolette tentatively links the theory to the thirteen-baktun cycle of the Maya, he told me that he disagrees with Kev Peacock's theory because "mag[netic] pole flips are induced by major coronal mass ejections unlike anything that occurs at present. The Sun has to first become transformed as a result of a cosmic dust incursion."[13] This implies that he doesn't really expect a major galactic superwave in the immediate future, but a minor one that would not bring a major change to Earth's geomagnetic field or climate, but would carry an electromagnetic pulse, or EMP, that would cause high voltage surges in power and telephone lines. In his online essay, quoted earlier, he points out that the minor superwave events have "EMP effects comparable to those of major superwaves,"[14] but for a geomagnetic inversion to occur, cosmic dust would have to cause a major increase in solar activity, which would in turn affect Earth's magnetic polarity.

LaViolette's comment to me, above, was made in the year 2000, before the solar maximum and subsequent unusual solar behavior. On November 4, 2003, came the largest solar flare ever recorded. The previous record was on April 2, 2001, when we had an X-22-level flare (classified then as an X-20 because at that point the gauge only went up that far; bigger events have since led to the development of higher scales, including the tentative use of a *Y* system). This time, estimates varied between X-28 and X-35 or more.[15] In fact, in March 2004 physicists from the University of Otago, in New Zealand, released the results of their ionospheric impact assessment,[16] in which they concluded that the November 2003 flare had been an X-45, which is more than twice as big as any previous flare. Fortunately, it wasn't pointing toward Earth, since the X-ray flare of March 6, 1989, which was pointing toward Earth—and shorted out the Canadian power grid, leaving six million people with no electricity (also burning out a New Jersey nuclear power transformer)—was somewhat less than an X-20.

The 2003 solar flare came after a record-breaking nine major solar eruptions in the space of twelve days. All this happened after solar maximum, in a period when the Sun should have quieted down.

Even before this event, on November 2, 2003, the online version of the magazine *New Scientist* declared that geophysicist Ilya Usoskin and colleagues from the University of Oulu in Finland, working with scientists from the Max Planck Institute for Aeronomy in Germany, had just revealed that the Sun is now more active than it has been for over a thousand years.[17] An announcement in July 2004, by Swiss scientists at the Institute for Astronomy, in Zurich, confirms the increased activity of the Sun via a study of Greenland ice cores. Significantly, in April 2006, NASA predicted that an immense solar storm was likely to occur in 2012.[18]

On August 11, 2003, it was announced on www.SpaceDaily.com[19] that the Ulysses satellite had detected a threefold increase in "stardust" flowing through the solar system since the last solar maximum, when the Sun's magnetic field weakened due to "increased solar activity." The stardust, which is usually kept out by the Sun's magnetic field, was predicted to "create more cosmic dust by collisions with asteroids and comets." However, a few days later, on August 20, 2003, another article appeared revising the reason for why the dust was increasing: "Instead of reversing completely, flipping north to south, the Sun's magnetic poles have only rotated at halfway and are now more or less lying sideways along the Sun's equator. This weaker configuration of the magnetic shield is letting in two to three times more stardust than at the end of the 1990s. Moreover, this influx could increase by as much as ten times until the end of the current solar cycle in 2012."[20]

ALCYONE EXAMINED

The concept of the photon band just won't go away, no matter how many times its flaws are pointed out. Internet discussion groups are indignant when the subject comes up, and then just when you think it is all settled, a new member starts the whole thing going again. It is as persistent as an archetype from the collective unconscious. Barbara Hand Clow describes the problem in *The Pleiadian Agenda,*[21] saying she wouldn't have given the matter much attention except that her Cherokee grandfather, Grandfather Wise Hand, had told her many years ago that Alcyone, the central star of the Pleiades, was her home. Then she noticed in *Calendrios Mayas Y Hunab K'U,* a book by Hunbatz Men in which he describes seventeen Maya calendars, that one of them is a 26,000-year cycle of the Sun's orbit around Alcyone. When she asked Hunbatz Men if the end of the Pleiadian cycle would coincide with the end of the thirteen-baktun cycle in 2012, he confirmed that it would.

As we have seen, the myth liberally mixes fact and fiction. Here are the facts:

1. The precession of the equinoxes takes between 25,600 and 26,000 years.
2. The solar system moves around a "central sun"—Galactic Center—but it takes 230–250 million years to complete one orbit around it.
3. The Maya had a thirteen-baktun cycle that ends in 2012, when the solstice Sun is close to Galactic Center and aligned with the center of a "belt of light" called the Milky Way.
4. The Maya used the movement of the Pleiades constellation to measure precession, and thus, with the Pyramid of Kukulkán, set up a Pleiades alarm clock set for 2012.

In the photon-belt myth, these facts are mixed up, so that the solar system moves around a central sun in the Pleiades and takes 26,000 years to do so, with the cycle ending in 2012.[22] The cycle used by Hunbatz Men encodes the important connecting concepts: that the Pleiades are used to check precessional movement toward galactic alignment in 2012. The fact that the mythologized version of Earth revolving around the Pleiades is astronomically incorrect does not alter the fact that the important concepts have been preserved in that myth.

Alexey Dmitriev's magnetized plasma band seems to be much closer to the photon-band concept than LaViolette's theory, and applies just as effectively to long-term solar cycles, geomagnetic reversals, and precession. It might also be encoded in ancient mythology, if anybody cared to spend enough time looking for it. However, the magnetized plasma band has some other fascinating implications.

PLASMATIC COMMUNICATION FROM THE NOOSPHERE?

William C. Levengood is a biophysicist based in Michigan who specializes in agriculture and seeds. Since 1991 he has been studying samples taken from crop formations across the world and has been publishing the results in peer-reviewed journals such as *Physiologia Plantarum*. Working with Nancy Talbott and John Burke, Levengood founded BLT Research and has been coordinating international efforts to establish scientific protocols for crop sample collecting. Samples are taken at predefined spacings and cross-sections of formations, and control samples (from measured distances away) are also collected and labeled, then sent off to Michigan for analysis. John Burke has also written articles based on the results of the investigations. They have found several anomalies and phenomena that cannot be explained, most of which had never been seen in Levengood's forty-year career before he started looking at crop circles.

Four of the main anomalies are:

1. The stalks are softened so they can bend without breaking.
2. Stalks sometimes have expulsion cavities (also known as blown nodes), holes at the stem nodes where the stalk has been rapidly heated from the inside, turning the moisture to steam so fast that a hole is blown and the sap comes out. This effect has been replicated in an industrial microwave oven (enlarged cell-wall pits are also due to internal heating).
3. An electrical charge has been found on the stalks, the amount of charge proportional to the amount of bending.

4. Bract tissue around the seed has been found to have an increased electrical conductivity.[23]

Additional effects include elongated plant stem nodes, stunted or malformed seed heads, and unusual seed germination, but the four anomalies listed above are the ones relevant to the point I am making here. John Burke says[24] that the team looked for an agent that could be responsible for these effects, and they realized that plasma (an electrically charged gas, or gas with the electrons stripped off) fits the bill nicely. Not only would it also move in a spiral when passing through the geomagnetic field, but it would also emit microwaves, heating the plants from the inside and causing the electrical changes. Meteorologist Terence Meaden had already suggested plasma vortices as the causative factor of crop circles, but the theory was abandoned because it was weather dependent, and the crop circles were found to be forming in a variety of weather conditions (and they were becoming more complex and thus less explicable as a natural phenomenon).

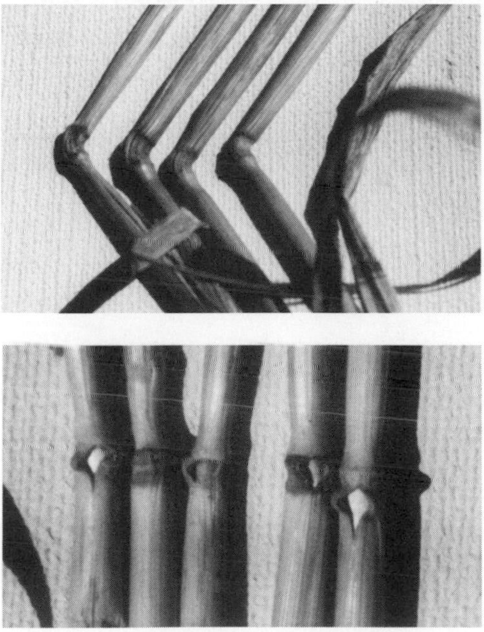

Fig. 20.4. Anomalies in stems from the crop formation at East Kennett, Wiltshire, on July 2, 2000 (see color plate 17). Above are stems bent and stretched at the nodal joint—a physical impossibility when laying down a crop manually (in a man-made formation). Below are "expulsion cavities," holes blown in the nodes where some kind of heat appears to have exploded the plant moisture outward. Photos: Melissa White, www.fairlyte.co.uk/

Looking around for an alternative source of plasma, the BLT team thought of lightning, which is a high-energy form of plasma but one that would cause charring of the crop. However, up in the ionosphere there are low-energy plasmas that glow when they become energized, in the form of the northern lights (aurora borealis) and the southern lights (aurora australis). The reflective layers of the ionosphere weaken at night, especially just before dawn (when the plasma is more likely to break through), the time many crop circles seem to appear. Also, the number of free electrons in the atmosphere increases one hundredfold at sunspot maximum, which coincided, from 1988 to 1989, with record numbers of crop formations up to that time. These observations led the team to take a closer look at the ionosphere. It turns out that there are forms of plasma now known as sprites that have recently been observed coming down from the ionosphere as far as cloud level. The BLT team deduced that a similar phenomenon might be responsible for crop formations, since this would also explain two other oddities about the crop designs.

There have been instances (in two thirds of the thirty-two crop formations where BLT had been able, by 1998, to obtain soil samples)[25] when high concentrations (twenty to one hundred times higher than the control samples) of micron-sized spheres of magnetite have been found in soil samples from crop circles. Despite claims of iron-filing misdemeanors by one hoaxer, this effect—found in twenty-one formations around the world—is said by BLT to be meteoritic dust that has been heated and melted into microscopic spheres. Most meteoritic dust can be found up in the ionosphere. A spiraling plasma tube would bring that down and concentrate it in the resulting crop formation. But BLT wondered why the plasma tubes would mostly discharge in a small area of England. When they looked at a geological map of the area, the answer was obvious: Wiltshire is mostly chalk, and chalk is an aquifer, a spongelike rock that holds a lot of water, trickling through and causing a negative electric charge. This would attract the plasma like a lightning conductor.

Researchers Glenn Broughton and Steve Page came to this same conclusion independently after checking geological maps of the UK and cross-referencing information from crop-circle databases. About 90

Fig. 20.5. The most important crop of the ancient Maya was maize. Here is Yum Kaax, Maya maize god, based on a drawing in the Dresden Codex.

percent of crop formations were found to lie on aquifers (which include limestone and greensand). The BLT team checked a field that regularly gets formations, using a fluxgate magnetometer, and found that the magnetic field that accompanied the charge vanished right after a formation appeared.

The only thing left unexplained was how the complex and apparently meaningful patterns could result from a natural process. This is where it gets complicated, involving magnetohydrodynamics, chaos theory, and waveguides, and the BLT team is still developing that part of its theory. However, if we look at photographs of sprites, we find that above a sprite is an expanding disk called an elf, both their namesakes (sprites and elves) being types of fairy folk—sometimes blamed for creating the first recorded crop circles. In *Passport to Magonia,* ufologist Jacques Vallee pointed out many parallels between fairies and today's extraterrestrials, such that he sees them as the same phenomenon. In fact, Alexey Dmitriev says of the vacuum domains, a category in which he includes elves and sprites, that he and his colleagues propose "a new non-meteorological approach to the study of these violent processes, now growing stronger."[26]

Extrapolating from Dmitriev's theory and combining the findings of BLT, crop formations seem to be caused by plasma from the ionosphere, where the plasma has been collecting, bleeding through from the magnetized plasma band that Dmitriev says the solar system is traveling into. If Teilhard's noosphere, the evolving mind layer that surrounds Earth, is located in the ionosphere, then perhaps the increasing number and complexity of crop formations is linked to the plasma buildup, and the noosphere is attempting to communicate with us, in the same way that Jung thought UFOs are projections from the collective unconscious. Jung also observed that psychic manifestations can be expected at the juncture of Platonic months, such as our current situation at the juncture of Pisces and Aquarius. If humanity is the evolving global neocortex—human neurons about to link into a unified component organ of the planet—then perhaps the fairy folk and some of the extraterrestrials are a life-form that collectively represents another, older part of the global brain.

THERMAL PLASMAS AT HESSDALEN

In the Hessdalen valley, in northeast Norway, unexplained lights have been under investigation since the early 1980s by a team led by the Norwegian engineer Professor Erling Strand. Since August 2000, Italian astrophysicist Dr. Massimo Teodorani and his team have joined this research effort, using radio-spectrum analyzers, spectroscopes, and other equipment.[27] In 2001, Teodorani published some results of the study, in which he found that the lights were 95 percent thermal plasmas (the other 5 percent were solid objects). The mysterious lights appear when there is a measured perturbation in the geomagnetic field, and the plasma is trapped in a magnetic cage that keeps it concentrated. They can shift out of visibility and become invisible (Teodorani says that at this point they shift from high-energy plasmas to low-energy plasmas) but are still evident on the radar. (When radar was first invented, these invisible blobs were known as "radar angels," and some radar equipment was modified to filter them out.) Linda Moulton Howe[28] has seen the very same thing in the crop formations of Wiltshire, using an infrared scope, and I was myself present on one of these occasions.[29] These plasmas are not ball lightning (which lasts only for a few minutes at most, whereas these plasmas lasted as long as two hours), and the physics behind them is not currently understood.

There are continual eyewitness accounts of balls of light (BOLs) in the Wiltshire area, and of these, many have been seen in association with crop formations. Bert Janssen's video documentary *Contact*[30] presents a fascinating compilation of sequences of these balls of light filmed over crop circles. In 2001, Nancy Talbott witnessed a crop circle being created by tubes of light in the Netherlands. The following night, Talbott returned to the formation and took flash photographs. Every film (four to five rolls) showed numerous light orbs. Some were quite large ("four–five times larger than a soccer ball") and almost transparent, while others were grapefruit size and more dense, but at the time the photos were taken none was visible to the naked eye. The photos and account can be seen at www.rense.com.[31]

In September 2003, it was announced at the online version of the magazine *New Scientist*[32] that Mircea Sanduloviciu and his colleagues at Cuza University, in Romania, had just succeeded in creating plasma balls in a laboratory that were able to "grow, replicate and communicate—fulfilling most of the traditional requirements for biological cells." Although the traditional view is that life can only be present if there is inherited genetic material, the Romanian team thinks it may have found the explanation for how life on Earth began, since the plasma spheres were created in microseconds when the researchers replicated early conditions on Earth by passing an electric arc through an argon plasma. Even more compelling is Sanduloviciu's conclusion that "the cell-like spheres we describe could be at the origin of other forms of life we have not yet considered."[33]

An article at the online version of the *Guardian* newspaper[34] specifically links this research to the separate studies of Paul Devereux and Andrew York in the UK and Michael Persinger and Gyslaine Lafrenière in the United States, in which the researchers correlated UFO sightings in Leicestershire and the United States to geological fault lines (where layers of rock grind against one another). Persinger and Lafrenière suggested that the air above the fault would be ionized in a piezoelectrical effect such as could be caused when crystals are compressed, resulting in a plasma light orb. So here we have some scientific evidence that the balls of light could actually be life-forms in themselves. The coverage of the Romanian announcement on the *Encyclopaedia of Astrobiology, Astronomy, and Spaceflight* website[35] says that the research implies the possibility that plasma life could be found in stars, in planetary magnetospheres, in areas of ionized hydrogen such as the Orion Nebula, and in ball lightning.

VACUUM DOMAINS

Let's return to the phenomenon Siberian geophysicist Alexey Dmitriev, in his tornado research,[36] identified as "vacuum domains," which we touched on earlier in chapter 12. He says these vacuum domains are increasingly being seen, and can rotate, emit light, and stop electronic devices, due to their strong magnetic field.

Vacuum domains[37] are only explicable in terms of the zero-point energy theory. *Zero point* refers to a vibrational energy retained by molecules even at the temperature of absolute zero. Zero-point energy also exists in a vacuum—there is enough in one cup of coffee to evaporate all the world's oceans, according to physicists Richard Feynman and John Wheeler. The category of vacuum domains includes ball lightning, earthquake lights, poltergeist lights, natural self-luminous objects (or plasmoids), and sprites. Of these, plasmoids, earthquake lights, and ball lightning are more frequently seen at the peak of solar activity. All the different forms of vacuum domains have the ability to penetrate through solid matter. Several types are associated with electrical or magnetic fields, and in 1995 Paul Devereux measured a huge jump in the surrounding geomagnetic field (eight hundredfold), coincident with the appearance of an anomalous light known in the Australian outback as a "min-min light." This light was about six miles away.[38]

Dmitriev notes that the unusual electromagnetic fields in a tornado can cause seemingly impossible things to happen: a maple leaf is pressed into hard stucco; the porous tip of a charred plank stays undamaged after puncturing a wall; and a pine stick doesn't break after puncturing an iron sheet.[39] David Wilcock cites cases where pebbles have gone through glass without breaking it and straw has been embedded in glass window panes.[40]

THE HUTCHISON EFFECT

The kinds of phenomena noted above are known collectively as the Hutchison Effect, after the Vancouver "wild scientist" John Hutchison,[41] who studied the longitudinal waves discovered by Nikola Tesla. Hutchison, who did not receive a conventional scientific education, but had a lifelong fascination with machines and electricity, had a collection of equipment in his lab, including Tesla coils, Van de Graaff generators, radio-frequency generators, RF transmitters, and many other devices. One day in 1979 he turned on all the machines at once, "to study possible field interactions," and was amazed when a steel bar levitated.[42] Continued experiments caused very similar effects as those associated with tornadoes, such as wood becoming embedded in metal without charring it or displacing any metal (cold melting); continuous levitations (including hovering fires, lights appearing and vanishing); melted steel and aluminium; and coins becoming embedded in metal. Video footage and photos of these supposed impossibilities can be seen on the Internet at www.youtube.com.[43]

These kinds of extraordinary events were sporadic at first, but by 1990 Hutchison had refined his approach so that he could produce such effects about five times per hour. His lab was raided twice, and some of the machinery was confiscated—regular scientists could not explain what was happening, and it was thought to be a "risk to national security." He eventually sold what was left of the equipment to avoid further hassles, but has recently rebuilt the lab and tried and failed to sell it on ebay for five million dollars. Wikipedia has a page covering the skeptical angle as well as the claims.[44]

You may recall from chapter 13 that Terence McKenna was told by the self-transforming "machine elves" that the "laws of physics will be transformed around 2012."[45] Dmitriev seems to echo this when he says that all the hundreds of thousands of self-luminous formations, or plasmoids, are having an ever-increasing influence on Earth's geophysical fields and biosphere: "The presence of these formations is the mainstream precedent to the transformation of Earth; an Earth which becomes more and more subject to the transitional physical processes which exist within the borderland between the physical vacuum and our material world."[46]

BIOCHEMISTRY OF GALACTIC ALIGNMENT

It was over forty-five years ago when he was a student that Russian biophysicist Simon Shnoll (now head of the Russian Academy of Sciences Biophysics Laboratory in Pushchino) was working on muscle protein and noticed an irregularity in biochemical reaction rates[47] (an example of this kind of reaction rate would be to measure the time that it takes for acetylcholine to build up in and dissipate from muscles). The results contradicted the theory, and when he asked his teacher, he was told that if he did fewer experiments, the anomaly would disappear.

Over the ensuing years, Schnoll investigated "anomalous statistical regularities in a wide range of physical, chemical and biological processes, from radioactive decay to the rates of biochemical reactions."[48] In 1998, in the Russian physics journal *Uspekhi Fizicheskikh Nauk* (Advances in Physical Sciences),[49] Professor Shnoll published the results of over thirty years of experiments in which he had made repeated histograms, or bar charts, showing rates of radioactive decay in which the decay never reached the ideal bell curve that should happen with increased trials. He found that the more readings he took, the more certain peaks were accentuated on the histogram. However, when he repeated the same experiment at the same time—in other words, had two or more identical experiments running at the same time—the histograms were almost identical. He then used a computer to compare the closeness of all the histograms, and a pattern finally emerged. There were clear peaks at intervals of just under twenty-four hours; 27.28 days; and "at three time intervals close to a year: 364.4, 365.2 and 366.6 days."[50] The first period corresponds "quite precisely" to a sidereal day (i.e., in relation to the stars), not a solar day. The second period is exactly one rotation of the sunspots—a "Carrington rotation."[51] Of the last three, one could either be a solar year of 365.2425 days or a sidereal year of 365.256363 days. The others are unknown.[52]

Thus biological processes, chemical reactions, and radioactive decay are all affected by astronomical cycles, including solar activity and sidereal cycles. This means that the orientation of Earth to the Milky Way has biological, chemical, and radioactive effects. This is exactly what the thirteen-baktun cycle is tracking: our orientation toward the equatorial plane of the Milky Way. As it now seems that the interaction of the geomagnetic field with the solar field and the galactic field can affect our biochemistry, this could be another factor in any forthcoming consciousness change. We have seen convincing evidence that the geomagnetic inversion has already started. Says Dmitriev, "The currently observed polar travel acceleration is not just a shift or digression from the norm, but is in fact an inversion of the magnetic poles; in full process."[53]

The results of an independent scientific study support Shnoll's findings. In 1997, parapsychologist James Spotiswoode, who is director of research at the Cognitive Sciences Laboratory in Palo Alto, California, published a paper titled "Apparent association between effect size in free response anomalous cognition experiments and local sidereal time," in which 2,500 trials were performed using remote viewing and ganzfeld experiments, a technique used in the field of parapsychology where a mild sensory isolation is induced by special goggles or half Ping-Pong balls combined with the wearing of headphones that emit white noise (combined frequencies) or pink noise (white noise without high frequencies). Anomalous cognition (AC) is another term for ESP, extrasensory perception. The idea of the experiment was to increase understanding of how ESP works by finding "a physical parameter which clearly modulated AC performance."

The results of the experiment showed not only that the time of the sidereal day (about four minutes shorter than a solar day) had a significant effect on the results, but specifically, that at a local sidereal time (LST) of about 13.5 hours plus or minus an hour or so, which corresponds to the rising time of Galactic Center on the horizon, there is a massive increase in human intuition or psychic ability. The implication is that the galactic alignment process could significantly boost human psychic abilities. To read about it further, see www.diagnosis2012.co.uk/new7.htm#intuit, and for the full range of Spotiswoode's work, go to www.jsasoc.com/library.html.

IMPLOSION POINT

John Major Jenkins has laid down a solid base with his comprehensive analysis of Maya archaeology and calendrics, showing that the galactic alignment process lies at the heart of the 2012 phenomenon. David Wilcock is constantly building from this base, synthesizing obscure findings as he goes. His approach of looking for scientific confirmation for the Ra material has produced some interesting results, including Dmitriev, Shnoll, and Smelyakov. Wilcock believes that as we travel farther into the magnetized plasma band, we pass through "ripples" that are spaced out by the phi ratio on either side of the band.[54] These are Smelyakov's bifurcation points. The plasma bands he sees as spaced out in a galactic-size version of the Parker spiral (the pinwheel in which plasma bands curve out from the Sun, as explained in chapter 12) about 25,920 years apart. This is how he combines Dmitriev's magnetized plasma band and Smelyakov's Auric Time Scale.

As we have seen, Smelyakov's Auric Time Scale did not reappear until October 2004, following the problem with the start-date. When it did reappear, only the start-date, end-date, and bifurcation points had been amended, but there were no amendments regarding the point I made about the fact that the series does not end in 2012, as implied by David Wilcock, but continues until 2037. When I told Wilcock about the 2037 implosion point,[55] his reply insinuated that he thought

I had misunderstood the Auric Time Scale.[56] He said, "I do not view this actually as a spiral formation in hyperspace."[57] However, he had already suggested that the series terminated in 2012 in his essay "The Ultimate Secret of The Mayan Calendar": "The entire cycle decreases to a certain final moment at the end of the cycle."[58] More specifically, in an interview with Wynn Free, Wilcock said, "Dr. Smelyakov was simply to take the beginning and ending point of the said Maya calendar cycle and to separate this cycle into partitions such that they are divided by phi with a spiral imploding toward 2012."[59]

When I simultaneously informed Sergey Smelyakov of the 2037 implosion point implied in his Auric Time Scale theory, he said I had raised some valid points that had not occurred to him at the time (see chapter 12). During the twenty-eight-month period I waited for the modified version of the Auric Time Scale to appear, I spent an afternoon working out a phi-based series with the same criteria as the Auric Time Scale that terminates on December 21, 2012. I present the result in figure 20.6. You can see that I have conceived of it as a spiral, like the spiral of Peter Russell, which converges on a white hole in time, or implosion point, in December 2012. If you prefer to see it as a series of ripples approaching the band of plasma, imagine viewing it looking into the spiral, centered on the part of the band that we shall meet on winter solstice 2012. As we cross the lines of the spiral, we meet the bifurcation points of the Smelyakov theory.

When I compared the new dates of the bifurcation points to the events listed by Smelyakov, they compared well, with some even closer to the events listed. Most impressive of all, though, was the fact that there is a new bifurcation point on July 18, 1945, which is only two days after the Trinity test, the first time (as far as we know) that a nuclear explosion ever occurred on Earth. Being within two days of this event makes this single new bifurcation point more significant than all the original points. The Auric Time Scale thus comes into a fairly close relationship with Timewave Zero, since that was locked onto history using the Hiroshima bomb—the first nuclear detonation

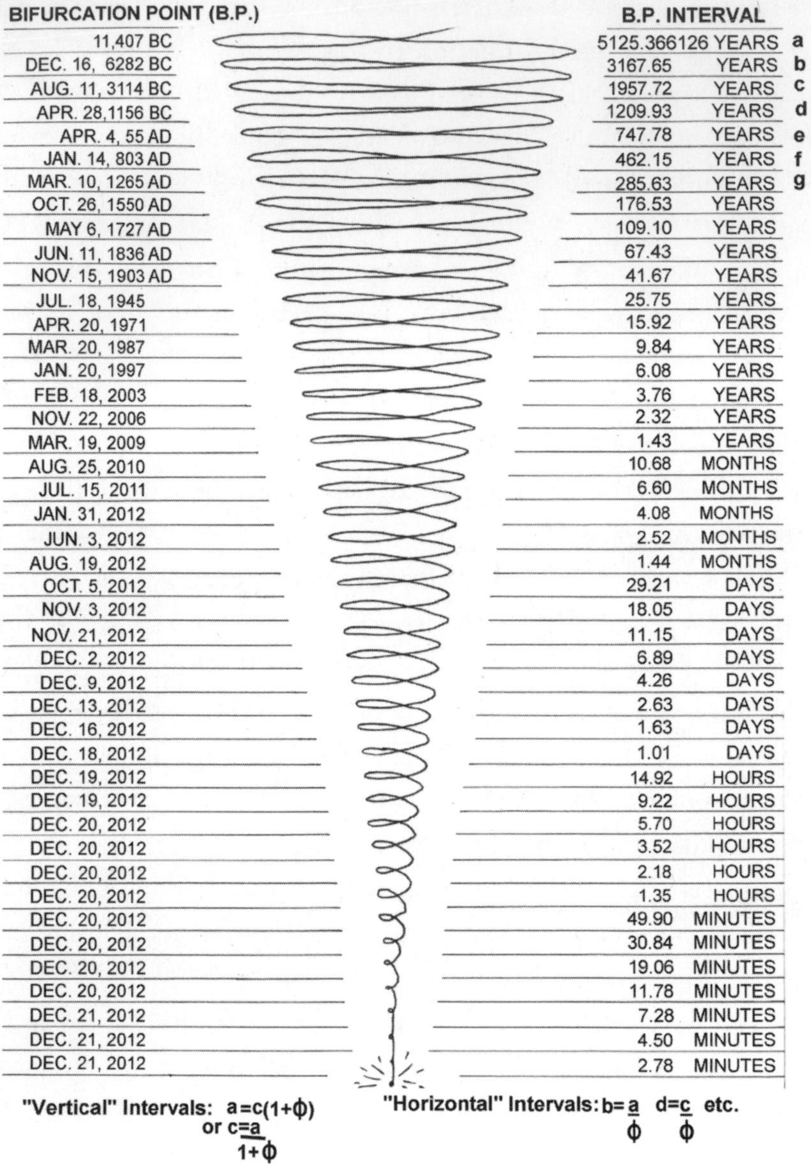

BIFURCATION POINT (B.P.)
B.P. INTERVAL
11,407 BC
5125.366126 YEARS a
DEC. 16, 6282 BC
3167.65 YEARS b
AUG. 11, 3114 BC
1957.72 YEARS c
APR. 28,1156 BC
1209.93 YEARS d
APR. 4, 55 AD
747.78 YEARS e
JAN. 14, 803 AD
462.15 YEARS f
MAR. 10, 1265 AD
285.63 YEARS g
OCT. 26, 1550 AD
176.53 YEARS
MAY 6, 1727 AD
109.10 YEARS
JUN. 11, 1836 AD
67.43 YEARS
NOV. 15, 1903 AD
41.67 YEARS
JUL. 18, 1945
25.75 YEARS
APR. 20, 1971
15.92 YEARS
MAR. 20, 1987
9.84 YEARS
JAN. 20, 1997
6.08 YEARS
FEB. 18, 2003
3.76 YEARS
NOV. 22, 2006
2.32 YEARS
MAR. 19, 2009
1.43 YEARS
AUG. 25, 2010
10.68 MONTHS
JUL. 15, 2011
6.60 MONTHS
JAN. 31, 2012
4.08 MONTHS
JUN. 3, 2012
2.52 MONTHS
AUG. 19, 2012
1.44 MONTHS
OCT. 5, 2012
29.21 DAYS
NOV. 3, 2012
18.05 DAYS
NOV. 21, 2012
11.15 DAYS
DEC. 2, 2012
6.89 DAYS
DEC. 9, 2012
4.26 DAYS
DEC. 13, 2012
2.63 DAYS
DEC. 16, 2012
1.63 DAYS
DEC. 18, 2012
1.01 DAYS
DEC. 19, 2012
14.92 HOURS
DEC. 19, 2012
9.22 HOURS
DEC. 20, 2012
5.70 HOURS
DEC. 20, 2012
3.52 HOURS
DEC. 20, 2012
2.18 HOURS
DEC. 20, 2012
1.35 HOURS
DEC. 20, 2012
49.90 MINUTES
DEC. 20, 2012
30.84 MINUTES
DEC. 20, 2012
19.06 MINUTES
DEC. 20, 2012
11.78 MINUTES
DEC. 21, 2012
7.28 MINUTES
DEC. 21, 2012
4.50 MINUTES
DEC. 21, 2012
2.78 MINUTES
"Vertical" Intervals: a=c(1+ϕ) or c=a/(1+ϕ)
"Horizontal" Intervals: b=a/ϕ d=c/ϕ etc.

Fig. 20.6. The projected Mach 2 version of Karpenko and Smelyakov's Auric Time Scale, with amended start point and recalculated with an improved fit to the thirteen-baktun cycle, reaching the center of the whorl on December 21, 2012—Peter Russell's "white hole in time." The same phi series as the original was used (see chapter 12, fig. 12.8) except that here, bifurcation points are all on the left and intervals are all on the right. Phi was taken as 1.6180339, and starting with the length of the thirteen-baktun cycle to six decimal places, the calculation was done with nine decimal places carried through, but with each interval rounded off on the diagram.

over a populated area. In November 2006, Smelyakov acknowledged the points I had made by incorporating them into a modified version of the Auric Time Scale theory entitled "The last multi-turns of the spiral of time before it rolls up to appear in a new reality," which can be found at www.astrotheos.com/Downloads/Spiral.zip.

21
SYNTHESIS

CONTAMINATED TRANCE?

It seems that everyone has a piece of the puzzle, whether small or large. So let's filter out the pieces that don't quite fit, or that have been force-fitted, to see if we can see the big picture coming into focus.

The McKenna brothers' book *The Invisible Landscape* was first published in 1975, when there was hardly anything in print about 2012. Frank Waters's *Mexico Mystique* was also published in 1975, but it put the end of the thirteen-baktun cycle at 2011. Terence McKenna modified the end of the timewave by thirty-four days in the second edition of his book, since by then he had heard about the thirteen-baktun cycle. If McKenna had heard of the Maya calendar ending in December 2012, he would have fixed the end point there, in the first edition, rather than at November 2012, so *The Invisible Landscape* is obviously an independent source on 2012, and it was revealed through the alteration of consciousness via DMT (the methylated tryptamines, psilocybin and psilocin, to be exact) and harmine.

Since Terence McKenna later became a kind of evangelist for the use of sacred plants, many of the people who got 2012 information via sacred plants would have already been familiar with the 2012 scenario, it seems. Yet the Syrian-rue user previously quoted in chapter 13 was surprised at his revelation, having given no credence to McKenna beforehand. Dannion Brinkley had his near-death experience in September 1975, so it is just possible he could have read *The Invisible Landscape*. Ken Kalb, on the other hand, had a near-death experience in Mexico in 1969, which led to the publication of *The Grand Catharsis,* about the history of consciousness between 1987 and 2012, a period in which he sees the entire planet having a near-death experience. But Kalb didn't write his book until 1995, so he was probably influenced by Argüelles's *The Mayan Factor*. This possibility of "2012 meme transfer" does not, however, apply to all puzzle parts. I have personally spoken to Jon King, and he told me that he knew nothing about the Maya, 2012, or Terence McKenna when he had his contact experience. I have also spoken to Robert Gilson, who assures me that he knew nothing about 2012 when he had his vision of a "period of justification."

In my opinion, there is enough independent corroboration here to show that this 2012 information is bubbling up from some deep, communal level of mind, and is emerging from various human transmitters in order to give us a spiritual kick up the Aristotle!*

We have seen how the Timewave-Zero concept sprang from a vision provided by the same mushroom that the Maya took. We have also seen how pure DMT allowed the "machine elves" (see chapter 13) to tell Terence McKenna of a forthcoming change in the laws of physics around 2012. Dr. Rick Strassman's research provides convincing evidence that DMT is involved in birth, death, and near-death experiences, as well as mystical experiences and close-encounter experiences. We have also mentioned Dr. Carl Callaway, of the University of Kuopio, Finland, whose research demonstrated that the near-death experience, out-of-body experience, and death itself are based on internal secretion of pinoline, DMT, and 5-meo-DMT.

---

*This is an example of a secondary cockney rhyming slang: Aristotle—bottle, bottle and glass—arse, as in the famous expression "Move yer aris!" The full tertiary slang would be "Move yer April!" where April in Paris = Aris. This is the approximate location (the muladhara, or root chakra), in which the kundalini serpent is coiled up, asleep, according to Indian philosophy.

When you also take into account the UFO wave that has been taking place in Mexico since 1991, and consider that Mexico is the land of the Aztecs and the earlier Maya who originated the calendar cycle that terminates in 2012, the pieces of the puzzle start to fall into place.

In other words, since many of the prophecies of Earth changes and change of consciousness in 2012 originate from shamans or medicine men using exogenous (outside the body) DMT and pinoline, and other 2012 prophecies originate from mystical experiences, contact experiences, near-death experiences, and out-of-body experiences, which are linked with endogenous (inside the body) DMT and pinoline release, we begin to see a massive connection between pineal hormones and a major event due in, or close to, 2012. Rick Strassman considered the possibility that a solar flare or some other astronomical event could trigger a mass turning-on of the *N*-methylating enzyme responsible for DMT production. Serena Roney-Dougal pointed out that the pineal responds to lunar and sunspot cycles via the changing geomagnetic field, and that sudden changes in the field correspond to poltergeist and ghost activity. A drop in the field, if you remember, increases the incidence of clairvoyance and telepathy, corresponding to secretion of a member of the telepathine family, pinoline.

ENTER THE SHEPHERDESS

We know that several people claim that *Salvia divinorum* has given them access to 2012 information, and also that this plant has a connection to preconquest Mexico. Some users refer to the deva, or consciousness, of the plant as the Shepherdess, or the Green Goddess. *Salvia divinorum* was not known to bind to any receptor sites—in other words, it was not understood how it produced a hallucinatory effect—until August 2002, when a paper was published[1] showing that Salvinorin A (now known to be one of two psychoactive constituents of *Salvia divinorum*) binds to the kappa-opioid receptors. Most other visionary drugs work on the serotonin (specifically the 5-HT-2a receptor) system, including the DMT family, so this enables us to cancel DMT out of the 2012-visionary equation for a moment and look from another angle.

After reading dozens of accounts of experiences people have had using DMT and *Salvia,*[2] I realized that one thing that is often described by users of both entheogens* is an out-of-body experience, or OBE, and that there are also well-known similarities between OBEs and NDEs (near-death experiences). There are equally well-known similarities between alien abduction experiences and OBEs. So if we refocus on the OBE rather than the external or internal means of achieving it, we can see that the source of information about 2012 seems to be the place that consciousness travels to when it vacates the body—what those in the esoteric world call the astral realm or astral plane. Remote viewers, or psychic spies, who can remote-view in time as well as space, work via the astral plane; we have already heard about the block around 2012 reported by some remote viewers.

I mentioned that the most interesting thing in Cotterell and Gilbert's *The Mayan Prophecies* was a reference to the effects of fluctuations of the geomagnetic field on the human endocrine system. They quoted a passage from a 1987 paper by Dr. Ross Adey, of Loma Linda University Medical Center, in Loma Linda, California: "About 20% of pineal cells in pigeons, guinea-pigs and rats respond to changes in both direction and intensity of the Earth's magnetic field (Semm, 1983). Experimental inversion of the horizontal component of the Earth's magnetic field significantly decreases synthesis and secretion of the peptide hormone melatonin, which powerfully influences circadian rhythms, and also reduces activity in its synthesizing enzymes (Walker et al., 1983)."[3]

So there is evidence that the secretion of pineal hormones can be affected by a change in the geomagnetic field, and we have already seen how not only Gregg Braden, but also mainstream scientists are saying that a geomagnetic reversal is imminent, and that the magnetic field

*An entheogen is a psychoactive sacrament, a plant or chemical substance taken to occasion spiritual or mystical experience: theo = god, spirit; gen = creation.

is approaching a 100,000-year peak that corresponds to a 100,000-year solar cycle. This 100,000-year cycle is about the same length as four cycles of the precession of the equinoxes, and has been linked to the Maya calendar by William Gaspar. Braden linked the forthcoming magnetic field change with the need for an extradimensional body. So is there any evidence linking out-of-body experiences with changing magnetic fields? Recall that in chapter 9 we reported the phenomenon of body-polarity reversal during OBEs, as discovered by the Monroe Institute,[4] along with its implication of a possible mass OBE induced by geomagnetic reversal or fluctuation circa 2012.

TEMPORAL LOBE EPILEPSY

Temporal lobe epilepsy (TLE) is caused by "unusual electrical activity in the brain's temporal lobes."[5] People suffering from TLE often report spiritual or religious feelings. In Clifford Pickover's article "Transcendent Experience and Temporal Lobe Epilepsy," he includes the testimony of one unnamed female sufferer who says:

> With TLE, I see things slightly different than before. I have visions and images that normal people don't have. Some of my seizures are like entering another dimension, the closest to religious or spiritual feelings I've ever had. Epilepsy has given me a rare vision and insight into myself, and sometimes beyond myself, and it has played to my creative side. Without TLE, I would not have begun to sculpt.[6]

The symptoms of an epileptic seizure have several points in common with feelings reported by alien abductees, according to TLE researcher Eve LaPlante in her book *Seized.* These include heat on one side of the face, a ringing in the ears, seeing flashes of light prior to abduction, an unnatural silence, a feeling of anxiety, a sensation of bugs crawling on the skin, and other symptoms. LaPlante says that religious prophets such as Mohammed, Moses, and St. Paul are all thought to have suffered from

TLE, since the descriptions of their experiences all show several signs of it. The Russian writer Dostoyevsky, who was himself a sufferer, recognized these signs in the stories of Mohammed. Other famous sufferers include the poet Tennyson and the artist Van Gogh, demonstrating that the condition can have positive, creative side effects. It is interesting to note here that the bifurcation points on Sergey Smelyakov's Auric Time Scale showed a correspondence to the coming of great religious teachers and philosophers, including St. Paul.

Very small fluctuations in the geomagnetic field have been found to affect people suffering from epilepsy (as mentioned in chapter 13), while the (disputably) psychedelic tryptamine bufotenine has been found in the urine of people suffering from epilepsy.[7] But tryptamines, including bufotenine, have also been found in smaller quantities in the urine of those not prone to epilepsy.[8] Steve Scott (a.k.a. Ian Hard) has temporal lobe epilepsy, and he brought my attention to the fact that since childhood he has sensed the importance of 2012.[9] A BBC-TV Horizon program called "God on the Brain," which aired April 17, 2003, dealt with the connection between TLE and religious experiences. It covered the work of Dr. Michael Persinger, of Laurentian University, in Canada, whose research indicates that people with TLE may be especially sensitive to magnetic fields. He has a specially designed helmet that directs controlled magnetic fields to the temporal lobes, and he has been able to trigger several types of religious experiences, including out-of-body sensations, physical feelings (such as being grabbed by the shoulders or leg), and even visions of alien abduction in some cases.[10]

Earth-mysteries researcher Paul Devereux was director of the Dragon Project, mentioned in chapter 13, which was started in 1977 as an attempt to monitor energy anomalies at sacred sites.[11] Many sites have been found to have magnetic anomalies, such as Carn Ingli, the Mount of Angels, in Wales, which has areas of reversed magnetic field. Devereux says, "Persinger and colleagues have produced some exciting statistical evidence that suggests that there is a link between some forms of psi activity in humans and the activity of the geomagnetic field."[12]

Devereux, who has tried Persinger's magnetic helmet, notes that a stone shaped like a seat in the Gors Fawr stone circle in Wales affects compass needles, and is also one of a pair of outlying stones that align to sunrise on the summer solstice.

> The sitter's head leans back against exactly that part of the stone that most disturbs the compass: the cerebral cortex is in close proximity to the natural magnetic effect. The stone is also involved in a midsummer-sunrise alignment. Would the effect be somehow enhanced then? Or does that clue tell us when the ritual was carried out at the magnetic "spirit" stone? Alas, we have much yet to learn in our fledgling study of geopsychedelics.[13]

Are magnetic effects at sacred sites enhanced at the solstices? Will the winter solstice in 2012 trigger an even more far-reaching effect? Should we be at such a sacred site for the galactic alignment (which is actually every winter solstice from 1980 up to 2016—see discussion later in this chapter)? Or will the changes in the background geomagnetic field affect the weak magnetic field that surrounds all our heads? Perhaps being present at a sacred site might protect us from changes in the background geomagnetic status. There is an Argentinean website (in Spanish, at www.argemto.com.ar) that explains how Russian

Fig. 21.1. Stone seat at Cairn T, Loughcrew Mountains, Ireland. The passage of the cairn is aligned to the equinox sunrise.

cosmonauts exposed to a magnetic field of 0 gauss (a gauss is a unit of magnetic induction equal to one ten-thousandth of a tesla, so 0 gauss therefore means no measurable magnetic field) experienced "confusion, aggression and madness," and postulates that Earth's magnetic field will drop to 0 gauss by 2012.[14]

ACUTE SCHIZOPHRENIC BREAK SYNDROME

At the moment of the summer solstice sunrise in 1972, Michael O'Callaghan was standing inside a Neolithic passage-cairn called Tibradden, on the hills above Dublin. He underwent an experience in which he saw the city below as a huge organism eating into the hillside, slowly poisoning itself with its own excretions. As the Sun rose over the horizon, shining directly up the ancient passageway onto him, O'Callaghan's vision metamorphosed into a classic mystical experience: "In this ecstatic condition, I had the delicious feeling of being utterly at one with nature, no longer a separate 'skin-encapsulated ego,' but a living extension of the biosphere in human form that could simultaneously look back upon—and identify with—the larger portion of its global Self, 'outside looking in,' as it were."[15]

O'Callaghan saw humankind emerging from the womb of history like a newborn baby, the flower at the end of a 4.6-billion-year germination process, with a huge potential—if only we could reach out and grab it. The Sun was the answer to the energy problem, he saw, with its free solar power. If we could stop our rampant consumerism in time, we could find ourselves at the beginning of the solar age.

O'Callaghan came to realize that an acute schizophrenic break was like an extended version of his solstice experience, and the world itself is on the brink of a schizophrenic break. He clears up one thing from the start: "Contrary to popular misunderstanding, the term 'schizophrenia' does not refer to the multiple personality syndrome; the Greek etymology of the word actually means 'broken soul' or 'broken heart.'"[16]

Cases of schizophrenia (often classed as "mental breakdown" by the mainstream psychiatric community) were seen by the psychologist Carl Jung as the start of a healing experience. The experience, if handled carefully through support, and not repressed or sedated, usually lasts about forty days (and forty nights). Very often it will include a death-and-rebirth experience, or "inner apocalypse." This is often projected outward and interpreted as the end of the world and an end of time. In fact, O'Callaghan points out that the word *apocalypse* means "a revealing," and this is a process of the unconscious mind revealing itself by archetypal imagery. R. D. Laing, who helped schizophrenic people through the healing process, says that the book of Revelation is obviously about an internal process of the destruction and reintegration of the ego.[17] St. John the Divine is said to have written it while he was imprisoned in a cave on the Greek island of Patmos (a darkroom retreat).

Jung said that schizophrenia can be defined as "when the dream becomes real,"[18] and this is the title of the brilliant four-chapter essay that Michael O'Callaghan has written on the subject.[19] Jung saw dreams as messages from our unconscious minds; when the dreams break into waking consciousness, we can no longer ignore them.

Schizophrenia is currently affecting approximately 2 percent of the population of the world, and we are not dealing with it properly. The current "treatment" is that people are just drugged to stop the whole process, while what they actually need is to be in a supportive environment for the forty days, by which time they will start to reintegrate—a process in which the three months following the forty-day schizophrenic break is spent at the facility, then another three months in a halfway house, then back to the outside world. This approach was tried in the 1970s by Dr. John Weir Perry at the Diabasis facility in San Francisco, and it was one-hundred-percent successful, with no relapses. In indigenous societies, anyone showing signs of an approaching "break" is taken aside and shown that the experience is a gift, and it is integrated into his or her subsequent training for a career as a shaman. The apprentice can then learn to navigate the otherworld, and reinvoke the experience at will by using trance techniques that often include sacred plants.[20]

As Ken Kalb sees it, the Earth herself is having a near-death experience. She is poised on the edge of an abyss where we must choose to cross the bridge of ecological sustainability or jump using our "Yamazaki" supercharged rocket plane with only half a pint of fuel left in the tank. Have you ever wondered why the (much flouted) Kyoto Protocol, an international agreement to cut pollution to target levels, has a deadline in the year 2012?

We have already seen the link between shamanism and the 2012 phenomenon, and that reintegration following the break can lead to a feeling of oneness with nature, so these messages seem to come from the collective unconscious, which is another concept that started with Jung (although the similar noosphere concept of Teihard de Chardin arose at roughly the same time). The same message is bubbling up through diverse terminals or human neurons in the developing global brain. The source could also be seen as the Gaian mind (some might call it the Galactic Federation, or the wee folk). It is being broadcast through the astral plane, or the fourth and fifth dimensions, or the next density level, or even via "voices in the head."

Whatever your favorite terminology, we are being told in myriad ways that Earth is about to enter a transition phase around the Gregorian year 2012, and that there will be geomagnetic effects that may open our pineal eyes. The first flowerings are the schizophrenics and psychics, those with TLE, channels, shamans, and visionaries, who are more sensitive to the first and smaller magnetic changes. As Earth's field gets nearer to reversal, it seems likely that there will be more pineal openings, and those of us who have not confronted our inner selves and our accumulated emotional baggage may find this experience to be more hellish. The fact that "tryptamines other than serotonin" (and different from those found in control subjects) and, later, DMT itself, were found in the urine of schizophrenic people[21] demonstrates the likelihood that the condition is connected with an altered pineal output—the first blurred opening of the pineal eye.

GNOSTIC SOULCRAFT

Graham Hancock, in *Heaven's Mirror: Quest for the Lost Civilization,* concludes that the survivors of the ice age civilization that ended around 10,500 BC passed on not only a knowledge of astronomy and precession that was related to the cyclic catastrophe that ended the ice age, but also a "science of immortality." This was a technology of the soul, in which techniques acquired over millennia of study and experimentation could be used in ways to preserve consciousness beyond the veil of death. Hancock says the knowledge was passed down through secret societies, and traces can be found in the gnostic writings found at Nag Hammadi.

In gnostic versions of the Genesis myth, the forbidden fruit gives

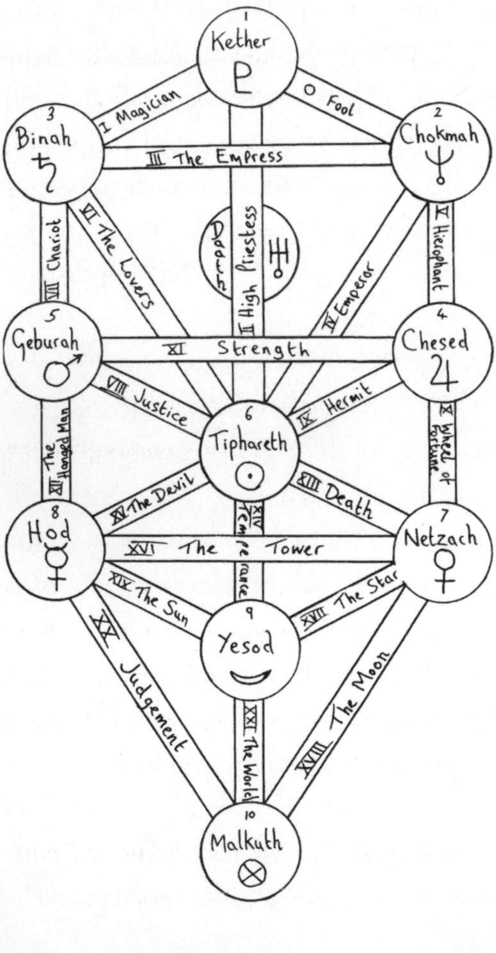

Fig. 21.2. The kabbalistic Tree of Life showing the eleven Sephiroth, or spheres, with their planetary attributions, and the twenty-two paths with their tarot attributions. This is the standard interpretation (the Crowleyan interpretation used by groups such as the AA [Astrum Argentum] and the OTO [Ordo Templi Orientis] reverses the position of the Emperor and the Star).

Key to Planetary Sigils

knowledge of light and darkness. The Tree of Knowledge can easily be seen as some sort of hallucinogenic plant that made those who had eaten it aware of other planes of existence and, via out-of-body experiences, they would have seen that consciousness can exist without the physical body. The Tree of Life thus became perceptible; the kabbalistic Tree of Life is a map of the paths of the netherworld, and the middle pillar relates directly to the chakras in our subtle body, situated along our own middle pillar, the spine. By performing certain exercises involving the chakras or power zones, and the snakelike kundalini energy, it seems that the astral body can be perfected or transformed into a vehicle with a much longer life span. Thus Hancock's study traced this science of soulcraft from a prediluvian race, through ancient civilizations in Mexico, Egypt, and other places, suggesting that it is linked to the precession cycle.

Some of the information can still be found in the Bible. In Matthew 22, the "kingdom of heaven" is compared to a marriage, in which those who are not wearing the "wedding garment" will not get to witness the wedding. In Luke and Matthew, we are informed about the pineal eye.

> Your eye is the lamp of your body; when your eye is sound, your whole body is full of light; but when it is not sound, your body is full of darkness. Therefore be careful lest the light in you be darkness. If then your whole body is full of light, having no part dark, it will be wholly bright, as when a lamp with its rays gives you light.[22]

This theme is expanded in Matthew 25:1-13, in which we are told to have our lamps fully fueled and burning, ready for the arrival of the bridegroom and his marriage feast. As to exactly what this means, John Major Jenkins, in *Galactic Alignment,* has provided us with a clue from a book he cites called *The Celestial Ship of the North,* by Valentia Straiton, in which someone named Edgar Conrow is quoted:

> The pineal gland is the "North Gate." This, in man, is the central spiritual creative center. Above in the heavens, it is found in the

> beginning of the sign Sagittarius, and is the point from which spiritual gifts are given. It is called "Vision of God," and is the Light within, a gift to the pure in heart, who verily may "see God," but to the impure, or those who abuse this great gift, the consequences are terrible. This North Gate, the creative center in man, the most interior center in the body, has become atrophied, and redemption or regeneration means its restoration to creative ability, by having the electrical, or positive, and the magnetic, or negative, forces restored in equal balance in man or woman.[23]

The alignment of the solstice Sun with the galactic equator, close to Galactic Center at the start of Sagittarius, is thus a time when the galactic pineal eye will radiate its light and restore our individual pineal eyes to their full potential of transmundane vision. However, not everyone will be ready. Luke's gospel implies that only the "unmarried" will qualify,[24] but it seems that he means male sexual energy should not be "wasted" in ejaculation, but used instead for the higher purpose of modifying our subtle body into its fully flowered form.

EGYPTIAN SHAMANISM

You can find an expanded essay on this subject, called "Soulcraft," at www.diagnosis2012.co.uk/soul.htm, in which all traces of this knowledge are exhumed and compared. Briefly, however, we can trace the earliest version back to Egyptian sources. The Egyptians had a shamanic religion in which the pharaoh would induce an astral flight to become Osiris (who was cut into pieces and then reassembled) in a death-rebirth experience. This was done by means of various techniques, which included hallucinogens, light, and sound[25] (see appendix 2); meditations timed by lunar phases; and possibly a darkroom retreat. Isis then reassembled the fourteen pieces of Osiris, having collected them (except for the phallus, which was lost) in the land of the phoenix (Phoenicia—now Lebanon) during the fourteen days of the waning Moon. She then bound him up like the pupa of a butterfly, and used a wooden phallus to conceive Horus, while Osiris ascended to heaven (possibly encoding the higher use of sexual energy). The phoenix was retained as the symbol of an immortal bird, but it is also connected with the recurrence of a long cycle of time.

The Egyptian winged disk, according to Charles Musès,[26] represents the "non-molecular body" that is developed as the vehicle for these flights. The timing of the whole seventy-day mummification process was synchronized with the seventy-day disappearance of Sirius into the Duat (the underworld period when Sirius is not visible). Musès points out the parallels between mummification and the process of metamorphosis from caterpillar, via cocoon and pupa (chrysalis), into butterfly. The wrapping of the body in bandages represents the silk-enswathed larva (caterpillar), and the folded wings embossed on the sarcophagus lids represented the wing shapes that are visible on the pupa or chrysalis case. The *mes-khent,* or "birth tent of skin," that was placed around the mummy represented the outer cocoon, while the ancient Egyptian word for the funeral chamber is literally "the birth chamber."[27] Humanity is seen as a larval stage in preparation for the final metamorphosis.[28] The seventy-day embalming period was a later metaphor for a metamorphic transformation—one that was commenced while the physical body was still alive, leading to the birth of the "winged form," or immortal body.

THE NONMOLECULAR BODY

Egyptian soul science (psyche-ology) is quite complex, with several nonphysical bodies being involved, but Jeremy Naydler has reconstructed the scheme in quite a convincing way in *Temple of the Cosmos.* The *khat* is the physical body (symbolized by a fish hieroglyph), and the *ka* (symbolized by upturned arms), or double, can be seen as "vital force"—an energy body that seems to be equivalent to an etheric body. The *ba* (symbolized by a bird with a human head), although often translated as "soul," is better understood as a seed-soul consciousness, since it is "but a preliminary to a yet more exalted state of consciousness."[29] It is consciousness externalized, and seems to be the equivalent of the astral body. "As a ba, a person had the experience of looking at his or her body as if from an outsider's standpoint. This experience was central to the Osirian initiation."[30] The *khaibit,* or shadow (symbolized by a dark human silhouette), represents "all the untransformed earthly appetites and obsessions that fetter the ba to the physical realm and prevent it from moving on"[31] (the astral body is also known as the emotional body). The *akh,* also called *akhu,* or *khu* (represented by a bird—the crested ibis), means "shining one" or "illuminated one." It is also called "the imperishable one" that returns to its source beyond the Duat.

Naydler says, "The akh may be understood as the ba divinized,"[32] but in order for the akh to be released from the body, a new spiritual body had to be germinated from the physical body as a vehicle for the akh. This spiritual body is the *sahu.* This sounds like the immortal man in Taoist yoga, who is gestated in the abdomen and expelled through a psychic opening in the top of the head to appear as a person sitting on a lotus within a golden sphere, as we shall see shortly. The lotus was also a symbol of rebirth in Egypt. Naydler says the re-membering following the dismemberment was an essential part of the Osirian initiation, which is a germination process in which the sahu is formed and the akh is attained, allowing access to the stellar realms beyond the Duat.

Naydler's interpretation implies that the conscious mind resides in the khaibit (shadow) and the unconscious mind in the ba (astral body). If the shadow can be conquered, then it can unite with the ba to form the akh (khu). At the same time, the sahu can be germinated from the ka, which now has access to more vital energy—the sexual energy, which is no longer being diverted into emotional blockages since the cleaning up of the shadow. This interpretation provides a scheme of the unification of male and female (conscious and unconscious), while retaining the evolution of the higher bodies.

Fig. 21.3. Above are the various Egyptian hieroglyphs and their later variations that represent the physical and nonphysical bodies that comprise a fully evolved human being.

All this seems to support the contention of Charles Musès that there were three postmortem paths that could be followed by the newly deceased Egyptian: the Hippo Path, in which the soul's vehicle disintegrated and the soul reincarnated; the Cow Path, in which the soul experienced a dreamlike existence in the Duat, or Underworld, from where it could either reincarnate or ascend; and the Lion Path, where a higher body allowed access to a higher dimension.

In the epilogue of Naydler's *Temple of the Cosmos,* the author recognizes the connection between Napoleon's invasion of Egypt and "his desire to discover the sources of Masonic Lore—which he believed lay in ancient Egypt."[33] He also recalls the Masonic Great Seal of America, with its implication that the "New Order of the Ages" was approaching and represented a "general spiritual awakening" linked to a reemergence of Egyptian knowledge (symbolized by the return of the capstone to the Great Pyramid). We have already seen how the Great Seal encodes the year 2012 on the top of the pyramid.

THE SPIRITUAL WEDDING

Just as feng shui, the Chinese system of geomancy, has an Earth ch'i and a Heaven ch'i, psychologist Mary Scott has found evidence that the fire serpent, or kundalini, which is said by the Hindus to lie coiled and sleeping at the base of the spine, is actually an Earth force. This kundalini is "the Shakti or divine cosmic energy with particular dominion over the physical world."[34] The evidence suggests that "some energies may enter the chakra at the base of the spine, from below, a point not stressed in Tantric texts"[35] (part of the evidence was from meditators who had a condition called physio-kundalini, the symptoms of which are concentrated in the pelvic area, but which start with sensations in the feet and legs). The kundalini is "coiled three and a half times around the Linga, the Shiva line of force which enters the body from above."[36] Shakti, including kundalini, is the feminine aspect of the divine in manifestation, while Shiva is the male aspect.

Kundalini yoga is a collection of techniques to raise the kundalini up the spine, cleansing the chakras on the way until finally the two aspects of the Divine, the male Shiva force and the female Shakti (kundalini), reunite in the *sahasrara* chakra at the crown of the head. This results in ego-annihilating *samadhi,* or trance of ecstasy. It is quite a dangerous practice and needs to be learned from a yogi. Kundalini energy is not *prana* (vital energy contained in the air we breathe), but works in close association with prana. Mary Scott says that "Kundalini is called the Logos in bodies, and this has to be taken to cover galaxies as embodiments of Brahman, as well as bodies in the usual sense of the word."[37] The prana and kundalini blend in the muladhara, or root chakra, at the base of the spine. In a similar way, Scott tells us, there must be a blending of forces in the Earth—terrestrial centers and channels through which cosmic energies pass, and in which they blend.

EARTH CH'I MEETS HEAVEN CH'I

This concept of a force from above meeting a force from below, and the meeting of the two taking place in the human body, was suggested over fifty years ago. In the Theosophical booklet *Kundalini: An Occult Experience,* published in 1947, G. S. Arundale suggests that there is an Earth kundalini and a Sun kundalini, which may be seen as two poles—negative and positive, respectively (similar to the Chinese Earth ch'i and Heaven ch'i). The booklet does not mention prana, but says that Earth kundalini flows up through the feet and limbs to the root, or muladhara, chakra at the base of the spine, where it is concentrated together with the Sun kundalini so that in the rare cases when the system works at its full evolutionary potential, the human acts as a rod connecting the two terminals and experiences a higher consciousness of union with the Divine, together with clairvoyance and other paranormal side effects. These higher states Arundale calls Buddhic and Nirvanic, which may perhaps be equivalent to global consciousness and solar consciousness.

Arundale also suggests that there are interplanetary and intersolar streams of kundalini, and that not only does Earth have chakras, but Earth and the Sun may themselves be chakras in larger systems. He also suggests that kundalini concentrates in hills and mountains, and places of repeated ceremonial gatherings such as churches, temples, and mosques, which he says are magnetized areas. He says that this kundalini "waxes and wanes" or "ebbs and flows," which suggests a lunar influence (though he doesn't actually say so). He hints at the

Fig. 21.4. The snake heads at the base of the stairway on the Pyramid of Kukulcan, Chichén Itzá, Mexico. The Maya daykeeper Hunbatz Men has demonstrated that the Maya had a spiritual technology utilizing the chakra and serpent/kundalini energy, in which one aspired to become Kukulcan (Quetzalcotal), the Feathered Serpent. He sees the pyramid as a reminder of the process (Men, *Secrets of Maya Science/Religion*, 109–45). Photo: Lee Anderson

connection between sexuality and kundalini in that "such remnants of sexual nature as there may have been seem to be transmuted and transformed into their true purpose . . . Godliness."[38] Since "Kundalini Fire is the essence of the Love of God,"[39] perhaps some kind of "love" in our sexual center is being converted to Love, or the Holy Spirit: eros transforming into agape.

Mary Scott does not mention Arundale, but has independently investigated the findings of dowser Guy Underwood and found many connections to kundalini lore. Underwood detected deep underground streams, called "water lines," which he sees as lines of force that cause discontinuities in a background force field coming from Earth. He also detected two other types of lines, called "track lines" and "aquastats." All these convoluted lines were found to have an intimate relationship with prehistoric sacred sites. In the course of his investigations, Underwood excavated to check his results and found that he had actually discovered several prehistoric circles, barrows, mazes, and stone-paved tracks, which have since been declared genuine sites. "Blind springs" are coils where lines converge, and which occur on all three types of lines. Water lines and aquastats, which often interweave with each other, converge on blind springs with seven coils, or with multiples of seven (forty-nine was the maximum number found by Underwood). Track lines converge on spirals of three and a half, or multiples of three and a half, just as kundalini is said to be resting in a coil of three and a half turns.

Scott says the lines correspond to the water veins and yin and yang dragon currents in the Chinese system, while the straight alignments known as ley lines correspond to the Chinese spirit paths. The snake/dragon theme is preserved in the names of many hills in Britain, and many of the churches built over sacred sites have been named after St. Michael, the one who slew the dragon (Michael was the biblical prototype for St. George, England's patron saint). Scott points out that Underwood found each blind spring had an associated "geospiral" and a pair of "halos" surrounding it, both of which altered according to the phases of the Moon. Since feng shui, the Chinese geomancy practice, places importance on planetary influences, and some earthworks trace planetary movements, she concludes that the ley lines are indeed spirit paths, with influences from the Sun and major planets, while the aquastats and track lines are the yang and yin dragon paths of feng shui, which are associated with energies of the Earth and Moon system. At sacred sites, the two types of paths meet in blind springs. "If this is so," says Scott, "the phases of the Moon and the lunar year would play an important role in the circulation of the intrinsic energies, while solar cycles and planetary movements would determine phases of activity and quiescence along ley lines. The priests and priestesses of the old pre-Christian religion in Britain must have been conversant with all the relevant astronomy, as we can gather from Sir Norman Lockyer's research into the subject."[40]

As I mentioned in chapter 1, John Major Jenkins, in *Galactic Alignment,* reasons that the process of galactic alignment will allow the kundalini shakti to flow through us, resulting in the shift to a more exalted state of consciousness. He sees the peak of the process occurring now, during the thirty-six-year "window of opportunity" that coincides with the period it takes for the winter solstice Sun to cross the galactic equator (approximately 1980 to 2016, centered on 1998, but flagged by the Maya at 2012).

According to Joan Parisi Wilcox's book about the Incas, *Keepers of the Ancient Knowledge,* we are in a window of opportunity between 1993 and 2012, during which time we can contribute to the coming transition by forming *saiwas,* or "columns of energy," that connect the upper world to the underworld via our bodies. This, again, sounds like Earth ch'i meeting Heaven ch'i.

Brian Stross, of the University of Texas, has written an essay titled "The Sacrum Bone: Doorway to the Otherworld,"[41] in which he shows that the Maya conceived of the sacrum (the fused vertebrae that lie in the center of the pelvis) as a kind of skull. The skull and the sacrum were seen as portals to the underworld, and it is a fact that our own language acknowledges this with the skull's "temples" and the word *sacrum,* which means "holy bone." Joining them together, we have the thirty-three vertebrae of the spine, which are related to the thirty-three degrees of Freemasonry.

Stross says the Maya saw the spine as a microcosm of the world tree, or axis (as did the Egyptians—the Djed columns represented the spine of Osiris/Orion and Earth's axis). However, the Maya associated the "portals" at each end of the spine with the celestial gates, where the ecliptic crosses the galactic equator. The galactic sacrum was the place of Creation in Orion, where the solar god One Hunahpu burst out of a turtle shell. The galactic temples were at the dark rift, where galactic alignment will occur through 2012, when the world tree aligns with the galactic tree and One Hunahpu emerges from the mouth of the jaguar-toad. Here we have a strong suggestion that galactic alignment will inaugurate a merging of Heaven and Earth (or the merging of the upper world, lower world, and physical world, as in the Peruvian prophecy), or the coming together of Rangi and Papa of the Maoris—Geb and Nuit of the Egyptians—in an ego-annihilating meeting of Shiva and Shakti.

HENDAYE REVISITED

Jay Weidner and Vincent Bridges have published an updated and expanded version of their book *A Monument to the End of Time,* titled *The Mysteries of the Great Cross of Hendaye: Alchemy and the End of Time.* The book shows in even more detail how the knowledge of a cyclic event connected with the cycle of precession was encoded into alchemy and gnosticism, along with a science of transformation. A Hermetic text, called Isis the Prophetess, indicates that this is a threefold transmutation, first of "the cerebrospinal energies and fluids";[42] second, using the new energies to perform physical transmutations; and last, the transmutation of time, converting the Iron Age to the Golden Age. Isis was in charge of the process, and it could not begin "until the stars were in the proper place."[43] Isis is connected with Venus, and we are reminded of the transits of Venus in 2004 and 2012, which Will Hart suggested would be some kind of trigger. However, Isis is also connected with Sirius, so this brings to mind Charles Musès's suggestion of Sirius as the trigger (the Maya version of the rebirth of First Father (One Hunahpu), which signifies that the galactic alignment, as the trigger, can be equated to the rebirth of Osiris, which was enabled by the wooden-phallus episode of Isis). Again, the secret is traced by Weidner and Bridges (but in even more detail in the new edition) through the Coptics, gnostics, Hebrews, Islam, the Knights Templar, Cathars, Priory of Sion, troubadours, and Grail romances, to the building of the cathedrals, where it was

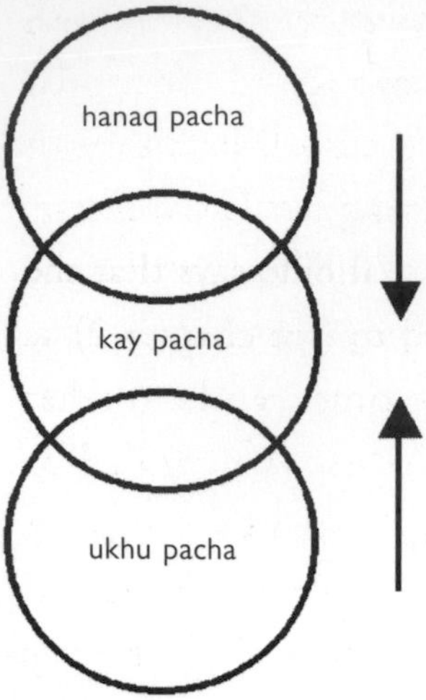

Fig. 21.5. According to Andean prophecy, *hanaq pacha,* the upper world, and *ukhu pacha,* the lower world, will converge on *kay pacha,* the everyday world, to form "one cohesive and paradisiacal world" around 2012. We can aid the process by performing *ayni*—digesting heavy energy—and by creating *saiwas,* columns of energy connecting the three worlds. See *Keepers of the Ancient Knowledge—the Mystical World of the Q'ero Indians of Peru,* by Joan Parisi Wilcox.

encoded in stone just in time before the church burned everyone at the stake.

As we saw in chapter 8, the authors conclude that a galactic core superwave, as predicted by Paul LaViolette, will bring a "double catastrophe," but they also say that "it could also be a window of opportunity for profound change, perhaps even the moment of ascension or the mass attainment of the Diamond Body, an immortal light body described in Tibetan Buddhism."[44] We also saw, in chapter 8, how the authors decoded the anagram in the inscription of the Cross of Hendaye as "Inca cave, Cuzco, Peru," and "Hail to the Cross at Urcos." Then, during a conversation with Dr. Juan Nunez del Prado, a professor (emeritus) of anthropology at Cuzco University (the National University of San Antonio Abad, at Cuzco), who was doing a lecture tour of the United States, they found out that there was a cross at Urcos. They later visited it, and also found caves near Cuzco, but the original cross had been replaced and the caves didn't give any answers, either. However, the Hendaye cross had predicted a catastrophe in the northern hemisphere, and intimated that Peru was a place of refuge. Cuzco and Urcos are just north of the Altiplano, a high plateau surrounded by even higher mountains, and would thus be well protected from tidal waves. It is also in the southern hemisphere. But perhaps the word *cave* should be taken in its Latin meaning of "beware," or, more aptly, "be aware"—in other words, be aware of the Incas near Cuzco, Peru.

In *The Mysteries of the Great Cross of Hendaye,* Weidner and Bridges mention another Peruvian link they have found that could be significant. They have spoken to Alberto Villoldo, a psychologist and medical anthropologist who has been studying the Q'ero priests and Quechua shamans of Peru for twenty years. Villoldo says that the word *pachakuti* that Joan Parisi Wilcox referred to (see chapter 2) as a cosmic transmutation, or overtuning of space-time, can be further analyzed. *Pacha* means both "Earth" and "time," while *kuti* means, "to turn upside down" or "to set right."[45] Therefore, the meaning is both a turning upside down of Earth and a turning upside down of time, but also setting them right—in other words, turning them the right way up again, according to Villoldo.

It is interesting that some people translate *pachakuti* as "earthquake," since this is the meaning of Ollin, the central symbol on the Aztec Sunstone that rules this era and its transition. The prophecies of the Q'ero speak of "a tear in the fabric of time itself," and "the end of time,"[46] according to Villoldo. However, both Joan Parisi Wilcox and Elizabeth Jenkins say that the pachakuti ended on August 1, 1993, while Villoldo uses the term to cover what they (Wilcox and Jenkins) call the first phase of the Taripay Pacha, which goes from 1993 to 2012 (the "full-blown form" is not expected until 2012, when the sixth-level priests emerge). Taripay Pacha means "the age of meeting ourselves again," which sounds just like the spiritual wedding.

THE PERUVIAN SOLUTION

Joan Parisi Wilcox spoke of the "engaging of the energy body" as a technique that will aid the spiritual evolution of humankind during the first phase of Taripay Pacha, from 1993 to 2012, following a pachakuti from 1990 to 1993. Villoldo has a slightly different version, in that he puts the pachakuti between the years 2000 and 2012,[47] but Wilcox and Elizabeth Jenkins (see below) have split the initial nineteen-year phase of the Taripay Pacha into two subphases:

the first is a seven-year period from 1993 to 2000 and the second is a twelve-year period from 2000 to 2012. Shaman and author James Arévalo Merejildo puts the start of the pachakuti in 1992,[48] which is also the start of the twenty-year period leading to 2012 signified by the Hendaye cross, though Weidner and Bridges say the pachakuti is between 2002 and 2012 (the cross flags 2002 as the halfway mark in the twenty-year period). In the material that follows we will go directly to the origin of these various versions. Suffice it to say, however, that whether we see the pachakuti as a three-year phase preceding a nineteen-year phase or simply as a twenty-year phase, the window of opportunity is open now.

Jay Weidner has recently made a film with Alberto Villoldo, *Healing the Luminous Body: The Way of the Shaman,*[49] and spoke about the film in a radio interview on Network X in January 2004. Here he implies that the pachakuti starts specifically in 2012:

> Villoldo tells us that the Incan shamans believe that 2012 heralds what they call a pachakuti, or period of upheaval and renewal that occurs at the end of time, in which the world will be turned right side up again. He tells us, "There will be a tear in the fabric of time itself, a window into the future through which a new human species will emerge. They call this new species *Homo luminous.*" I don't know if this is what is really happening, but if the pursuit of alchemy is to turn lead into gold, so perhaps the real goal of time and history is to turn the lead of the human body into the gold of *Homo luminous.* Perhaps, like the ancient Egyptians believed, we will become stars in the heavens being born in the Bardo dreams of Orion. To be honest, though, I don't really know.[50]

Villoldo says that the prophecy of the Return of the Inka, or Return of Pachakuteq, will herald the end of time.[51] The Inka (I use this spelling to differentiate between the people, Incas, and the god-men, Inka) Pachakuteq was the ninth man-god and was the one who built Machu

Picchu, and after him there were four more, if you count the one-year reign of Atahualpa (but he was very cruel and so does not seem to qualify). Atahualpa succeeded his brother Huaskar, but both are said to have failed in their duties as Inkas, and are said to currently be in the underworld, helping the beings there to prepare for the forthcoming merging of the three planes. The twelfth true Inka will emerge between 2000 and 2012, Villoldo says:

> The Star Rites are held at the Snow Star, at Mt. Ausangate. They are believed to summon the return of the Inka. Before Spanish rule, the Inka were the political and spiritual leaders of a mighty empire that spanned a large part of South America. But the shamans are not speaking about the return of a ruler that will re-establish the now-forgotten glory of the empire. They speak of the possibility of Luminous Ones emerging among us.[52]

This statement contradicts what Dr. Juan Nunez del Prado says, but before commenting, let's look at another extract from Villoldo's www.thefourwinds.com website:

> The processes the Inca shamans have been the keepers of are known as Karpay, the great rites of passage. The Karpay connect the person to an ancient lineage of knowledge and power that cannot be accessed individually—it can only be summoned as a village. This power can ultimately provide the fuel for an individual to leap into the body of an Inka, a luminous one, someone who is directly connected to the source that fuels stars. A person who completes the cycle of ceremony of the Karpay is able to emerge into the Fifth Sun, similar to the Hopi prophecies of emerging into the Fifth World. They can become the Luminous One, also known as "Man (or Woman) who walks in Peace."[53]

STRAIGHT FROM THE HORSE'S MOUTH

While attending Juan Nunez del Prado's two-day workshop in Richmond, London, on May 25–26, 2004, I had an opportunity to talk to Dr. del Prado and to ask him about the prophecies. He confirmed the versions written in Joan Parisi Wilcox and Elizabeth Jenkins's books. The return of the Inka means the process by which the first twelve fifth-level priests emerge, with the first one emerging from the Mt. Ausangate (Qoyllur Riti, or Snow Star) festival and eventually a group of twelve doing a ceremony in which two sixth-level priests emerge. These are the "shining ones" who will lead Peru and the world into a new era of peace; so, presumably, the fifth level of consciousness will

Fig. 21.6. Juan Nunez del Prado prepares a *despacho* to Pachamama, Mother Earth. The despacho is an offering consisting of several items such as flowers, leaves, and sweets, infused with *sami*, or refined energy, then folded into a bundle.

have to become established before the Luminous Ones of the sixth level begin to increase.

In Villoldo's statement above, he says that the Karpay process cannot be accessed by individuals, whereas del Prado provides a Hatun Karpay, fourth-level initiation, for individuals all over the world. The reason for this disagreement seems to be that Villoldo has not found any Andean priests who know about the Hatun Karpay initiation process. However, del Prado's teachers were Don Andreas Espinosa and Don Benito Qoriwaman, who were the last of the great *paqos* (Andean priests) and who died in 1981 and 1988, respectively. Juan Nunez del Prado initiated Alberto Villoldo in 1993.

It was Don Benito Qoriwaman who gave del Prado the formula for the pachakuti (this information was passed on sometime before 1987). He said that the pachakuti would last three years, followed by a seven-year period of preparation, followed by a twelve-year period of manifestation. Then the Taripay Pacha would begin. On January 1, 1990, two masters came to del Prado's house to tell him that at a meeting of paqos they had all agreed that the pachakuti would start on August 1, 1990. Then del Prado applied the date to Don Benito's formula and ascertained that in August 1993* the pachakuti would end and the seven-year period of preparation would start. The seven-year period would then end, and the twelve-year period of manifestation would therefore start on August 1, 2000. The end of the whole transitional process would end on the first day of the Taripay Pacha: August 1, 2012 (August 1 is the Q'ero New Year's Day).

Juan Nunez del Prado confirmed that at the time he got the second part of the prophecy in 1990, he had heard nothing about the Maya calendar, nor had he even traveled outside of Cuzco.

*Joan Parisi Wilcox and Elizabeth Jenkins both have August 1, 1993, for the end of the pachakuti, but del Prado said August 3, 1993, when I spoke to him. However, although I double-checked this with him, in light of the fact that August 1 is the New Year's ceremony for the Q'ero, and the other four dates in the sequence are all August 1, it may be that this should also be August 1.

THE URCOS CONNECTION

There is another connection between the Cross of Hendaye and Peru. Weidner and Bridges decoded "Urcos" from the inscription on the cross, but since the original cross at Urcos had been replaced, they remained mystified about its connection to the enigma. The answer lies in a book by Elizabeth Jenkins, *Initiation: A Woman's Spiritual Adventure in the Heart of the Andes.* It was published in 1997, two years earlier than Joan Parisi Wilcox's book, and is a similar story of a woman who undertook the fourth-level initiation under the guidance of Juan Nunez del Prado, who explains that "attaining the fourth level has to do with developing the astral body."[54]

Wilcox tells a story about the emergence of twelve fifth-level paqos. In Elizabeth Jenkins's version, she goes into a little more detail. She had just taken her Hatun Karpay, or initiation, into the fourth level, when she was told the prophecy of the twelve fifth-level paqos. The fifth level is a level of consciousness that allows miraculous healing abilities. There will be six males and six females—the six male fifth-level priests are called *Inka mallku* and the six female fifth level priestesses are called *nust'a.* The first Inka mallku will emerge at the aforementioned festival of Qoyllur Riti, an annual festival at a very high altitude, at the base of a glacier. At this festival, 30,000 to 70,000 people (sources vary) produce a huge bubble of living psychic energy, visible to those initiates who have developed their clairvoyance.

> [The first Inka mallku will] follow a specific travel route to the village of Urcos, and there, at the door of the church, he must meet and recognize another Inka mallku who has risen simultaneously here, near the Temple of Wiraqocha. These two must then travel together to Cuzco, where they will meet and recognize the third Inka mallku, who will rise at the public ritual of Corpus Christi on the main square in front of the central cathedral of Cuzco, the ancient Temple of Wiraqocha in Cuzco.[55]

Juan Nunez del Prado told Elizabeth Jenkins this when she was at the Temple of Wiraqocha at Raqchi (not the one in Cuzco).

The story continues that the three Inka mallkus travel on to Lima, where they meet the fourth and the first nust'a. The five continue on, gathering the rest of the twelve at specific times and places, and they all return to the Temple of Wiraqocha to perform a "coronation ritual." This will cause the emergence of the new *sapa Inka* and *qoya*—the supreme Inka ruler and his sister-wife—the first initiates of sixth-level consciousness. The sixth-level beings will have a visible shining aura. Then the mythical hidden city of Paytiti will materialize, just like the prophecies of the New Jerusalem and Shambhala, and the golden age will begin. After this, the seventh and highest level of consciousness is possible. This level is called *titantis ramji,* or deity incarnate. These seven levels seem to correspond exactly to the seven man types of Christian esotericist Boris Mouravieff (see the online essay "Soulcraft" for more details on this)[56] and the seven Worlds of the Hopi.

So the church at Urcos, where there must surely be another cross of some form, will play a key part in the coming shift, when the first two fifth-level priests meet there to begin the process leading up to the emergence of the sixth level. The emergence of the sixth-level priests is expected to be "around 2012 or so,"[57] according to Elizabeth Jenkins:

> This time period, from 1993 to 2012, represents a "critical period" in the development of human collective consciousness. These nineteen years mark the time when a significant percentage of humanity can and must pass from the third to the fourth level. We must be able to leave fear behind and learn to share our cultural gifts and achievements. If we can truly learn to live ayni—sacred reciprocity—and to share all of our accumulated knowledge without fear of each other, then we can discover our wholeness, like putting together pieces of a puzzle that make up our human family, our ayllu. It is up to the people of the Earth, us, to maximize this

> critical period in order to bring in the Taripay Pacha. We must not waste this opportunity![58]

The first step in preparing people for the coming change is to cleanse. In the Karpay ceremonies, the paqos use their medicine bundles to "extract any huchas, or heavy energies associated with our past—the pain, grief and discontent that we all carry from our personal histories, as well as the violence we carry within us."[59] Alternatively, fourth-level paqos can perform Hucha Mikhuy, in which they digest the *hucha,* or heavy energy of other people, absorbing it through the *qosqo* (literally, navel—see below). The fourth-level initiation, as Elizabeth Jenkins described it, also involves the use of several caves, some of which are close to Cuzco, to connect with Pachamama (Mother Earth or Gaia) in order to cleanse the energy bubble (subtle body) of heavy energy and absorb finer energies (sami). Could this be the reason that the Hendaye cross points to the caves around Cuzco? The caves are places of living energy that have been in continuous ceremonial use for centuries, and are thus potent tools for developing the subtle body.

As we have seen, emotional blockages divert the vital energy from its true purpose of feeding our energy bodies, and thus from being ready to "meet ourselves again." There are plenty of people who specialize in this very technique of helping people to release all attachments, fears, complexes, and other baggage, so a trip to Peru should not be necessary for this. The process can be continued by practicing the exercises in the back of Wilcox's book. Elizabeth Jenkins's Wiracocha Foundation (www.inka-online.com) also runs courses and expeditions to initiate students into the fourth level, Hatun Karpay.

Laurence Lucas, author of *The 21st Century Book of the Dead,* provides another route, one that combines the knowledge of the Maya, Egyptians, Tibetans, Chinese qi gong, yoga, and astrology to form a "liquid lightbody" or "holographic energy field" specifically to prepare for 2012.

Also with 2012 in mind, Ananda Bosman (mentioned in chapter

13) does a Vortexijah, or Diamond Body, training, based around the "secret Tibetan Buddhist teachings of the Vajra Chakra: the Diamond Body," cross-referenced to findings from superconductive physics (see http://phoenix.akasha.de/~aton/DIAMOND1.html). Although his site says he wouldn't be doing any more after 2001, he did run another course in Switzerland in May 2004 and there is a video available.

THE CORICANCHA AND THE GOLDEN SUN DISK

The Incas had a similar concept to that of the Hindu chakras, in which the Incan "chakra" is seen as an energy belt with a focal point. There are four of these belts and they can perceive as kinds of eyes; the fifth belt consists of the two eyes and the pineal eye, or some say that the two eyes and pineal eye are the fifth, sixth, and seventh eyes. The energy belt at the navel or solar plexus position is the one through which most of the energy exchange is performed, and is called the qosqo, mentioned above. This corresponds to the place where many people report their astral body connecting to their physical body via a silver cord.

The city of Cuzco thus means "navel," and is seen by some as the navel, or solar plexus chakra, of the world. The center—the navel—of the city was the Coricancha, or Temple of the Sun, which stood where the cathedral now stands, and in the cathedral grounds the original egg-shaped *omphalos* (navel) stone still survives. The Coricancha was the center point of a system of forty-one *ceques,* or alignments, of sacred places (or *huacas*), which totaled 328. The system was a giant lunar calendar of twelve months of an average 27.3 days each (a sidereal month of 27.3 days is the passage of time for the Moon to return to the position of a given star), according to anthropologist Dr. Tom Zuidema of the University of Illinois.[60]

This is also very close to the 27.28-day Carrington rotation of the sunspots discovered by Simon Shnoll to affect biochemical reaction rates. Knowing that the Incas had a soulcraft technology, and that Charles Musès has found a direct link between soulcraft and the

Fig. 21.7. A replica of the golden Sun disk of Coricancha, seen here on display in the Museum of Anthropology, Lima, Peru.

lunar calendar, perhaps one use of the ceque system was for the timing of those operations. According to Paul Devereux, the ceques were essentially "the physical correlates of the routes of shamanic flight in the spirit landscape; lines either marking the designated ways over the landscape for the disembodied mind of the shaman, or recording where the shaman's spirit had travelled, or both."[61] Juan Nunez del Prado has been working with Q'ero paqos to reconnect the sacred sites of South America by laying down "energy filaments." At the same time, they cleanse and reactivate the *chunpis,* or energy belts, of Pachamama, Earth Mother.

In the Coricancha, which still has intact solstice alignments, there once stood a massive golden Sun disk that is at the center of several myths. (However, some of these myths have their source in a 1961 book called *Secret of the Andes* by Brother Philip, who was actually a ufologist-

Theosophist called George Hunt Williamson.) It was said to have reflected the morning sunlight onto the mummies of the Inka rulers. Other myths connect it with the pachakuti (but many Internet sources say that the pachakuti is a five-hundred-year period, and two patchakutis make one Inti, or Sun) and say it also represented the central Sun of the galaxy.[62] According to psychotherapist and workshop facilitator Bonita Luz, "It has been prophesied that during the time of the tenth pachakuti, which began in 1992, the Solar Disk would be reactivated . . . 2002 was the opening and activation of the Temple of Illumination on the Ethereal Plane. With this opening, the energy frequency of the Solar Disc will finally be restored to its original power."[63] Note that the 1992 and 2002 dates also connect to the Cross of Hendaye. However, the location of the disk is not yet known, since it was removed by the Incas to prevent the Spanish from getting it, and it is said to be either in Lake Titicaca or in the hidden city of Paytiti.

The cover of this book shows a full-size reconstruction of the Solar Disk, which is displayed in a museum in Lima, Peru (there is also a miniature version five and a half inches in diameter in the National Museum of the American Indian, in New York). It shows a central deity with protruding tongue, like the Aztec Sunstone. It represents the fulfillment of the prophecy, the time, Joan Parisi Wilcox says, when the three realms—the upper realm, Earth realm, and underworld (hanaq pacha, kay pacha, and ukhu pacha)—will merge or coalesce into "one paradisiacal world here on Earth."[64] This is very similar to the "dissolving of the veil" predicted by the Maoris for 2012, in which the sky and Earth will meet.

22
PROGNOSIS

TAOIST YOGA

There is an important factor to consider, if we are to find ourselves navigating the "invisible landscape" of the astral realms. Knowledge of this factor has arisen from a study of neolithic sacred sites, combined with certain clues from Taoist yoga, so we shall follow the clues in the next few paragraphs.

In Taoist yoga,[1] breathing and visualization exercises are practiced in which the generative force is diverted from its normal route (ejaculation) and is circulated around a system of power zones that closely resemble the Hindu chakras. A leaflet from the occult bookshop The Sorcerer's Apprentice[2] claims the system is similar to kundalini yoga, but safer. During a long process of internal alchemy, the original spirit cavity is located and transformed into the "precious cauldron" (which is equivalent to the activation of the pineal gland), and an "immortal foetus" starts to develop. Eventually, the spirit body separates from the physical body, then a golden light emerges from the physical body (up from the precious cauldron, through the top of the head, or "heavenly gate") and combines with the spirit body. In the version given by Lu K'uan Yu, the immortal man or spirit body sits upright within "a mass of golden light, the size of a large wheel."[3]

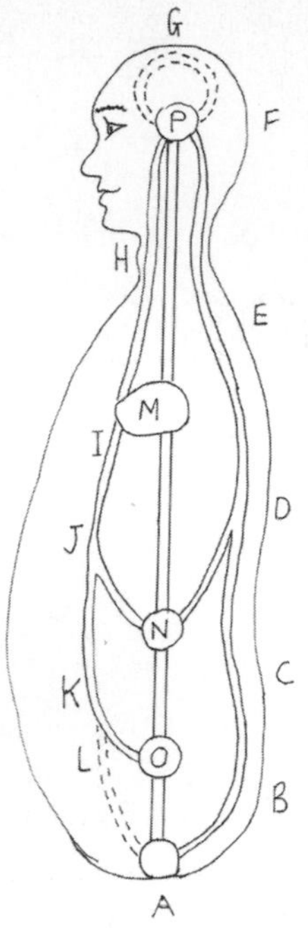

Fig. 22.1. The Taoist system conceives of three channels, and on the central channel we have **A:** cardinal point North—the mortal gate; **N:** the solar plexus; **M:** the heart; **P:** the center of the brain, and **G:** cardinal point South—the heavenly gate. After *Taoist Yoga,* by Charles Luk.

> When spirit manifests for the first time, it should only be allowed to leave the physical body in fine weather and it should be well looked after, like a baby just born. Its egress should on no account take place when there is thick fog, heavy rain, gale, thunder and lightning.[4]

The spirit body is "very sensitive to fear and awe, which should be avoided at all costs."[5] This immortal man within a golden wheel is reminiscent of the original Merkabah[6] (Hebrew for "chariot," discussed in chapter 9) that was seen by Ezekiel, enlarged on by Hurtak, developed further by Drunvalo Melchizedek, and connected by Gregg Braden with the zero point in 2012. Braden said that when the component tetrahedrons were counterrotating,* the whole Mer-Ka-Ba—he spells it like this to emphasize that the concept includes the Egyptian ka, or etheric

*But Drunvalo Melchizedek sent out a statement in January 2003 saying that the tetrahedrons should not spin in opposite directions. J. J. Hurtak, on the other hand, has implied that Melchizedek has the whole Merkabah concept totally wrong. See chapter 9, note 3.

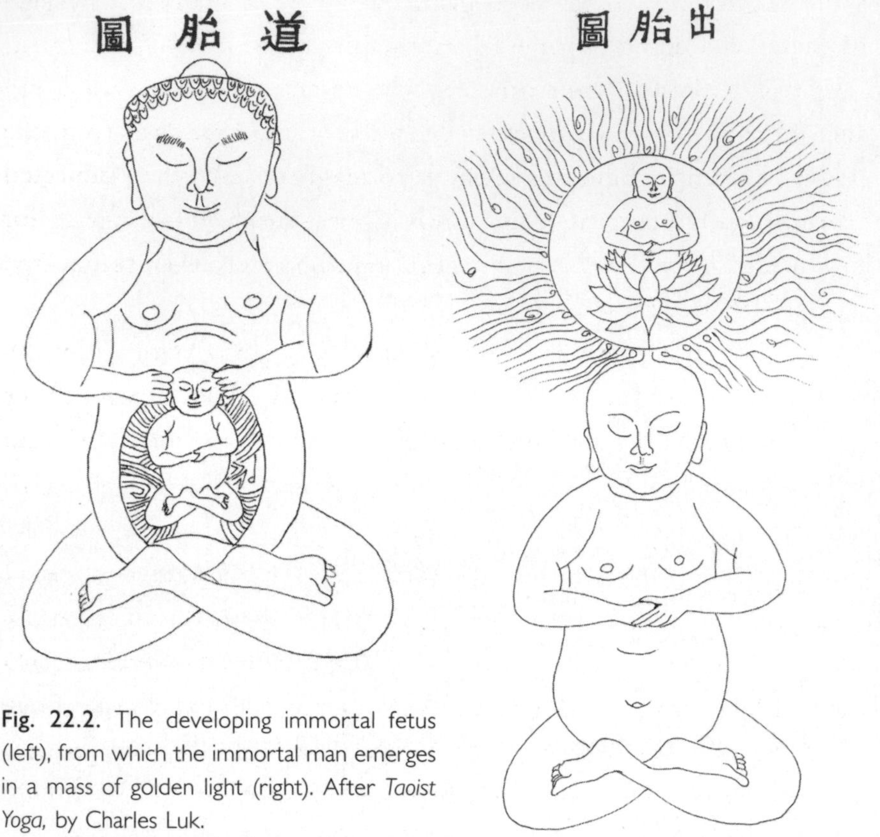

Fig. 22.2. The developing immortal fetus (left), from which the immortal man emerges in a mass of golden light (right). After *Taoist Yoga,* by Charles Luk.

body, and ba, or astral body—would look like a flying disk. Lu K'uan Yu also describes the spirit body as a "golden ball of light."

KILNER

In 1911, Dr. Walter J. Kilner, of St. Thomas's Hospital in London, published the results of experiments prompted by Baron Karl von Reichenbach's earlier conclusions about the existence of an aura around the body that could theoretically be used as an aid to medical diagnosis. After Kilner's eyes were sensitized by looking at daylight through a dark screen, he looked at human subjects against a black background, through a lighter screen made from dicyanin dyes sandwiched between sheets of glass.[7] What he saw was an envelope of energy composed of three layers, the innermost of which he called "the etheric double." The other two he named the "inner aura" and the "outer aura," which together extend at least a foot away from the body: "The entire aura was sensitive to magnets, and when subjected to the negative current from a Wilmshurst machine—a device for producing an electric charge—it vanished completely, then reappeared with an increased intensity."[8]

In 1970, biophysicists and biochemists from Kirov State University of Kazakhstan, in Russia, published the results of a study in which they had been examining the human energy body by means of an electron microscope, following up on the discoveries of Semyon Kirlian, a Russian electronics engineer who, in 1939, discovered that a photographic plate placed between human skin and the electrode of a high-frequency electrotherapy machine recorded a luminescent shape surrounding the body. The Kirov State University team called this the "biological plasma body," because they found it to be "some sort of elementary plasma-like constellation made of ionised, excited electrons, protons and possibly other particles . . . It is not a chaotic system. It's a whole unified organism in itself." This is the "non-molecular body" of Charles Musès.[9]

The Russian scientists showed that the plasma body is charged up by the oxygen we breathe—particularly by ionized air—and that it is affected by the eleven-year-plus sunspot cycle, the Moon, thunderstorms, strong winds, illnesses, emotions, the thoughts of others, and changes in the planets. A plasma is a gas that can be created in electric storms or thermonuclear reactions, causing all the electrons to be stripped from the nuclei of the atoms. In *Supernature,* biologist Lyall Watson points out that not much is known about the plasma, or how it could be created at body temperature, but it is known that "the only thing that will contain its energy effectively is a magnetic field—and we know that the body has one of these."[10] But maybe it is possible that the plasma is created outside the body, and invades it in the manner of Philip K. Dick's plasmate.

PLASMATIC INVADER

Philip K. Dick was a science-fiction writer, now famous as the author whose stories have been made into the films *Bladerunner, Total Recall, Minority Report*, and *A Scanner Darkly*. His writings were thought-provoking mixes of psychology, philosophy, and weirdness, and he already had forty books in print when in February 1974 he had a mystical experience that changed his life.

Dick was under a lot of stress at the time, and in pain with an impacted wisdom tooth, waiting for a delivery of prescribed painkillers. The young woman who delivered the medication was wearing a necklace with a golden fish pendant, like the fish symbol worn by the early Christians. Dick was hypnotized by the pendant initially; then it reportedly fired a beam of pink light at his head, causing anamnesis, which comes from the Greek word for "loss of forgetfulness." In this case, an existence as a second-century secret Christian superimposed itself onto his contemporary existence. Dick interpreted this as an invasion of his mind by a "transcendentally rational mind," which he called Valis (Vast Active Living Intelligence System). It "was equipped with tremendous technical knowledge. It had memories dating back over 2,000 years, it spoke Greek, Hebrew, Sanskrit; there wasn't anything that it didn't seem to know."[11]

Like Carl Jung, Dick found that his visions had much in common with the philosophy of the Gnostics, fragments of whose scriptures were unearthed in 1945 when forty-nine codices, or gnostic books, were found near the ancient town of Chenoboskion, in Egypt. They are known as the Nag Hammadi library, as that is the name of the nearest modern town. Dick played with various interpretations of the experience in his remaining books, but in *Valis,* in an appendix, he presents his most enduring take on it.

He saw his experience as an invasion by the "immortal one," which he called a plasmate, since it is a form of energy. In *Valis,* the plasmate had been slumbering in the form of the buried gnostic scrolls, until it awoke when they were unearthed and read: "As living information, the plasmate travels up the optic nerve to the pineal body. It uses the human brain as a female host in which to replicate itself to its active form. This is an interspecies symbiosis."[12]

The Gnostics have a notion that the soul of each of us is a divine spark—a fragment of the Godhead scattered throughout creation—and that every fragment will eventually be reconnected into the Godhead. This will occur at the end of time.

NEOLITHIC EARTH-NET

The Taoist warning against thunderstorms may have wider relevance. Lightning invariably strikes at the crossing points of water lines (underground streams), according to dowsers, and sacred sites have been found by dowsers to invariably be positioned at these points, on what dowsers call blind springs. However, in his book *Earth Lights,* Paul Devereux observes that nearly all the stone circles in the UK are in areas of low thunderstorm activity. These facts seem at first to contradict each other, until you hear the traditional belief about a thunderstorm happening if a barrow is desecrated. This is not just a reference to the wrath of the gods, as we shall see, but it certainly must have crossed the minds of a certain research group, as reported by dowser Tom Graves.[13] The group were surveying a stone circle, and when they hammered a stake into the ground, it triggered an immediate two-minute rainstorm.

In his book *Needles of Stone,*[14] Graves presented evidence that the neolithic network of alignments that join up sacred sites was designed—at least partially—as a weather-control system. Standing stones next to barrows acted like lightning conductors, or, to be more accurate, like Wilhelm Reich's cloudbuster. Reich was a psychologist who claimed to have discovered evidence that Sigmund Freud's hypothetical libido—the life energy that he said was essentially sexual energy—was real and could be captured and stored in a specially designed box. Reich called the energy "orgone," since it powers the orgasm, and the box was called an "orgone accumulator," and was composed of alternate layers of organic and inorganic materials. He designed a machine that looked like an antiaircraft gun, a collection of various lengths of metal tubes on a swiveling base, connected by a steel cable to a body of water. This was the cloudbuster. When pointed at clouds, it dispersed them, and when pointed at a cloudless sky, it could create clouds, and even storms on some occasions. It is only in recent years that Reich's ideas have been taken seriously and researchers have successfully repeated his experiments.

In the case of a barrow and standing stone, Graves explained that a negative charge from an underground stream was passed up the stone and into a positively charged cloud, thus neutralizing it and causing rain; or the positive charge from the cloud was passed down via the stone, along the underground stream and into the barrow, which stored it (a barrow is made up from alternate layers of organic and inorganic layers, like an orgone accumulator, says Graves). This also neutralized the cloud and avoided lightning. The charge then slowly leaked back into the sky via stream and stone, or was neutralized by the negative charge from the underground stream (see fig. 22.3 on page 380). The two polarities involved would be orgone and DOR (or deadly orgone radiation, the stagnant, negative version of orgone life energy), which are ch'i and sha in the Chinese system. The system is still actually working after all these years, since Graves points out that the Isle of Jura has a lot of megalithic sites, and thunderstorms are very rare there.

Why would neolithic people have wanted to avoid lightning at all costs? This is a question that has gone unanswered until now. Two facts mentioned by Graves have a bearing on this. First, several of Graves's correspondents told him that the telepathic transmission of complete mental images was easier between sacred sites than elsewhere, and we have already seen that telepathy works better when geomagnetic activity is quiet. It so happens that in *Places of Power,* Paul Devereux reports that stone circles are areas where the geomagnetic field is lower than the background level.[15] This suggests that sacred sites were used for telepathic-message relaying—but that is only part of the story.

The second fact from Graves is that his "magician friends" told him that they had found both forms of astral traveling—"projection of the imagination, and projection of the entire personality"—to be much easier "by hitching a ride on the energy that passes along ley lines or overgrounds."[16] However, the interpretation of leys as energy lines is no longer very viable following Devereux's work (though this does depend on one's definition of the word *energy*), and it might be better to follow Mary Scott's lead and describe convoluted lines as dragon currents and straight lines, or "overgrounds," as spirit paths.

Devereux, as editor of *Ley Hunter* magazine for sixteen years and one of the founders of the Dragon Project, makes a strong case that the ley-line network is the remnant of a system of shamanic flight alignments, where ancient barrows, standing stones, stone circles, and even pre-Reformation churches (built on sacred sites) were aligned in straight lines. The surviving beliefs of the Kogi tribe of Colombia support the theory. Their shamans, or seers, are called Mamas, and they are trained from birth and kept in the dark for their first nine years, so they can tune in to the *aluna,* or Spirit Earth (here we have some evidence that Ananda's dark-room retreats are a genuine ancient technique of shamanism). Interestingly, as reported in Alan Ereira's 1990 BBC-TV documentary on the Kogi, and quoted by Devereux, they also refer to the aluna as the life force and the Mother.

> [The Mama walks spirit paths] in a world visible only to the mind's eye, the world of aluna . . . they were lines of thought, not lines on the ground. But there were points where the two worlds used to meet. . . . The Mamas used to travel in aluna and meet other people from other places there. But as the years passed, there were fewer and fewer people who reached the spirit world to talk with them. Now there is no one. It is deserted.[17]

Devereux also traces the use of hallucinogens back through medieval times, when our modern stereotype of the witch had its origin. Wise old women who had accumulated knowledge of medicinal plants over a longer-than-average life brewed up tropane-containing plants like henbane, mandrake, datura (thorn apple), and belladonna in their cauldrons. Tropanes are known for causing sensations of flight, and also for their ability to be absorbed through semipermeable membranes. The wise old women lived alone at the edge of town as outcasts among people who mostly died around thirty years of age (the classic witch appearance is simply an exaggerated old person, whose teeth have fallen out and the nose appears to touch the chin, with long white hair, wrinkles, and hairy growths). The flying ointment was stirred up in the cauldron with a broomstick, and the broomstick was then "ridden" (rubbed between the labia).[18] Devereux shows that the tradition of out-of-body flight is rooted in prehistory,[19] and it seems that the neolithic shamans would use the low-magnetic environment of stone circles as a sacred space in which to shamanize, or achieve trance. Remember that Earth's magnetic-field strength is dropping as we approach 2012 . . .

ECSTATIC FUSE BOX

Finally, we can perhaps answer the question of why neolithic peoples needed weather control. First, remember what was said about the immortal body in Taoist yoga—there was a warning to the student never to come out of the body during thunderstorms or lightning. Then recall Kilner's dicyanin aura filter. The aura was visibly affected by a magnet and the electric current from the Wilmshurst machine caused it to completely disappear. Could this be the reason that the megalithic priesthood feared lightning? If they were out of their body on an astral flight, a sudden bolt of lightning would at best send their astral form back to their body in a state of shock, and at worst cause death, and possibly even the loss of their astral body. Since lightning is naturally attracted to the nodal points of their system, the priesthood would have needed a lightning-prevention system, like a kind of integral fuse box. This is exactly what Graves described as the function of a barrow and standing stone (see fig. 22.3).

The two functions we have associated with the ancient grid are the very same two abilities enhanced by our natural pineal hallucinogens: the beta-carbolines (like pinoline) triggering telepathy and the tryptamines (like DMT) triggering out-of-body experiences (literally, ecstasy, from the Greek *ekstasis,* "standing outside oneself"). What all this means is that the changing geomagnetic environment is probably going to cause the place the Kogi call aluna to become populated again, as more people find they are "standing outside themselves" in a normally invisible landscape.

Like the witchcraft stereotypes of old hags, broomsticks, and cauldrons, the pointed hats that are still used in our caricatures of witches and wizards conceal another secret. In 2003, the National Germanic Museum in Nuremberg held an exhibition in which, for the first time, four golden hats from the Bronze Age were displayed together. The most recent find, which was bought by the Museum for Prehistory and Early History in Berlin in 1996, has been subjected to a detailed study by Professor Wilfried Menghin, director of the museum.

The hat is about three thousand years old, thirty inches high, and was found in Switzerland in 1995. It is covered with a total of 1,739 Sun and Moon symbols, and when compared with the other hats, it was found that the embossed symbols on it represented a quarter of a Metonic cycle, which is a nineteen-year cycle discovered by Greek astronomer Meton in the fifth century BC. Meton found that 235 lunar months are equal to nineteen solar years, after which the lunar phases repeat on the same solar calendar date. The Ezelsdorfer hat, one of the four hats on display in Nuremberg, which was found in Germany, has 1,737 symbols, and the four hats together show that the 1,735 days in fifty-seven lunations (a quarter of the 235 lunations of the Metonic cycle) had been recorded as a logarithmic table. According to Professor

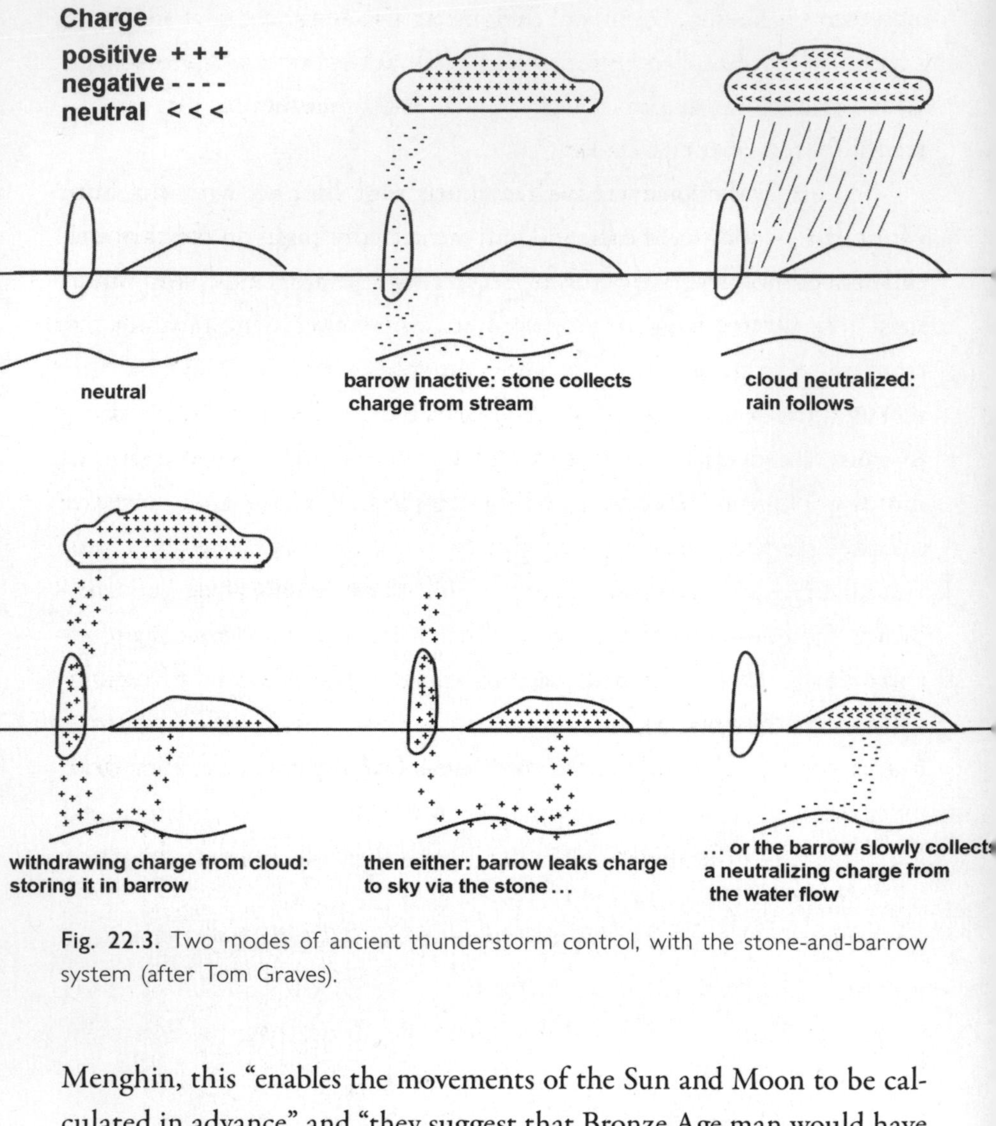
Charge
positive + + +
negative - - - -
neutral < < <
neutral
barrow inactive: stone collects charge from stream
cloud neutralized: rain follows
withdrawing charge from cloud: storing it in barrow
then either: barrow leaks charge to sky via the stone . . .
. . . or the barrow slowly collects a neutralizing charge from the water flow

Fig. 22.3. Two modes of ancient thunderstorm control, with the stone-and-barrow system (after Tom Graves).

Menghin, this "enables the movements of the Sun and Moon to be calculated in advance" and "they suggest that Bronze Age man would have been able to make long-term empirical astrological observations."[20]

A third golden cone hat, found in 1835 near Schifferstadt, in Germany, still had a chin strap attached, and at 1300 BC is dated the earliest of the four hats. The fourth surviving hat was found in France, but there are reports of five similar cone hats found by peat diggers in Ireland in the seventeenth and eighteenth centuries that have now disappeared, according to the research of Professor Sabine Gerloff, of

Erlangen University, Germany. Gerloff has also suggested that the Mold gold cape (a kind of low-slung, solid-gold-embossed collar designed to attach to a cloak), found in Wales in 1831, was another part of the ceremonial dress of the priest-kings.[21]

So our Halloween fancy-dress outfits of Merlin, with his Sun, Moon, and star-covered cape and hat, are actually based on the calendrically encoded ceremonial dress of Bronze Age "king-priests" who would have been revered as Lords of Time. (See color plate 22 for a demonstration of the wearing of these hats and collars.)

In his twenty-year study of Thornborough Henge[22] (see chapter 19), senior lecturer in late prehistory at Newcastle University Dr. Jan Harding has concluded that the complex included alignments to the summer solstice, lunar standstills, and the rising of the constellation of Orion. In 2003,[23] he announced that the entire complex was a reflection of Orion's Belt, and the aerial shots seem to confirm this, with the three henges in a row, one slightly offset. The third henge is not immediately visible, since there is a copse of trees now growing on it, but it has been well preserved within the trees. The site was inhabited from

Fig. 22.4. The Pyramid of the Moon, at Teotihuacán, Mexico, as seen from the Pyramid of the Sun, with the author in the foreground. Like those of several other ancient monuments, the layout of the complex appears to represent the Belt in the constellation Orion.

3500 BC to 2500 BC, so it looks as though this Orion ground plan preceded the one at Giza (according to the officially accepted chronology, in which the Great Pyramid was built around 2500 BC). The pyramid layout at Teotihuacán in Mexico also appears to be a representation of Orion's Belt. We have seen evidence that both ancient Egypt[24] and neolithic England had shamanic priesthoods, whose members traveled out-of-body and were thus able to get an aerial view of the terrestrial Orion's Belt they had constructed.

In *Shamanism and the Mystery Lines,* Paul Devereux reveals that according to tribal peoples all over the world and to modern investigators of out-of-body travel, the astral body can sometimes appear as a light ball. Although he said in *Earth Lights* that there is no evidence to associate ley lines with UFOs, he does admit the following about a random sample of leys selected for his book *The Ley Hunters Companion:* "37.5% of the leys we were selecting showed some evidence of UFO events occurring on them or in their immediate vicinity. It was impossible to determine whether this UFO incidence was associated with the alignments as such, or with the sites forming the alignments."[25] It could be that the electromagnetic conditions are only conducive to the physical manifestation of the light balls at the nodes—the sacred sites with their unusual energy fields (they are also commonly witnessed at crop formations). Balls of light recorded at Hessdalen could still be tracked on radar after passing out of the visible spectrum.

ENOCH'S GLORIOUS DEVICE

The book of Enoch (1 Enoch, or the Ethiopic Enoch, as opposed to the later book, the Secrets of Enoch, also called 2 Enoch or the Slavonic Enoch)[26] has been dated to the years between 200 BC and 100 BC (between the Old Testament and the New Testament). Specialists in Jewish religious texts have agreed that much of the New Testament used Enoch as source material. It is an apocalypse (revealing) similar in some ways to St. John's book of Revelation.* Its attribution to Enoch as the author is thought to be a calendrical reference, since Enoch is stated in Genesis[27] to have been seventh in line from Adam, and to have lived to the age of 365 years.

There is a lot of calendrical information in the book, and when Robin Heath, author of several books on megalithic sites, analyzed it using a simple computer program,[28] he found that the observations were taken from somewhere far to the north of Israel—in fact, from the latitude of Stonehenge. In *Sun, Moon and Stonehenge,* Heath further showed that the descriptions of the Sun- and Moon-rises within six portals in the east, and the settings within six portals in the west, refer

*In the Apocalypse of Weeks (Enoch chapter 93 and part of 91, replaced to original order, as in the SPCK (Society for Promoting Christian Knowledge) edition, the whole Creation is predicted, from Adam to the Last Judgment. There is a series of weeks, starting with the week from Adam to Enoch, in which Enoch is the seventh born. This is not the same as the seven days of Creation that we covered in part 1. There are ten of these periods called weeks, and the first five are easily identified from the Jewish patriarch mentioned in it (by description, not by name). I have looked at the Genesis genealogies and there does not seem to be a regular period encoded, but the "weeks" seem to be between 600 and 1,000 years long. The patriarchs of the first five weeks are Enoch, Noah, Abraham, Moses, and Solomon. The next week is the one in which the book of Enoch was officially written down, and it predicts that in it, "a man shall ascend," and at its close, "the house of dominion shall be burnt with fire." These could well refer to the crucifixion in 33 AD and the destruction of the temple in 68 AD (70 AD according to some sources). This is a pretty amazing prediction, considering that this is the oldest part of the book of Enoch, at around 200 BC, regardless of whether you take the ascension story literally. The seventh week describes the rise of an "apostate generation," which could refer to the Arab takeover of Jerusalem and the building of the al-Aqsa Mosque on the site of the Temple in Jerusalem in 691 AD. At its close was to be elected "the elect righteous of the eternal plant of righteousness," possibly the nine original knights of the Knights Templar in 1118. In the eighth week, "a righteous judgement [would] be executed on the oppressors," and at its close, the oppressed would be rehoused. This seems to be the Jewish reclamation of Israel in 1947–48. This means we are now in the "ninth week," during which the Judgment will be revealed to the world, and the works of the godless will vanish and the world will be "written up for destruction." The Last Judgment of the fallen angels and the New Heaven and New Earth are not to be expected until the end of the next week, which will be between fifteen hundred and two thousand years in the future (St. John interprets it as a thousand years).

specifically to observations made from a location known as The Grave (at the "entrance" to the trilithon horseshoe), through the "portals" between the sarsen stones at Stonehenge. This fits perfectly with Enoch chapter 34, in which he is taken by the angel Uriel "towards the north to the ends of the earth, and there I saw a great and glorious device . . ."[29] Authors Christopher Knight and Robert Lomas (see below) point out that Enoch also seems to have visited Newgrange, in Ireland, a passage tomb built around 3000 BC that is covered in brilliant white quartzite, when he (Enoch) visited a "wall built of crystals."[30]

Robin Heath remarks that in the Celtic book *The Tales of Taliesin,* the author states, "I was instructor to Eli and Enoch." He is also mentioned in *The Mabinogion,* another Celtic book dated around 1300 AD; there is even a hill called Carn Enoch, near Preselli, in Wales, where the bluestones of Stonehenge originated.[31] I have also discovered that the original ancient name of the county that borders the Boyne Valley in Ireland, where Newgrange is located, was Uriel.[32]

Enoch was the great-grandfather of Noah, and in the book of Enoch, not only is Enoch given calendrical information, but he is also told about the forthcoming flood, and passes this information down to Noah. Robin Heath points out that before 3000 BC, astronomical alignments were mostly stellar. He suggests that just prior to 3000 BC, Earth had a close encounter with an asteroid, which is the origin of the worldwide Flood myth. He suggests an approximate date of 3300 BC. Knight and Lomas suggest 3150 BC, and also suggest that the grooved-ware people of neolithic Britain actually predicted this, after making observations following the previous impact with a comet that they put at 7640 BC. We have already seen that other researchers put the Flood at 3114 BC, along with the start of the thirteen-baktun cycle.

Knight and Lomas, who are Masons, have researched ancient lodge documents in an attempt to trace the roots of Masonic tradition. They started looking at Enoch, since he is mentioned in several Masonic degrees. They have traced back Masonic ceremonies to the Knights Templar (so convincingly that many previously incredulous Masons now concur); the original nine knights were direct descendants of Jewish temple priests. The Templars excavated under the Temple of Jerusalem and found the huge stash of treasure described in the Copper Scroll of Qumran, plus accompanying secret texts, contemporary to the Dead Sea Scrolls. These were then buried under Rosslyn Chapel in Scotland, which is a copy of Solomon's Temple, complete with nine vaults. The information in the scrolls was encoded into cathedrals sponsored by the Templars. This is when the Masonic guilds started, which laid out the stonework.

The grooved-ware people were trading in the Mediterranean area (e.g., tin for use in bronze production), and their beliefs were passed on to the Caananites and Phoenicians, who bequeathed them to the Israelites when they arrived from Egypt. Knight and Lomas pointed out (after Martin Brennan) that if a lens tracks the midday Sun for 366 days and the dots are joined, the resulting figure is two spirals around the solstices, with flattened curves at the equinoxes.[33] Presumably, if the shadow from a post or gnomon is marked at midday every day for 366 days, the same result would be achieved. Thus a spiral represents a quarter of the year. There are many spirals at Newgrange, which is aligned so that the winter solstice sunrise illuminates the chamber each year. There is a conspicuous triple spiral in the inner chamber at Newgrange. According to the Knight and Lomas theory, a triple spiral would represent three seasons: nine months, or the gestation period of humans. This is the period around which the Maya tzolkin of 260 days is based, and, as we know, the thirteen-baktun cycle that ends in 2012 is structured like a giant tzolkin of 260 katuns, possibly representing a planetary gestation that precedes the birth of the next Sun (rebirth of One Hunahpu) on the winter solstice of 2012.

Fig. 22.5. The triple spiral engraved in the central chamber at Newgrange, Ireland.

Enoch compares all the righteous and elect at the time of the Judgment to "fiery lights"[34]: they shall have been "clothed with the garments of glory."[35] He says, "Ye shall shine as the lights of heaven and the portals of heaven shall be opened to you,"[36] and they "shall go where the days and seasons are prescribed for them."[37] Is this a description of the golden light-ball bodies in the timeless astral dimension?

Despite the fact that the Flood myth borrows heavily from the Sumerian myth of Utnapishtim, perhaps the whole apocalyptic genre of Judaism and Christianity has its roots in Britain and the megalithic attempt to predict the next catastrophe. But perhaps it is in essence a prediction of the next geomagnetic fluctuation, which will cause the mass opening of the pineal eye, and exosomatic (outside the body) transition. A clue here is the "forty days and forty nights" that rain fell upon Earth while Noah was in the Ark.[38] We have already seen that this is the period of the acute schizophrenic break, with its altered pineal output. Other biblical instances of it are Moses' forty days and forty nights on Mt. Horeb, in the Sinai, talking to God;[39] Elijah's escape to the desert, where he sees an angel and continues for forty days and forty nights to Mt. Horeb and hears God speaking to him;[40] and Jesus' forty days and forty nights in the desert, being tempted by the Devil, then visited by angels.[41]

Having seen the connection between the book of Enoch and Stonehenge and Newgrange, we might notice that Newgrange was designed to identify the winter solstice sunrise, and that the current orthodox position (after Professor John North's 1996 book, *Stonehenge: Neolithic Man and the* Cosmos) is that the primary alignment at Stonehenge is toward the midwinter (winter solstice) sunset. The summer solstice sunrise, attended by thousands at Stonehenge each year, in which the first gleam of the Sun appears just next to the Heel Stone, and when fully risen sits exactly on it, is an alignment that did not exist when the monument was constructed. The change in the obliquity of the ecliptic has caused the Sun to move between one and a half and two Sun-widths toward the Heel Stone.

This changing obliquity cycle of 41,000 years is exactly divisible by the thirteen-baktun cycle, as William Gaspar pointed out. Earth's axis

Fig. 22.6. The passage and light box at Newgrange, Ireland, showing a close-up of the quartz blocks that form most of the perimeter wall.

takes 20,500 years to go from minimum to maximum tilt, and another 20,500 years to go back. This means it takes exactly four thirteen-baktun cycles for a half cycle, and we are just at the end of the Fourth World, or Sun, of the Hopi and Maya, which ends on a winter solstice. At Newgrange, the changing obliquity has moved the light beam back, so it no longer fully illuminates the triple spiral of gestation,[42] but falls in the center of the chamber, where it would (if the basin was still in its original position) hit the central-stone reflecting basin, lighting up the whole chamber—signifying an emergence of humanity from its techno-chrysalis into fully metamorphosed imago forms.*

Crop formations, as we have seen, are attracted to aquiferous geology, and these are the same areas where most sacred sites are preserved. The Avebury area of Wiltshire is the focus of the activity in the crops, and is also the location of the world's largest stone circle, which encircles the whole village of Avebury, which was subsequently built within it. The crop circle phenomenon is undoubtedly interacting with something at our sacred sites.

This was especially noticeable in 2003. On July 5 of that year, a line of Bronze Age barrows at North Down, Wiltshire, had a formation that appeared as an extra barrow in the sequence, but with the appearance of a Bronze Age shield (see figs. 22.7 and 22.8). Three sacred-site formations also appeared in Yorkshire in 2003; the first of these three appeared next to Thornborough Henge on June 21 (see the Ecstatic Fuse Box section, above, where we discussed Thornborough), alerting many people to the existence of this little-known earthwork and attracting attention to the fact that it is under threat from quarrying, thus prompting a campaign to save the site. It has since been featured in a BBC-TV documentary about the Orion connection discovered by Dr. Jan Harding. Another crop formation, which appeared on July 9, 2003, was visible from the Rudston Monolith, the highest standing stone in the country (twenty-six feet); and yet another design arrived

---

**Meta-morphe* is Greek for "change of form."

several weeks later, on August 3, at Boroughbridge, in the same field as the Devil's Arrows standing stones.

Not only are the crop formations focusing our attention on sacred sites, but they may also be affecting our endocrine systems. Circle researcher Lucy Pringle conducted an experiment in which hormone levels were ascertained using an ESR (electrical skin resistance) meter on a woman aged seventy-two and a man aged fifty-four, before and after entering a crop formation. After visiting the circle, their melatonin levels were found to have increased considerably."[43] Melatonin is made in the pineal gland and is the raw material for producing internal hallucinogens. There are several women who have reported either postmenopausal bleeding or unexpected menstruation during visits to crop formations (two of these are friends of mine; others have been reported by Lucy Pringle).[44] This could implicate other endocrine glands, along with the pineal, since they are involved in the feedback loops of human sexual hormone production. These menstrual and postmenopausal effects may be

Fig. 22.7. A crop formation resembling a Celtic shield joins a line of burial mounds at North Down, near Avebury, Wiltshire, July 6, 2003. Photo: Nick Nicholson

Fig. 22.8. Closer view of the Celtic shield crop formation. Photo: Nick Nicholson

connected to magnetic fields. Jerry Decker, who runs the website www.keelynet.com, a clearinghouse for information about free energy, gravity control, and alternative health, says, "A steady state magnetic north pole field will stimulate the production of melatonin and eliminate PMS [premenstrual syndrome], as long as field strength in the vicinity of the pineal is in the 400 mg. range."[45]

Tom Graves compared the sacred sites to a system of acupuncture points that show a measurable difference in skin resistance from surrounding areas. These have been measured with a device called a Tobiscope[46] and photographed using Kirlian photography. The acupuncture points are exactly where the Chinese always said they were. Similarly, Earth's acupuncture points—sacred sites—show a measurable difference in geomagnetic field from surrounding areas, and would seem to be nodal points in a network of meridians on the planetary body, where we can access higher dimensions of mind. We are being shown the relevance of the neolithic mind-set—what Terence McKenna called "the archaic revival."

THIRD EYE MAGNETIZED

Jerry Decker, reporting on KeelyNet,[47] says that in a telephone conversation with Walter Rawls, coauthor, with Albert Roy Davis, of four books on magnetism,[48] Rawls told him about an experiment in which a mask[49] was made to hold a long cylindrical magnet over the pineal gland. This was worn for between ten and thirty minutes per day, for three to four weeks[50] (this must be done with a long bar magnet, to separate the poles, and it must be the north pole, since the effects of a south pole could be dangerous to the health):

> Within the first week, he was sitting at his desk reading documents when he noticed something move, out of the corner of his eye. As he looked up, the ghostly figure of a man had walked through one wall, moved across the room and disappeared through another wall. The figure was totally unaware of Walter. Further exposures to this north pole field took place over a second and third week. The second week, the same ghostly figure moved through the room and glanced toward Walter as he passed through. This time, the figure appeared to have slightly more detail, not quite so ghostly. The third week, busy working on documents, Walter noticed a change in the room. When he looked up, the wall had dissolved away and he was looking at a small hill where a man and woman sat beneath a tree. It was the same ghostly male figure who he'd seen on the other occasions. He sat quite still, watching this pastoral scene for several minutes. The man looked over toward Walter and appeared startled. It was as if he clearly SAW Walter this time and possibly recognized Walter as the ghost HE had seen the previous week! The image faded away and the wall restored to its normal condition. From that moment on, Walter never used the pineal stimulator again.
>
> In conversations about this, with Walter and other interested people, it was mentioned that there is a theory that we have multiple energy bodies, much like the Ka and the Ba of ancient Egypt. Each

> energy body lives in another reality, yet communicates with our consciousness here in this reality. Another comment was that consciousness simply creates an energy body in whatever reality it VISITS. Prolonged presence in a given reality increases the density of the energy body, moving from a phantom, ghostly form that is at first not easily perceptible to the inhabitants of the other reality until the intruding energy body has become sufficiently dense to trigger their senses. That could explain why repeated exposures would add density to Walter's other reality body, allowing its denizens to perceive him, thus the startled reaction from the male figure.[51]

The American Society for Psychical Research has commented that ghost sightings and paranormal activity are more frequent during times of high solar emissions;[52] there were also increased sightings of the "American Yeti," Bigfoot, at the time of the last solar maximum, and there have also been reports of increased admissions to mental hospitals during these periods. This illustrates that as the Sun gets nearer to its megacycle peak, more people's magnetic cerebral switches are activated.

We all have significant deposits of magnetite (a naturally occurring magnetic oxide of iron) in the pineal gland, and also layers of magnetite in the sphenoid and ethnoid bones of our sinuses, with which we detect magnetic fields, according to the findings of Manchester zoologists Baker, Mather, and Kennaugh.[53] Anemic people have an absence of these deposits. There are two types of people who are even more sensitive to these fields, according to the work of paranormal researcher Albert Budden.[54] What is known as electromagnetic hypersensitivity (EMH) occurs when the magnetism of the pineal (and sphenoid/ethnoid) magnetite is increased by environmental factors. In EMH type A, this happens by "prolonged irradiation at a hot spot, often where the fields involved emanate from a natural geological source, i.e., tectonic activity at faults."[55] Stone circles, as Devereux found, lie mostly on or near geological faults (UFO sightings are also common in these areas).[56]

Electromagnetic hypersensitivity is acquired in B types by exposure to a major electrical event, or MEE. These include being close to a lightning strike, major electrocution, proximity to ball lightning or earthlights, ECT (electroconvulsive therapy), and defibrillation (resuscitation following heart attacks, etc.). People with type A hypersensitivity tend to see angels, guides, and friendly apparitions, while those with type B hypersensitivity tend to have "alien" abduction experiences, says Budden, whose wife is a type B. Whitley Streiber, who has written several books about his abduction encounters, was not only exposed to "multiple MEEs" in his earlier life, but he also experienced nosebleeds and the feeling of "knots" in his sinuses.[57] Streiber recognizes that electromagnetic hypersensitivity plays a part in his experiences,[58] but still sees his experiences as real. It could be that sudden magnetization of the internal magnetite tunes us in to a different part of the astral plane—what is known in the esoteric world as the lower astral—whereas a slow, steady magnetization gives access to the higher astral plane.

Carn Ingli is a peak on the Preselli Hills overlooking Newport, on the coast of Wales, and it is an example of the kind of environment that would slowly increase pineal magnetic sensitivity. The name Carn Ingli means "the Peak of Angels" (or "Mount of Angels") and is so named after the experiences of St. Brynach, who lived the life of a hermit on the peak, where he communicated with angels. In *Places of Power,* Paul Devereux reports that the hill has some unusual magnetic anomalies, including a few places where the field is fully reversed by 180 degrees, which can be checked with a compass. In 1987, a woman felt she was "being subjected to waves of some kind of force"[59] coming from the peak, and the closer she got, the stronger the feelings became. This is how the reversed field came to the notice of Dragon Project investigators.

TELEPATHIC INTERPLAY

We have seen that many catastrophe theories associated with 2012 are seriously flawed; yet Earth changes are increasing by the month. Either the field-effect energy reversal suggested by John Major Jenkins or our entry into the magnetized plasma band may be the mechanism anticipated by ancient cultures that will trigger not only increased solar activity and weather extremes but also a geomagnetic reversal, during which time itself may seem to slow down. (I know exactly what this is like from a motorcycle accident in 1976, in which I demolished a small Italian car with my elbow—it was like a slow-motion film.) This also happens in dreams, where you can have a whole dream of falling for thousands of feet, based on a physical fall from the bed to the floor. This is Ken Carey's "moment of Quantum Awakening": an out-of body experience into a timeless realm caused by a magnetically induced pineal-eye opening. Some have even suggested that our memories will be wiped out by a magnetic reversal. If this interpretation is correct, then it is certainly an interesting time to be alive, to be part of an evolutionary quantum leap that may be seen as the boundary between *Homo sapiens* and our descendant, *Homo spiritus.*

Preparation for an event such as this would involve confronting our shadows, a clearing-out of the subconscious or unconscious mind of all repressed feelings, and forgiveness of all those who have caused us problems in the past; a letting-go of all bitterness, resentment, negative emotions, self-doubt, and ego problems, a dissolving of learned fear and anger responses; and getting ready to treat others as ourselves, as in the Maya greeting *"In lake'ch"*—"I am another yourself."[60] This process is *metanoia,* a Greek word meaning "mind change," but which generations of biblical scholars have mistranslated as "repentance."

These are exactly the issues that are sorted out by years of meditation, or contemplative prayer, but since we need to get a move on, there are all kinds of workshops available to help us with this, as well as practicing techniques of out-of-body travel and familiarizing ourselves with the geography of the otherworld. We might even try taking iron pills, as recommended by dowsers, to improve our magnetic sensitivity (or drink iron-rich springwater, such as that from Chalice Well, in Glastonbury). We can spend more time at sacred sites, to gently increase pineal and sinusoidal magnetic sensitivity. Then perhaps we will be ready for what esoteric and religious writer and teacher Alice Bailey called "a telepathic interplay which will eventually annihilate time," in which we can cooperate as human neurons in a global brain.

Archaeologist Laurette Séjourné has interpreted the end of the Fifth Sun of the Aztecs in a similar way.

> We know that the Fifth Sun appears after four others have come and gone, and that this Fifth Sun is itself destined to be superseded by another. We cannot believe that these myths of solar cataclysms were inspired only by the eternal cosmic renewal manifest in natural cycles. Apart from the fact that Nahuatl religious phenomena, which are so spiritual, could never be so simply explained—in terms that is, that have no inner meaning—there are various indications that the Fifth Sun is the creator of a great and indestructible work: that of freeing creation from duality . . . The essence of the Nahuatl religion is contained in the revelation of the secret which enables mortals to escape destruction and to resolve the contradiction inherent in their natures by becoming converted into luminous bodies.[61]

Thus we could look upon Earth as a large hibernating creature, like a bear that is recycling its body waste but can only keep it up for a certain amount of time before it either has to wake up or poison itself. The unconscious mind of the bear knows when spring is coming, the trigger

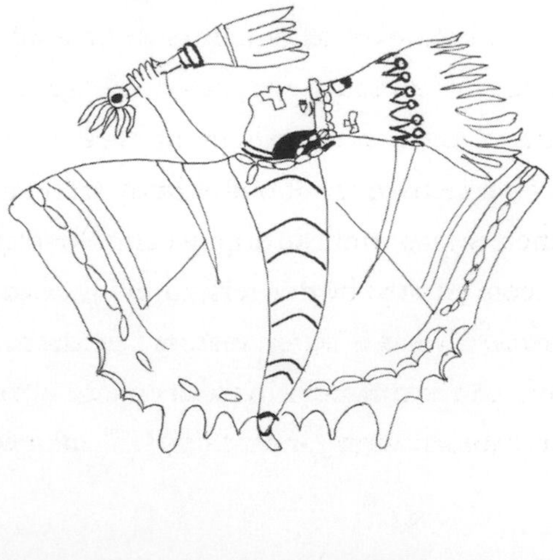

Fig. 22.9. "Human emerging as butterfly." A shield design from the Tepatlaoztoc Codex. After *The Book of the Fourth World: Reading the Native Americas Through Their Literature*, by Gordon Brotherston.

that will arouse it from its slumber—just as the year 2012 is emerging from the collective unconscious as the end of time, or year of ascension, or the year of catastrophe, or the deadline for cleaning up the planet through the Kyoto Protocol. As we get nearer, there should be more outbreaks of telepathy and clairvoyance, like a pan of maize on a stove that starts with a few pops and speeds up until the whole pan goes crazy.

CONCLUSION

In our quest to investigate the meaning behind 2012, we have seen that the tribal elders are divided in their expectations of Earth changes at the end of the thirteen-baktun cycle. We have looked at catastrophic theories of axis shift, crustal displacement, rogue asteroids, and wandering planets, but most of these theories are unconvincing because of errors and inconsistencies. Paul LaViolette's galactic-core-explosion theory is quite convincing, but not firmly linked to 2012 . . . until we consider Weidner and Bridges's study of the Cross of Hendaye. The enciphered message on the cross seems not only to be warning of a cyclic catastrophe, but the alchemists behind the message have given clues that this event might also act as the trigger of an evolutionary leap to a new kind of human with a functional, shamanic, luminous body, or vehicle.

The magnetized-plasma-band theory of Dmitriev and Wilcock is also intriguing, explaining all the changes going on in the solar system, and also implying that there is a link to a forthcoming evolutionary leap and to the increase of "self-luminous formations" in the atmosphere, which some branches of science are on the verge of accepting as a new kind of life-form. Both these theories, of Weidner and Bridges and of Dmitriev and Wilcock, would allow an explanation of the 100,000-year solar cycle that now seems to be emerging, since it is about four cycles of precession in length. Both these theories are linked to the precession cycle, which is, of course, what is being measured by the Long Count calendar of the Maya.

A third fascinating idea emerges from the findings of Simon Shnoll, combined with John Major Jenkins's work, and implies that the galactic alignment process itself may cause biochemical or neurochemical changes in humanity, which may be part of a process of a mass awakening of kundalini.

Gregg Braden's observation about the falling geomagnetic-field strength being the sign of a forthcoming magnetic reversal is borne out by several scientists; there is also a trend for greater bursts of geomagnetic activity at the time of the solar maximums (and during the progressively more frequent and more intense solar storms), as discovered by Kev Peacock.

When we consider the findings of Michael Persinger and Serena Roney-Dougal, it seems that we can expect a general increase in telepathy, interspersed with more bouts of poltergeists, UFOs, and other paranormal phenomena that may correspond to increased output of pinoline or DMT, respectively. Rick Strassman's suggestion of a sudden switching-on of the *N*-methylating enzyme that controls DMT production is a similar scenario.

Persinger's findings regarding those suffering from TLE, along with the story of Michael O'Callaghan's mystical experience of oneness, suggest that the geomagnetic fluctuations could prompt a mass "schizophrenic" break for humanity, an inner apocalypse or revealing, in which our suppressed complexes will emerge in visible form. This may be a healing process and part of a near-death experience for Earth herself, as her megamind is congealed from its constituent human particles. There is a similar kabbalistic doctrine of the Heavenly Man, which says that throughout the cycles of history, certain individuals manage to achieve perfection and become part of an evolving megabeing: "This process is repeated, so says the tradition, through seven great epochs of evolution, until at the end there is to be found one Heavenly Man, which consciously mirrors the mind and consciousness of the Eternal, that mind and consciousness which is immanent in all manifestation, and which in the Qabalah is known as Adam Kadmon."[62]

In Persinger and Lafrenière's book *Space-Time Transients and Unusual Events,* they suggest the concept of a geopsyche.

> Our concept of the geopsyche essentially involves the interaction between large numbers of biological systems and the geomagnetic environment within which they are immersed. It contends that at certain critical numbers of biological units (of a species), a matrix is formed with the capacity to be energized by the intense geophysical forces of nature. When energized, this matrix acquires the potential to display behaviours and patterns of its own.[63]

The authors go on to explain how this transformation could be brought about, using the analogy of thin wires that are all in the same relatively small space but together constitute a cable.

> Now, suppose a moving magnetic field is applied to the entire population of the little wires. The result would be the induction of an electric current in each of those wires and the production of an electric field over the entire volume. Within limits, increasing the number of wires in the cable would amplify the intensity of the field. Although the electric field would in principle be produced by the individual wires, it would take on the properties of a matrix with properties of its own.
>
> In the principle of the geopsyche, the elements are not thin wires, but the electronic units of human organisms—the brains of individual people, millions of people. They are not packaged into cables, but are clustered into areas of high population density. The geopsyche emerges when a changing magnetic field is applied to the population of individual electronic units.[64]

In the geopsyche concept, assuming the "energizer" to be consistent with previous geomagnetic storms or electrical disturbances, Persinger and Lafrenière say the duration of the phenomenon would be anywhere between a few seconds to a week. In other words, it is only transient. This is not really what Teilhard de Chardin was talking about when he predicted a "planetized humanity" that has "a communication of mind and spirit that will make the phenomenon of telepathy, still sporadic and haphazard, both general and normal."[65] But if an unusually powerful energizer causes a permanent change by triggering mass electromagnetic hypersensitivity, or by turning on the *N*-methylating enzyme, then we have a model for the mechanism of transformation.

Although we have been unable to reach a definite conclusion at this point as to what we can expect around 2012, hopefully we have been able to sift out some of the misinformation and have increased our understanding, so that we can pool our knowledge more effectively and work together to manifest a golden age. In the words of Terence McKenna:

> We are the inheritors of a million years of striving for the unspeakable, and now with the engines of technology in our hands, we ought to be able to reach out and actually exteriorize the human soul at the end of time; invoke it into existence like a UFO and open the violet doorway into hyperspace and walk through it; out of profane history and into the world beyond the grave, beyond shamanism, beyond the end of history, into the Galactic Millennium that has beckoned to us for millions of years across space and time . . . This is the moment! A planet brings forth an opportunity like this only once in its lifetime and we are ready and we are poised and as a community we are ready to move into it, to claim it, to make it our own. It's there! Go for it! And thank you.[66]

Good luck, readers!

Afterword

PS: PUSHING UP THE CRAZIES

A fortune-teller is so sure he is going to die in 2012, he has bought himself a grave and a cross, inscribed with the date. The healthy 70-year-old claims to have had visions which allowed him to calculate the year of his demise. In 1991, Alexandru Marin [the fortune-teller] saw a "big star" in an unusual position and, a year later, a "strong light" shone in his house. "In 2012 we will see if I am right," he said. A priest said [Marin's] purchase [of the grave and the cross] had shocked residents in the village of Branistea, Romania.

LONDON METRO NEWSPAPER,
OCTOBER 3, 2001 (THANKS TO LAURENCE LUCAS
AND NEIL KILBURN)

Appendices & References

Appendix 1

TWO ACCOUNTS OF OUT-OF-BODY EXPERIENCES

This account of an out-of-body experience, by Robert Anton Wilson, appeared in his introduction to a book by David Jay Brown titled *Brainchild*. The experience was induced in 1986 by a prototype "brain machine" made by retired NASA engineer Mike Hercules. Hercules already had a machine on the market called Pulstar, and the new one was called Pulstar 2, but unfortunately it never went into production because Hercules died the following year. During Wilson's second "trip," in addition to Pulstar 2, Hercules hooked him up to a "second generation" electroencephalograph (EEG) that shows separate readings for all four types of brainwave—beta (waking), alpha (relaxed, light meditation), theta (dreaming, deep meditation), and delta (deep sleep). The Pulstar 2 had four electrodes that were attached to Wilson's head at the "right and left, front and back," and Wilson had control of the control knob.

He started sending electromagnetic waves of 6.5 hertz into his brain. He went rapidly into a state of deep meditation, which he recognized as being normally very difficult to attain. Hercules then took over the control knob, and turned the Pulstar 2 down to 4.0 hertz, and Wilson left his body and had "an interesting tour" of the Rocky Mountains.

He later heard that 4.0 hertz is the frequency used by Robert Monroe's Hemi-Sync tapes, which are designed to produce OBEs by "binaural beat" sounds. Hercules asked Wilson if he'd like to repeat the experience, with the electroencephalograph attached to his skull as well. Wilson agreed, and this is his account:

> Mike turned the frequency down to 4.0 hertz and then lower, and lower, and I went deep, deep into the quiet, serene place of hypnosis and then away from the bodily sensation-and-emotion circuits, and then up and out, over the Rockies, over the North Pole, down past Iceland to Ireland, to Dublin Bay, and I found the hill of Howth on the north rim of the bay and swooped gently down on the old street, curiously called The Haggard, where I was living at that time. I entered my house and found my wife reading in the living room and became very concerned with not doing anything to frighten her, and then I browsed a bit around Howth Hill and went back over the North Pole and found the Rockies and pursued the Continental Divide and found Boulder and Mike's house, and went back into my body in his laboratory.
>
> I was feeling as happy as a cat in a tuna factory. It had been a glorious trip, like Lilacs in the Sky with Delphiniums. Then Mike showed me the EEG for the period of my Far Journey. The lines were flat. My brain showed no beta, no alpha, no theta, no delta. Nothing. Void. Clinically, it looked like the graph of a dead brain. Whatever was going on during my flight over the pole, I was not using my imagination, which would have produced high delta and some beta, at least.[1]

Notice that, to get to Ireland from Boulder, Colorado, and to return again, Wilson went via the North Pole, suggesting an interaction between the astral body and the geomagnetic field.

The second account is from Terence McKenna's *True Hallucinations,* in which, while in Kathmandu, Nepal, in 1969, he produced an effect similar to Wilson's experience, but via exoegenous DMT, following an earlier dose of LSD, combined with a sexual encounter (this was not an uncommon kind of activity in Kathmandu in the 1960s). After finding himself among the "machine elves" (the morphing, multicolored beings mentioned in chapter 13, which told McKenna the laws of physics would change in 2012), at the moment when the whole vision would normally have faded, the scene changed.

> I suddenly found myself flying hundreds of miles above the Earth and in the company of silver disks. I could not tell how many. I was fixated on the spectacle of the Earth below and realized that I was moving south, apparently in polar orbit, over Siberia. Ahead of me I could see the Great Plain of Shang and the mass of the Himalayas rising up in front of the red-yellow waste of India. The Sun would rise in about two hours. In a series of telescoping leaps, I went from orbit to a point where I could specifically pick out the circular depression that is the Kathmandu Valley. Then, in the next leap, the valley filled my field of vision. I seemed to be approaching it at great speed. I could see the Hindu temple and the houses of Kathmandu, the Temple of Svayambhunath to the west of the city and the Stupa at Boudanath, gleaming white and a few miles to the east. Then Boudanath was a mandala of houses and circular streets filling my vision. Among the several hundred rooftops I found my own. In the next moment I slammed into my body and was refocused on the rooftop and the woman in front of me.[2]

Again, the astral traveler was moving in polar orbit, showing the relationship between the geomagnetic field and the astral body, and he was flying in the company of silver disks, giving the impression that not only do astral bodies have the appearance of UFOs, but as astral forms they are also subject to the geomagnetic field. This is all very relevant to the essay that follows, "The Astral Transition."

Appendix 2

THE ASTRAL TRANSITION

Clues from Egypt and Siberia Suggest a Shamanic Experience in 2012

ECDYSONE AND THE NONMOLECULAR BODY

As mentioned in chapter 21, Dr. Charles Musès, a.k.a. Musaios, suggested in his 1985 book, *The Lion Path,* that the Egyptians had developed a technology in which tones, lights, and a then unidentified plant (later identified as khat) were used to "open a rusty valve," or trigger the production of hormones similar to the ecdysone (ecdysterone) produced by larval forms of insects, which allows the adult form to emerge. In this way, they would encourage the gestation of a nonmolecular body that would allow the survival of consciousness beyond physical death. As already mentioned, every external mummy case has folded wings on it, in imitation of the embossed wings found on a chrysalis. The process was started well before death and completed during the seventy-day embalming period, connected to the seventy-day disappearance of Sirius.

Musès, a mathematician, philosopher, and computer scientist, was convinced that the synchronous perihelion (closest point of planetary orbit to the Sun) of Pluto, with the periastron (closest approach of two stars to each other) of Sirius B and Sirius A, which both occurred in 1994 and only happen concurrently every 90,000 years, allowed a flow of resonant energies, and the possibility of an evolutionary jump for those prepared to follow the clues left by the Egyptians. He produced a series of audiocassettes of sounds consisting of special frequency combinations, designed to be used at astrologically predetermined times while meditating in order to produce "the higher human analogue of the lepidopteran metamorphic hormone, ecdysterone." This would then "activate certain genes whose functions would otherwise remain inaccessible." The transformative energies would start slowing in January 1999 and stop by May 2000, and Musès says that those unable to complete their development before then would have to "reincarnate on the life-bearing planets of other solar systems that are on a later (and non-Plutonic) Meta-Cycle." He expected a culmination of the whole process on November 12, 2004, when he said that Earth changes would start.[1]

THE PYRAMID TEXTS

In William R. Fix's 1979 book *Star Maps,*[2] the author shows the correspondence between shamanic flight and the initiation rituals of Egypt. In the 4,300-year-old Pyramid Texts, on the walls of the Pyramid of Unas, it is stated repeatedly that "he is not dead, this Unas is not dead." In fact, Fix makes it plain that the reason why no bodies have ever been found in any pyramids—even those that were sealed—is because the pyramids were designed for initiation, to facilitate an OBE (out-of-body experience), in which the pharaoh would be gone for about three days, first orbiting the planet and then going to the circumpolar stars, to become a purified spirit (hence the term *astral projection*). Upon his return, the pharaoh, Unas, was told, "Put on thy body."

Shamans, too, traditionally fly to the polar stars. The axis mundi, or World Tree, which represents the Earth axis (and connects with the underworld below and heaven above), is shown on some of their drums as the route taken to the polar stars. Shamans usually employ drumming, fasting, and powerful plants to access the other planes, and typically experience dismemberment, where they are torn to pieces, then put back together again, as a kind of rebirth. This is exactly what happened to Osiris when his brother Set (plus seventy-two conspirators) tricked Osiris into getting into a coffin and then threw it into the Nile. The coffin became embedded in a tamarisk tree and was eventually used as a pillar in a palace. Isis found the coffin and hid it, but Set discovered it and cut Osiris into fourteen pieces. With the help of Thoth, Isis found all the pieces except one, the phallus, and reassembled Osiris. With Thoth's magic and a wooden phallus, Isis conceived Horus. Then Osiris ascended.

MYSTERY RELIGIONS

These shamanic themes formed the Osiris cult, then found their way into other mystery religions, such as the Greek mysteries of Dionysus at Eleusis, Attis in Asia Minor, Bacchus in Italy, and Mithras in Persia. The life stories of all these god-men involved a voluntary death, a flight from the body, a descent into the underworld, an ascent to heaven, then a resurrection. They also had a sacrament, which in later versions was wine but in the Mysteries of Mithras (or Mithraism, a mystery religion popular among the Roman military), "developed from older rites which used consecrated bread and water mixed with the intoxicating juice of a psychedelic plant called Haoma."[3] The various cultures all adapted one of their gods to take on the role of the resurrecting god-man. But when a Jewish sect wanted their own mystery religion, it was a bit more difficult, since they only had one god, so they based the mysteries around the figure of the Jewish Messiah, resulting in Christianity. Timothy Freke and Peter Gandy, in their book *The Jesus Mysteries,* have pointed out thirty correspondences between the Dionysus/Osiris Mysteries and Christianity. I have also found Mithraism/Christianity correspondences in *Pears Cyclopedia,* and these can be seen in the chart opposite.

DIONYSUS/OSIRIS MYSTERIES	MYSTERY OF MITHRAS (MITHRAISM)	MYSTERY OF JESUS (CHRISTIANITY)
1. Virgin birth of Son of God	Miraculous/Virgin birth of sun-god	Virgin birth of Son of God
2. Born in cave on December 25	Born on December 25—originally the winter solstice	Born in "stable" (original = "cave") on December 25
3. Crucifixion (or stuck to/in a tree) to atone for sins of humanity	Mithras sacrifices a bull to save humanity from sin	Crucifixion to atone for sins of humanity
4. Birth prophesied by star		Birth prophesied by star
5. Three shepherds visit the birth	"Magi" are astrologer-priests of Mithras	"Magi" (Matt.) or three shepherds (Luke) visit birth
6. Baptism of initiates	Baptism of initiates	Baptism of Jesus and initiates
7. Water into wine		Water into wine
8. Miracles		Miracles
9. Transfiguration		Transfiguration
10. Twelve disciples	Mithras surrounded by twelve signs of zodiac	Twelve disciples
11. Eating of bread and wine (= flesh and blood) to commune with the god	Initiates take "Holy Communion" of watered-down wine with bread or wafers embossed with the cross	Jesus and disciples share wine and bread; initiates take Holy Communion of wine and bread or wafers embossed with the cross
12. Death of savior redeems sins of the world	Death of bull redeems sins of the world	Death of savior redeems sins fo the world
13. Descends into Hell and resurrects after three days	Mithras (Sun) "dies" at winter solstice and is "reborn" after a three-day standstill	Descends into Hell and resurrects after three days
14. Ascends to Heaven to appear as judge at the end of time	Mithras waits in heaven for the End of Time, when he will return to Earth to awaken the dead and pass judgment	Jesus ascends to heaven for the End of Time, when he will return to Earth to awaken the dead and pass judgment
15. Three women followers visit empty tomb		Three women followers visit empty tomb
16.	Sunday is the holy day	Sunday is the holy day
17.	Use of bell, holy water, and candles	Use of bell, holy water, and candles

This chart illustrates the striking similarities between the stories of Dionysus, Mithras, and Jesus and the methods of worship adhered to by followers of the religions that sprang up around each of them.

PRECESSION ENCODED

Giorgio Santillana and Hertha von Dechend, in their 1969 book *Hamlet's Mill,* have shown that there has been knowledge of the precession of the equinoxes for millennia, and that it has been encoded into mythology all over the world. This concerns the slow movement of Earth's axis in a circle, which takes about 26,000 years to complete. Plato called this the Great Year of 25,920 years. The vernal (spring) equinox slowly moves through all twelve zodiac constellations over the 25,920 years, until it comes back to its starting point. Actually, it is the background stars that are moving—the equinox stays in roughly the same place. Each constellation takes 2,160 years to cross the equinoctial point, and it takes seventy-two years for each of the 360 degrees of the sky to rotate. The constellations also oscillate up and down over the precession. In *The Orion Mystery,* Robert Bauval and his coauthors have shown that the Egyptians were measuring this movement from the First Time of Osiris (Zep Tepi), which they put at 10,500 BC, when Orion was at its lowest point, to the Last Time of Osiris, which is coming up soon (half a precessional cycle later), when Orion reaches its highest position.

These precessional numbers, 72, 360, 2,160, 4,320 (2 × 2,160), and 25,920, have been found encoded into Egyptian myths, such as the one where Ra, upon discovering his wife's infidelity, said she was not to bear children on any of the 360 days of the year. Thoth intervened and played a game of draughts (senet) with the Moon, and won one seventy-secondth part of the Moon's light, creating the extra five epagomenal days, on which were born Isis, Osiris, Set, Nephthys, and Horus the elder (360/72 = 5). Encoded here is 72 × 360 = 25,920. Remember also the seventy-two conspirators of Set, who caused Osiris to become part of the tree!

Every year, the Egyptians held a festival for "Raising the Djed." The Djed was a pillar with four rings on it, which represented the backbone of Osiris, the tree he was entombed in, and the axis of Earth. In her book *Catastrophobia,* Barbara Hand Clow has pointed out that the relief showing the Raising of the Djed in the Temple at Abydos shows the Djed at an initial angle of "20 to 25° off vertical"—the same as the angle of tilt on Earth's axis. When measuring the image with a protractor, it is actually within a degree of the 23.5-degree tilt angle. Clow says that around 9500 BC, at the end of the ice age, the catastrophe pinpointed by Allan and Delair in *When the Earth Nearly Died,* when supernova debris passed close to Earth, caused the start of the precession cycle by knocking the axis away from vertical.

DJED = *AXIS MUNDI* = TREE OF LIFE = TREE OF KNOWLEDGE

Some Egyptian papyri show the World Tree administering spiritual nourishment, thereby combining the concepts of the World Tree, Tree of Life, and Tree of Knowledge. This connection is also emphasized on Stela 25 from Izapa, Mexico (see chapter 1, fig. 1.6), which shows "Hunahpu with Seven Macaw in his polar perch." The polar axis is shown looking very similar to a Djed pillar, with Seven Macaw representing the Big Dipper (Plow/Great Bear) constellation. In his book *Maya Cosmogenesis 2012,* John Major Jenkins has shown that the Maya were tracking precession, and that this is what is behind the Long Count calendar. The Crocodile Tree on the stela represents the Milky Way, and Hunahpu is shaking Seven Macaw off his perch, representing the precessional movement of the Big Dipper away from the celestial pole (and also signifying the change in preferred shamanic destination to visual Galactic Center, via the Crocodile/Caiman Tree). Notice that there is a snake wrapped around the two trees!

In Plaincourault, France, there is a mural on an old chapel showing the Tree of Knowledge with the snake wrapped around it, and Eve talking to the snake. The tree is a fly agaric mushroom. The fly agaric is, of course, a hallucinogenic, and John Allegro, who translated the Dead Sea Scrolls, went so far as to suggest that the early Christians used it as a sacrament. Bishop Jim Pike (a friend of sci-fi author Philip K. Dick) found the argument so convincing that he went to the desert near Qumran, on the Dead Sea, in search of fly agaric, but unfortunately died of dehydration while there. Gordon Wasson's book *Soma: Divine Mushroom of Immortality* identifies fly agaric with *soma,* the sacrament of the Hindus (meaning "body," which gives us the word *somatic*), but Ananda thinks that soma was like an alternative ayahuasca, made from Syrian rue and acacia.

TAKING THE PEE?

Every year at Christmas, we reenact our own mystery play, which has somehow survived to remind us of all this: December 25 used to be winter solstice. It is the birthday of the god-man Osiris/Dionysus/Mithras/Jesus, who, as Osiris, represents the polar axis (Djed) and its 72 × 360–year precession cycle. We now pretend that a wise old man named Santa Claus flies through the sky, pulled by reindeer (having frozen time, so he can visit everyone), and enters our home via the chimney. He is depicted as wearing a red and white outfit (though this is a more contemporary depiction, after St. Nicholas, bishop of Myra, who wore those colors; originally he was shown wearing green or other variations, though the gifts were always traditionally wrapped in red and white), leaving presents under a pine tree that we have brought indoors, or in socks hanging over the fireplace. Then he flies off to the North Pole.

In Siberia, since time immemorial, around winter solstice, shamans have gathered the red and white fly agaric mushrooms from under the very same pine trees we put in our living rooms, and distributed them from a sack. In Siberia, the entrances to homes are in the roof, and double as a smoke outlet, so they really are entering via the chimney. The mushrooms are left to dry out by hanging them over the hearth. The mushrooms have to be gathered quickly, because the reindeer love them. However, it is a fact that the psychoactive ingredient can be recycled several times, and unpleasant side effects such as vomiting are thus avoided. It is also a fact that the reindeer love eating yellow snow. If a reindeer herder wants to gather his herd, he only has to urinate, and they all come running.

That is why the reindeer all fly along in frozen time. Even the reindeer names reinforce the encoding.[4] The Christmas tree also represents the *axis mundi,* with the candles or lights and baubles representing stars and the star on the top representing the Pole Star. It is usually a five-pointed star, and therefore encodes seventy-two, since a pentacle is formed by 5 × 72-degree angles. Santa lives at the North Pole, which encodes the OBE destination, and Santa's helpers, the elves, are either the fourth-dimensional astral beings encountered or even our higher-dimensional selves—also represented by the angel/fairy on the tree.

What crazy kind of world is this? Is Jack's trip up the beanstalk to get the golden eggs another version? Is Humpty Dumpty an exploded planet?

WHAT DOES IT ALL MEAN?

As we have seen, Jenkins has shown that the Maya were tracking precession with the Long Count calendar, but rather than the vernal equinox, they were tracking the movement of stars (the Milky Way) against the winter solstice. He also showed that the Maya thirteen-baktun cycle (5,125 years) ends on winter solstice 2012. What is more, the Maya used to take hallucinogens, including toad venom and psilocybin mushrooms. Jenkins has also found some evidence that they may have taken *Amanita muscaria* (fly agaric),[5] but they mainly used the psilocybin mushroom. This contains a close relative of DMT, which is also produced in the pineal gland, or third eye, and which, when young, is shaped like a pine cone. It was after ingesting these mushrooms that the McKenna brothers were inspired to investigate the I Ching–encoded timewave that ends in 2012.

Some have suggested that the Egyptians also took psilocybin mushrooms. They did take the blue water lily, which, says Paul Devereux in

The Long Trip: A Prehistory of Psychedelia, "was the form in which Isis restored the murdered Osiris, and was thus a symbol for him."[6] The lily has hypnotic effects. Devereux says the Egyptians also had access to opium (which gave Errol Flynn an out-of-body experience) and khat, which produces a dreamlike state. Also, in the Americas, a different species of tobacco was taken, and in huge doses. A near-lethal dose can give an OBE. We have heard recently of the "cocaine mummies" found in Egypt (examination of mummified tissues showed traces of cocaine, which must have come from the coca plant, which grows only in South America). Well, they also contained large amounts of nicotine, another plant that should not have been in Egypt—very upsetting for Egyptologists!

Devereux has shown that most of these power plants can cause OBEs when taken in large enough doses, or in combination (though they can be very dangerous or unpleasant in some cases), and that they have been used for just such purposes for thousands of years. In another book, *Shamanism and the Mystery Lines,* Devereux makes a very strong case that these magical shamanic flights were associated with the alignments we know as ley lines, where ancient barrows, standing stones, stone circles, and even pre-Reformation churches (built on sacred sites) are aligned in straight lines.

Will the galactic winter solstice in 2012 allow a mass out-of-body experience? Will celestial and terrestrial grids align? Will a passing celestial body realign the *axis mundi* to its pre-Fall state? Or maybe the falling geomagnetic field will lead to a magnetic reversal and switch our polarity (OBE researcher Robert Monroe found that the body's polarity reverses while the consciousness is out of body). Maybe the solar megacycle or galactic alignment will trigger it . . . or perhaps it will be stimulated by the magnetized band of plasma through which the solar system is passing. Better start practicing those OBEs.

—First posted on Diagnosis 2012 website, August 2001

Appendix 3

CROPGNOSIS

Crop Circle Research from a Gnostic Perspective

In the "Plasmatic Communication from the Noosphere" section in chapter 20, we found that there may be a connection between Alexey Dmitriev's theory of our entry into a magnetized band of plasma and the mechanism behind crop formations as proposed by John Burke of BLT Research. It is a logical step from this position to wonder whether the effects observed in crop circles could be a foretaste of the physical effects that we will experience as the solar system becomes further embedded in the plasma band. So does this mean our kneecaps will explode like the crop-stem nodes? Let's hope that on a global, less concentrated level, the more severe effects will not be triggered—after all, Earth has passed through the plasma bands many times before without wiping out all life-forms (once every 25,920-year cycle of precession, according to Wilcock's theory, but there are arguments for other lengths of interval, such as four cycles of precession, around 100,000 years).

PLANT EFFECTS AS PRECURSOR AND METAPHOR OF HUMAN EFFECTS

It was while reading through the "Stunted, Malformed Seeds and Germination Effects" section of the "Plant Abnormalities" page[1] on the BLT website that the possibility of similar plasma effects on humans jumped out at me. BLT has discovered four levels of effects on seeds and germination, according to the stage of crop maturity at the time of exposure:

1. Immature crop (before flowering): the plant develops normally, but is totally devoid of any seeds.
2. Young crop where seed is forming: seeds will be smaller, with reduced germination.
3. More mature plants with fully formed embryo: seeds small, with no growth variability. All germinating seeds exhibit synchronized growth.
4. Mature plants with fully formed seeds: massive increase in growth rate and vigor. Seedlings can tolerate extreme stress without harm.

So if humanity is to be affected in a similar way in the near future, what exactly could this mean? There are several possible interpretations. Here is the first one that occurred to me:

If "crop maturity" represents spiritual preparation and seed germination represents spiritual transformation:

1. Those totally unprepared will be unable to make a dimensional transition.
2. Those who have started preparation will make a partial dimensional transition.
3. Those who have almost completed preparation will make a low-level transition to the next dimensional or evolutionary phase.

4. Those who have completed their spiritual development will make a complete transition to a comparatively godlike state.

The second interpretation. If "crop maturity" represents spiritual preparation and "seed germination " represents fertility:

1. Those totally unprepared will be totally sterilized.
2. Those who have started preparation will be partially sterilized. Reproduction may produce a race of mutants or dwarves.
3. Those who have almost completed preparation may reproduce a race of clones.
4. Those who have completed their spiritual development will produce a next generation that will be a race of gods.

As a possibly corroborative factor to what may seem like crazy speculation, above, the BLT "Other Facts" page[2] tells us that after visiting the 1996 Julia Set fractal crop formation at Stonehenge, "more than a dozen women reported menstrual abnormalities, the most unusual of which was the re-occurrence in several post-menopausal women of their previous normal menstrual cycles." As already mentioned, according to Lucy Pringle's experiments,[3] this is probably connected with an increased secretion of melatonin from the pineal gland that sometimes occurs in the crop-circle environment. So perhaps plasma can have effects on human reproduction. This pineal activation could also be involved in the godlike developments, with the rising of kundalini and the spiritual wedding that we covered in part 4 of this book.

THE SOWER

I hesitate to continue to the next part, since I am not an advocate of modern Christianity. However, it is now accepted by some researchers that there were inner teachings used by the early followers of The

Way, and that these early followers were actually Gnostics. Anyway, I remembered at this point the parable in the Bible called The Sower, so I located it in three versions: Matthew 13, Mark 4, and Luke 8. Incredibly, it is a parable of four kinds of seed germination:

1. Seeds that fell on the path were eaten by birds.
2. Seeds that fell on rocky ground had a brief germination, then withered.
3. Seeds that fell among thorns germinated, then grew a little, but became choked.
4. Seeds that fell on good soil grew to full size and produced a huge amount of grain—"some a hundredfold, some sixty, some thirty."

Even more incredibly, the parable is explained (for the benefit of the disciples, or inner gnostic brotherhood) as representing four types of spiritual development:

1. Those who misunderstand "the word"
2. Those who accept the word for a short time (until stressed)
3. Those who accept the word for a longer time (until distracted)
4. Those who accept the word and understand it, who bear fruit and yield a hundred-, sixty-, or thirtyfold.

Matthew 13 continues with another wheat parable, about weeds choking the crop, and then gets very sinister, talking about harvesttime, when the grain is collected into the barn and the weeds are burned. The harvest is "at the close of the age" and the reapers are angels. The theme is then expanded on in Revelation 14, the "Harvest of the World," which is very nasty. However, some Bible scholars think that the author added these parts to an original text (called proto-Matthew, or the Q-gospel).

The "Word," of course, is Logos, in Greek, and is a gnostic term that has been universally misunderstood, so perhaps a good place to start would be a website called The Outlawed Logos Teachings of Jesus, at www.gnosticchristianity.com,[4] and in particular the page "Details of the Five Stages of Consciousness."[5] This finds many models from psychology, mythology, paleontology, and several religions, all of which agree on the five stages of development in human society and in individuals as an evolution of consciousness, or "intellectual metamorphosis." The mass of humanity is stuck in stages two and three—stage five is, in Teilhard de Chardin's model, "'the convergence' or 'mega-synthesis' of hyper-personal consciousness, oriented toward God, the Omega Point." Also, the parable of the sower is seen from the same gnostic perspective as the very same five-stage model, where the fifth stage is a result of the "100-fold yield."

In summary, here are the five stages of gnostic consciousness:

Stage I refers to nonrational beings.
Stage II refers to rational beings without a conscience.
Stage III refers to rational beings with a conscience.
Stage IV refers to the expanded consciousness (enlightenment) achieved by using both judgmental and nonjudgmental laws of logic.
Stage V refers to an amplified form of consciousness in which the whole is more than the sum of the parts.

So concealed in the results of scientific research, we find a clue that the side effects in humans of forthcoming plasmatic exposure (as the solar system becomes more embedded in a magnetized beam of plasma) may depend on the evolved state of each individual at the time of maximum immersion. Those who are "ripe" will become fully functioning neurons in the planetary brain, as Gaia completes her gestation. This will allow all functioning human neurons to communicate telepathically, so ripeness would signify a readiness to have every thought telepathically broadcast. The timing of maximum immersion seems likely to be coincident with the end point of the

Maya thirteen-baktun cycle, which is a 260-katun macroscopic version of the 260-day tzolkin. While the tzolkin governs individual gestation, the thirteen-baktun cycle seems to govern planetary gestation, and the due date is winter solstice 2012.

—First posted at www.swirlednews.com, October 2002

NOTES

If you find that a Web page referenced here is no longer online, try looking in the Internet Archive at www.archive.org or http://web.archive.org.

CHAPTER 1. THE MAYA CALENDAR

1. Jenkins, *Tzolkin,* 110.
2. Aveni, *Stairway to the Stars,* 42.
3. Jenkins, *Tzolkin,* 110, 113.
4. Ibid., 111. Barbara and Dennis Tedlock were apprenticed to a daykeeper who confirms this.
5. Scofield, *Day-Signs,* 5.
6. Ibid.
7. Jenkins, *Maya Cosmogenesis 2012,* 328.
8. Ibid., 330.
9. Ibid.
10. Ibid., 212.
11. Ibid.
12. Brotherston, 299. As well as the Aztecs and Toltecs, the Zuni, Navaho, and some Incas say that we are in the Fifth Sun or World, while the Maya, Hopi, and some other Incas say we are in the Fourth Sun or World. Encoded into these two versions we have the precession of the equinoxes, where five cycles of 5,200 tuns (or 5,125 years) add up to 26,000 tuns (25,627 years), and a half cycle of the change in the obliquity of the ecliptic or axis tilt cycle, where four

cycles of 5,200 tuns (5,125 years) add to 20,800 tuns (20,500 years)—half of the 41,000-year cycle. www.diagnosis2012.co.uk/new4.htm#ages.

13. Quoted in Gilbert and Cotterell, *The Mayan Prophecies,* 70: ". . . an anonymous manuscript from 1558 entitled Leyenda de los Soles, seemingly derived from one or other of two earlier documents, the Chimalpopoca Codex and the Cuauhtitlan Annals . . ."
14. Séjourné, *Burning Water,* 138.
15. For example, José Argüelles's *The Transformative Vision,* 14: "For this fifth age, Ollin, means 'Earth-Shaking,' change or movement. It spans at the most a little more than a millennium, consisting of thirteen 52-year Heaven cycles, from 843 to 1519, and nine 52-year Hell cycles, from 1519 to 1987." The Cherokee webmaster and self-professed authority on Cherokee teachings, Dan Troxell, agrees.
16. Jenkins, *Tzolkin,* 288–89.

CHAPTER 2. TRIBAL PROPHECIES

1. http://web.archive.org/web/20030218184249/http://www.greatmystery.org/mayaspring2002.html; also http://web.archive.org/web/20030404171024/http://www.greatmystery.org/mayaspring2002initiation.html.
2. www.galacticmaya.com/English/Interview.htm.
3. http://latinola.com/story.php?story=144.
4. www.chiron-communications.com/communique%207-10.html.
5. See footnote on page 23. Barrios uses August 1987, taken from Shearer/Argüelles, but this is an error.
6. www.kachina.net/~alunajoy/timekeeper.html.
7. Skinner, *Millennium Prophecies,* 157. See also www.kachina.net/~alunajoy/awake.html.
8. Waters, *Book of the Hopi*, 30.
9. Ibid., 178.
10. Ibid., 11.
11. Ibid., 33.
12. Ibid., 408.
13. Quoted all over the Internet—here is an example: http://web.archive.org/web/20040601110224/http://seventhmoon3.tripod.com/hopi.htm.
14. Timms, *Prophecies to Take You into the Twenty-First Century,* 157.

15. Waters, *Book of the Hopi,* 408.
16. www.dreamscape.com/morgana/pan.htm.
17. http://nativenet.uthscsa.edu/archive/nl/9201/0113.html and http://nativenet.uthscsa.edu/archive/nl/9201/0113.html.
18. Waters, *Book of the Hopi,* 143–44.
19. Waters, *Mexico Mystique,* 272–73.
20. See Coe, *The Maya,* 149; and the 1971 Penguin paperback edition, 174.
21. Hope, *Ancient Egypt,* 191.
22. Temple, *The Sirius Mystery,* 324–25.
23. Ibid., 298.
24. http://web.archive.org/web/20040531200539/http://lightnews.org/December+98fifthworldofpeaceand1.htm.
25. www.amazon.com/exec/obidos/tg/detail/-/0892818042/qid=1080309050/sr=1-1/ref=sr_1_1/102-5288386-9508124?v=glance&s=books.
26. http://web.archive.org/web/20071118143811/http://trox11.tripod.com/snake/snake.htm.

    See also same web address with alternative endings: .com/native/wheels.htm and .com/native/calendar.htm.
27. www.2012unlimited.net/hitler.html.
28. Grandmother Twylah Nitsch is the author of several books: *Creature Teachers; Entering Into the Silence; The Seneca Way;* also coauthor of *Other Council Fires Were Here Before Ours: A Native American Creation Story as Retold by a Seneca Elder and Her Granddaughter.*
29. http://web.archive.org/web/20031210172812/http://alternativeapproaches.com/magick/amerind/amerind3.htm.
30. http://web.archive.org/web/20070220172012/http://www.livinglakes.org/stlucia/credomutwa.htm.
31. www.wam.umd.edu/~tlaloc/archastro/ae32.html.
32. http://web.archive.org/web/20030414030848/http://forums.abovetopsecret.com/viewthread.php?tid=558.

CHAPTER 3. ASIAN CALENDARS

1. Godwin, *Mystery Religions in the Ancient World*, quoted in Murry Hope, *Time,* 108–9.
2. Santillana and von Dechend, *Hamlet's Mill,* 242, 290, 409.

3. Jenkins, *Galactic Alignment,* 244–45.
4. Ibid., 268.
5. Ibid., 74–85.
6. Jenkins has 18.51 years on pp. 77, 273–74, of *Galactic Alignment* but this is incorrect. See www.alignment2012.com/Table%20of%20Contents-2.htm.
7. Jenkins, *Galactic Alignment,* 84–85.
8. Bernbaum, *The Way to Shambhala.* Edwin Bernbaum's book *The Way to Shambhala,* originally published in 1980 (reprinted in 2001), was my source for the Kalachakra information. It clarifies and completes various versions circulating on the Internet and in books that have only part of the story.
9. Argüelles, *The Transformative Vision,* 302.
10. Bernbaum, *The Way to Shambhala,* 83.
11. I used CyberSky for this, which can be downloaded as a free demo here: http://cybersky.com.
12. www.montclair.edu/RISA/d-kalki.html.
13. http://web.archive.org/web/20010418000605/http://www.skyboom.com/ddasa/index12.html.
14. http://web.archive.org/web/20030302215931/http://www.livinginjoy.com/gc/booklet_e.html.
15. Grasse, *Signs of the Times,* 17.

CHAPTER 4. I CHING: ANCIENT CHINESE LUNAR CALENDAR

1. The article is preserved on Sheliak's new website, as the last section, Standard, Revised, and Random Generated TimeWave Results, on this page: www.johnsheliak.com/SubSpace_Bridge_Domain/Sheliak_Formalization.pdf.
2. McKenna, Terence, and Dennis McKenna, *The Invisible Landscape,* 189; 1993 edition, 196.
3. Ibid.
4. Ibid.
5. http://groups.yahoo.com/group/timewavezero2012/.
6. Michell, *The View Over Atlantis,* 110–11.
7. To see an animation of the ways that the Franklin Magic Square adds to 260, see: http://grapevine.net.au/~grunwald/une/KLAs/maths/Benfranklin_magicsquare.html.

8. Peter Meyer became disillusioned with the Timewave for a while, but as a result of the Huang Ti–9/11 discovery, he now has renewed his interest and has set up a new website from which it can be purchased: www.hermetic.ch/frt/frt.htm.

CHAPTER 5. ABRAHAMIC RELIGIONS

1. www.magicmusicmyth.com.
2. www.godsweb.com.
3. http://web.archive.org/web/20041027102824/http://www.geocities.com/bob_hunter/jesusreturn.html.
4. http://web.archive.org/web/20030713224114/http://www.jvim.com/catalog/leftbehind.html.
5. http://web.archive.org/web/20070322234707/http://www.siscom.net/~direct/revelation/. The book of Revelations website, now available as a book, *The Final Days of Mankind.* See Howard, *The Final Days of Mankind.*
6. Daniel 9:24.
7. See also Psalms 90:4 and Ezekiel 4:6.
8. www.be-ready.org/when.html.
9. Matthew 24:32–35, Mark 13:28–31, Luke 21:29–33.
10. Luke 21:29–33.
11. Psalms 90:10.
12. Revelation 11:2, 13:5.
13. Revelation 12:14.
14. Daniel 7:25, 12:7.
15. Revelation 12:6, 11:3. The trampling in Daniel is 2,300 "days," which would seem to be affected by the day-equals-year rule, but in Revelation it is 1,260 days, whereas the 1,260-day figure in Daniel seems to refer to actual days—three and a half years—so we can presume that the five instances of 1,260 days in Revelation also refer to actual days, two of these periods adding to the seven-year period usually called the Tribulation.
16. www.authorsden.com/visit/msgboard.asp?AuthorID=10788.
17. www.christiantrumpetsounding.com/7_yr_come_again.htm. Aparently completely in ignorance of the other connections, October 4, 2005, is listed on a web page called "Pre-wrath rapture" as being 14,000 days after the Israeli recapture of Jerusalem on June 7 1967, where 14,000 days is also the interval between the

triumphal entry of Christ to Jerusalem (April 6, 32 AD) and the destruction of Jerusalem by Rome (August 5, 70 AD). The reliability of the dates is open to discussion. www.whatsaiththescripture.com/Timeline/IIndex.Entries/16C.html.

18. http://web.archive.org/web/20031004022228/http://aplus-software.com/thglory/7000year.htm.
19. http://victorian.fortunecity.com/byzantium/56/y6k.html.
20. http://web.archive.org/web/20030405022140/http://aplus-software.com/thglory/7000year.htm. The fourth part of the article is not linked to it, but can be found here:http://web.archive.org/web/20021230192522/aplus-software.com/thglory/rapture5.htm.
21. www.teshuvah.com/#T.
22. www.sichosinenglish.org/books/timeless-patterns/06.htm.
23. http://weeklywire.com/ww/03-15-99/austin_xtra_feature2.html. John Whalen's original article: http://web.archive.org/web/19990203120253/http://www.sfweekly.com/extra/apocalypse.html. Abdal Hakim Murad's article: www.sunnah.org/audio/millen.htm.
24. http://thetruth.hypermart.net/wrath/13_when_shall_it_be.htm.
25. Keter Publishing House Ltd.: Israel, 1973; now available as a CD–ROM.
26. Alford, *Gods of the New Millennium,* 237. Alford refers to a papyrus in the Cairo Museum, ref. 30646, that was found in a tomb in Thebes and mentions a "Game of 52" that Thoth challenged mortals to play, and which they usually lost. Alford sourced this information from Zecharia Sitchen's *The Lost Realms* (Sitchen, *The Lost Realms,* 82–83).
27. http://exodus2006.com.

CHAPTER 6. SUNSPOT CYCLES

1. www.alignment2012.com/mproph.htm.
2. www.alignment2012.com/mproph.htm.
3. www.gsfc.nasa.gov/gsfc/spacesci/sunearth/solarmax.htm.
4. The e-book *Something Is Wrong with Our Sun* is now available at Kev Peacock's website, http://users.spin.net.au/~aeutiwer/CDBook. The original solar magnetic fields theory page is here: http//web.archive.org/web/20020202170326/http://people.mail2me.com.au/~vk4vkd/Geomagnetic.htm.
5. Jazz Rasool's website: www.bodymind.co.uk/.

6. Telephone conversation January 2, 2004.
7. Poynder, *Pi in the Sky,* 131.
8. Jenkins, *Maya Cosmogenesis 2012,* 366, and www.alignment2012.com/mproph.htm.
9. Brennan, *The Stones of Time,* 7.
10. *Kindred Spirit,* vol. 2, no. 1 (1990), 24; at www.kindredspirit.co.uk.

CHAPTER 7. ASTRONOMICAL CLAIMS

1. Mardyks and Alana-Leah, *Maya Calendar,* 106–07.
2. www.fieldwerks.com/galactic_alignment.htm.
3. http://edj.net/mc2012/truezone.htm.
4. http://hem.passagen.se/alkemi/gaia2012.htm.
5. http://sunearth.gsfc.nasa.gov/eclipse/eclipse.html.
6. www.diagnosis2012.co.uk/5thsun.htm.
7. You can download the software at: http://cybersky.com.
8. Velikovsky, *Worlds in Collision;* and Bourbourg, *Histoire des Nations Civilisées du Mexique,* I, 120, 311, paraphrased in Velikovsky, 158–59.
9. Jenkins's full, unabridged review of *The Mayan Prophecies* can be found at www.alignment2012.com/mproph.htm.
10. Download shareware online at www.angelfire.com/zine/meso/meso/skyglo36.zip.
11. Excellent free Maya calendar software can be downloaded from http://members.shaw.ca/mjfinley/Maya.htm. It is not obvious from most astronomical software screen shots when exactly Venus becomes visible/invisible due to solar glare—this is when you need the Burden of Time software—but don't forget to set the start-date to August 11 every time you use it.
12. Appleby, *Hall of the Gods,* 211.
13. Ibid., 60. Appleby is quoting the *Washington Post* of December 30, 1983. You can see a photocopy of the newspaper clipping at www.enterprisemission.com/images/pulsar-Wash-post1.gif. See Appleby, *Hall of the Gods,* 61, for the *Newsweek* July 13, 1987, quote.
14. Appleby, *Hall of the Gods,* 58.
15. Hope, *Time,* 221.
16. Temple, *The Sirius Mystery* (1998), 239.
17. Santillana and von Dechend, *Hamlet's Mill,* 430–37.

18. http://nibiruancouncil.com/html/nibirureturnarticle.html.
19. Clow, *The Pleiadian Agenda*, 30.
20. www.yowusa.com/authors/jvdworp/0300/phobos/phobos1.html.
21. www.yowusa.com/authors/jvdworp/0300/phobos/phobos2.html.
22. http://groups.yahoo.com/group/astro-revelation/message/1265.
23. Rudisill, *E-Din*, 210.
24. Ibid., 282–86.
25. Ibid., 344.
26. Ibid., 267.
27. Ibid., 268.
28. Ibid., 329. Something the author overlooked is the possible connection to the forty-nine-year Sigui period of the Dogon (based around the orbit of Sirius B), whom Robert Temple traced to Egypt, and who also base their calendar on Sirius, where the heliacal rise (with the Sun) of Sirius coincides with the annual flooding of the Nile.
29. Ibid., 328.
30. Ibid., 329.
31. Ibid., 344.
32. Ibid., 333.
33. http://web.archive.org/web/20050216172224/http://www.edinnet.com/CommentsonDiagnosis2012.htm.
34. Ibid., 285.
35. Ibid., 329.
36. http://web.archive.org/web/20011215071914/2near.com/edge/maya/hunter.html.
37. http://web.archive.org/web/20041010063106/http://www.2near.com/edge/maya/hunter-2.html.
38. Keith Hunter, e-mail to author, September 20, 2003.
39. Ibid.
40. A more thorough critique, including the figures with seven decimal places, can be found at www.diagnosis2012.co.uk/new3.htm#kh.
41. LaViolette, *Earth Under Fire*, 301.
42. Ibid., 314.
43. Ibid., 107.
44. Ibid., 187.

45. *Galactic Tsunami,* by David L. Souers, at http://members.tripod.com/galactic tsunami/chapbychap.html.

46. http://web.archive.org/web/20021009094440/http://therealmidori.tripod .com/disaster.html.

CHAPTER 8. ALTERNATIVE ARCHAEOLOGY

1. Collins, *Gods of Eden,* 209.
2. Ibid., 200.
3. Bauval and Gilbert, *The Orion Mystery.*
4. *The Return of the Phoenix* was self-published in 1997, but can be found in updated form online at: www.michaelmandeville.com/phoenix/phoenix.htm.
5. Appleby, *Hall of the Gods,* 360.
6. www.michaelmandeville.com/phoenix/phoenix.htm.
7. West, *Serpent in the Sky.*
8. In *Hall of the Gods,* Appleby gets the dates right on p. 362, but on pp. 363–64 he says 12,500 BC, when he meant 12,500 BP—Before Present.
9. Appleby, *Hall of the Gods,* 296.
10. Bauval and Hancock, *Keeper of Genesis,* 283.
11. Ibid., 70.
12. Appleby, *Hall of the Gods,* 317. This is when he says the causeway of the Pyramid of Khafre was pointing to cross-quarter sunrise.
13. Ibid., 359.
14. Jenkins, *Maya Cosmogenesis 2012,* 76.
15. Ibid., 81–87.
16. Collins, *Gateway to Atlantis,* 354.
17. Ibid., 355.
18. Ibid.
19. Ibid., 354.
20. Dick, *Valis. Valis* means "Vast Active Living Intelligence System." See his appendix (p. 229 in the Vintage edition).
21. Found at Raymond Mardyks's website, Galactic Astrology Centre (www .geocities.com/Area51/Nebula/9172/gac.html), which is no longer available. This website was accessed between 1997 and 1999. The archived site was accessed until April 2007, when it was blocked.

22. www.rense.com/ufo5/undergiza.htm. This page on Rense.com is the first version, as far as I know. No references are given for this statement, but I have traced it to a book called *Tracing Our White Ancestors,* by Frederick Haberman, pp. 22–23, which in turn cites source statement from a book called *History of Phoenicia,* by George Rawlinson. The Haberman paragraph is quoted online here (but I have corrected the title of Rawlinson's book): "Let us now analyse the word 'Phoenician' and 'Phoenicia.' Professor George Rawlinson, in *History of Phoenicia,* tells us that Phoenicia derived its name from the forests of date or Phoenix palms which grew there in great luxuriance. So far so good; but whence did the Phoenix palm derive its name? Horapollo says: 'A palm branch was the symbol of the Phoenix.' Yes, but what or who was the Phoenix? Sanchoniathon, the Phoenician writer, states that 'Phoenix was the first Phoenician.' Phoenix, then, was a man. Now, the word *Phoenix* is the Greek form of the Egyptian term *Pa-Hanok,* the house of Enoch. In Hebrew, Enoch also is Hanok. Thus the mystery of that ancient race is solved: they were the sons and descendants of Enoch and of Noah and his three sons, who after the Flood started their westward march." http://web.archive.org/web/20031230194238/http://www.childrenofyahweh.com/Teaching+Letters/letter_27.htm
23. Hurtak, *The Keys of Enoch,* p. xiii. Hurtak confirms that he saw the next "great cycle" starting in the year 2004. Hurtak, *The Keys of Enoch,* 549.
24. Appleby, *Hall of the Gods,* 360.
25. http://web.archive.org/web/20000511015737/http://cob250.dn.net/members/panspermia/quickening.html.
26. Sitchin, *Genesis Revisited,* 328. "It can be simply illustrated that when the planet's perihelion was in Cancer, its first appearance had to be from the direction of Sagittarius."
27. http://ambilac-uk.tripod.com/index.html.
28. See http://web.archive.org/web/20000614172854/www.freeyellow.com/members7/howardmj/page1.html.
29. See the article "Crystal Clear" (if only it was!) at http://web.archive.org/web/20000615025422/www.freeyellow.com/members7/howardmj/page5.html.
30. Lehner, *The Complete Pyramids,* 108.
31. www.siloam.net/jenkins/ambiliac.html.
32. Ibid.
33. Weidner and Bridges, *A Monument to the End of Time,* 177.

34. http://portalmarket.com/timms.html.
35. Weidner and Bridges, *A Monument to the End of Time,* 164.
36. Ibid., *A Monument to the End of Time,* 196.
37. Ibid., 231.
38. In *Stalking the Wild Pendulum,* Bantam edition, 1979–81, pp. 33–36 and 54–56, Bentov explains that in meditative states the breathing is so gentle that it no longer interferes with the feedback of the aorta, allowing the system to become resonant so that the whole body resonates at the same frequency as Earth, 7.8 hertz. How this is affected by the increasing rate of Schumann resonance remains to be researched.
39. For more on this, see the essay by Vincent Bridges, "Star Birth Bardo in the Body of Orion": www.sangraal.com/bardorion.htm.
40. Weidner and Bridges, *A Monument to the End of Time,* 234.
41. http://vincentbridges.com/?s=raising+the+djed.
42. http://web.archive.org/web/20030429113301/http://www.vincentbridges.com/millennium/update.html.
43. Revelation 21:16.
44. Weidner and Bridges, *A Monument to the End of Time.* See appendix G, pp. 274–79, and www.sangraal.com/library/gnomon.htm.
45. http://web.archive.org/web/20030429113301/http://www.vincentbridges.com/millennium/update.html.
46. Ibid.
47. http://web.archive.org/web/20030609053833/http://vincentbridges.com/update.html.
48. Ian Crane, e-mail to author, September 26, 2002. Ian said he was positioned approximately one hundred meters due east of the paws of the Sphinx at dawn on September 22, 23, and 24, from 5.30 AM, and saw nobody on the causeway up to the Great Pyramid, or in front of the Sphinx, so he assumes that the Nicki Scully group was positioned on the plateau. He added, "To be on the plateau at that hour of the morning would have either required a permit from the Egyptian Antiquities Department (Zahi Hawass and co), or necessitated a large amount of baksheesh to have changed hands!"
49. Gilbert, *Signs in the Sky,* 262.
50. Ibid., 265. Or 2500 AD—the date of its highest altitude as stated in *The Orion Mystery.* Both these dates were calculated with SkyGlobe software. To

see why they are wrong and why the end time is now, see the second footnote at the bottom of page 128.

51. Gilbert, *Signs in the Sky*, 232. See Matthew 22 for the wedding garment parable, and also Revelation 19:7–9 for the Wedding of the Lamb.
52. Jenkins, *Maya Copsmogenesis 2012*, 389. Astronomers vary between 1998 and 1999 in their calculations of the exact time of the galactic alignment, so the end of the process varies between 2016 and 2017, with 2018 being the first definite year when the process is over.
53. Gilbert, *Signs in the Sky*, 269–70.
54. www.diagnosis2012.co.uk/orp.htm.
55. http://ourworld.compuserve.com/homepages/dp5/pole1.htm.

CHAPTER 9. THE GEOMAGNETIC FIELD

1. Braden, *Awakening to Zero Point*, 19.
2. Various scientific websites confirm these claims of a declining magnetic field, though they differ in details, e.g., www.britannica.com/bcom/eb/article/6/0,5716,108976+16+106190,00.html, which says the average time between reversals is 300,000 to a million years, and takes 5,000 years to reverse, but it does shrink to zero during reversal, allowing entrance of cosmic rays that would cause mutations

   The Answer Man (http://web.archive.org/web/20010108211800/http://www-spof.gsfc.nasa.gov/Education/FAQ.html) says the average time between geomagnetic reversals is every 500,000 years, and the last one was 700,000 years ago, and that the field strength has dropped 8 percent in the last 150 years.

   Ask the Space Scientist (http://image.gsfc.nasa.gov/poetry/ask/amag.html) says it reverses every 250,000 years, is decreasing at 0.07 percent per year, and will be at zero in 4,000 years; the reversal can take less than 100 years, and the last one was 250,000 years ago; *American Scientist* (http://web.archive.org/web/19980224015411/http://www.sigmaxi.org/amsci/articles/96articles/Fuller.html) says it was 780,000 years ago; *Science Frontiers* (www.science-frontiers.com/sf041/sf041p13.htm and www.science-frontiers.com/sf101/sf101g11.htm) says the Steens mountain volcanic record indicates that a 180-degree reversal took 4,500 years, but included three periods of very rapid change; in *Earth Under Fire*, Dr. Paul LaViolette says that the Steens mountain information shows that a reversal would take only two months; geologist Andrew P. Roberts (http://earth.agu.org/revgeophys/robera01/robera01.html) suggests that confusion arises because people fail to differentiate between "polarity transitions and excursions of the geomagnetic field." More recently, sources have been agreeing that the last reversal was 780,000 years ago, that the average time between reversals is 250,000 years, and that at the current rate it will be 1,000–2,000 years before a reversal, though some sources now admit that certain researchers say it could happen in a few weeks. The latter would be preferable, since that means we would not be without our protection from the solar wind for long.

3. Braden, *Awakening to Zero Point,* 51. The Merkabah was the divine vehicle of light described by the Hebrew prophet Ezekiel as having the appearance of "wheels within wheels," and made its first appearance in New Age literature in 1973, in J. J. Hurtak's *The Keys of Enoch,* as combinations of interlocking pyramids: bipyramids or tripyramids, for example. Hurtak describes an Order of Melchizedek, who administer to Man through the Merkabah in order to prepare humankind for the "fear-inspiring day of Jehovah, when the Sons of Light will appear" (p. 590). Merkabah meditation has now become a widespread practice, but in April 2000 Hurtak said in an interview that "sacred teaching like the Merkabah . . . is not given to those who wish 'instant nirvana' nor experiences based on multi-level marketing. In order to work universally through Merkabah, it requires the rigors of inner development, for some, over a whole lifetime, or many lifetimes, and is NOT in the realm of simple meditation and dream time experiences, which operate on a global as opposed to universal level" (www.lightnet.co.uk/informer/interviews/hurtak0004.htm).

Hurtak's *The Keys of Enoch* describes five bodies that must be fully developed before the Christ body, or Overself body, can be worn. Only then can a Merkabah take his "physical body and consciousness body" off the planet for the next phase. See Hurtak, *The Keys of Enoch,* 446 for an explanation of the interrelationship of the five bodies, Overself body and Merkabah. Bernard Perona, who changed his name to Drunvalo Melchizedek, claims to be the current authority on the "Merkaba," as he spells it, and it is his version to which Braden is referring, but Melchizedek sent out a "sequel to an open letter" in January 2003 that said, "The Melchizedek tradition does not teach, nor has it ever, that we spin the top tetrahedron one way and the bottom one the other way. Anyone who has studied this work carefully knows this is not true. This is wrong, and I clearly say so."

4. Braden, *Awakening to Zero Point,* 32.
5. There is some evidence that the pyramid interacts with the geomagnetic field, since there is a patent, registered in 1949 (www.amasci.com/freenrg/tors/drbl.html), with the patent office in Prague, Czechoslovakia, for a cardboard replica of the Great Pyramid that will sharpen razor blades if they are put inside with the sharp edge facing east–west, while the pyramid is aligned north–south. However, this suggests an enhanced magnetic field rather than a reduced one. There are rumors that magnetometer readings in the King's Chamber of the Great Pyramid show a higher than background level (www.sacredsites.com/africa/egypt/great_pyramid.html) but I have been unable to trace the source of this information as yet. There is, however, a very high radiation measurement in the King's Chamber, since the granite walls under the weight of all the limestone exude radon gas, giving it a similar saturation to some stone chambers in Cornwall, England (Devereux, *Places of Power,* 194–97). Dr. Paul Brunton experienced a stunning spontaneous projection of the astral body when he spent the night in the King's Chamber (see Brunton, *A Search in Secret Egypt,* 57–78). Paul Devereux, who coordinated the Dragon Project, a study of energies at sacred sites, confirms that at the Rollright stone circle in Oxfordshire, England, there was "some evidence growing of a fluctuating field within the circle relative to outside, changing over a period of hours." Devereux, *Places of Power,* 91.
6. Hutchison, *Mega Brain Power.*
7. http://ebeltz.net/classes/histgeo3.html.
8. http://web.archive.org/web/20040306161359/http://www.telegraph.co.uk/connected/main.jhml?xml=/connected/2002/04/17/ecfhead.xml.
9. http://observer.guardian.co.uk/international/story/0,6903,837058,00.html.
10. Ibid.
11. http://news.bbc.co.uk/1/hi/sci/tech/2889127.stm.
12. LaViolette, *Earth Under Fire,* 92.
13. J. R. Petit, J. Jouzel, et al., "Climate and atmospheric history of the past 420,000 years from the Vostok ice core in Antarctica," *Nature* 399 (June 3, pp. 429–36, 1999); www.nature.com/nature/journal/v399/n6735/pdf/399429a0.pdf.
14. www.aist.go.jp/aiste/latestresearch/2002/20020329/20020329.html.

15. Published in the June 10 issue of *Earth and Planetary Science Letters* (Elsevier, volume 199, issues 3–4).
16. www.eurekalert.org/pub_releases/2002-06/dc-1c060602.php.

CHAPTER 10. ICE CYCLES

1. Gaspar, *The Celestial Clock,* 80.
2. Ibid.
3. www.homepage.montana.edu/~geol445/hyperglac/time1/milankov.htm.
4. Ibid.
5. www.ngdc.noaa.gov/paleo/milankovitch.html.
6. http://aa.usno.navy.mil/faq/docs/seasons_orbit.php.
7. Ibid.
8. There is a good explanation of the Milankovitch cycles at: http://web.archive.org/web/20030314175647/http://www.d.umn.edu/~hmooers/geology/GlacialGeology/Notes2.pdf.
9. www.atmos.washington.edu/1998Q4/211/project2/moana.htm.
10. The title of the article is "New evidence for imminent change in global temperature patterns," by Ernest C. Njau: www.sciencedirect.com.

CHAPTER 11. O.O.O.O.O

1. Published in the *Express* on March 13, 1999. http://web.archive.org/web/20030421005031/http://www.nunki.net/PerDud/TheWorks/Express/MountainArk.html.
2. Ibid.
3. Brotherston, *The Book of the Fourth World,* 115.
4. Furlong's article (posted on his website) "What on Earth Was Happening in 3100 BC?" sums up the cogent points: www.kch42.dial.pipex.com/3100bc.htm.
5. http://personal.eunet.fi/pp/tilmari/.
6. Baillie, *Exodus to Arthur,* 178.
7. Ibid., 179.
8. http://abob.libs.uga.edu/bobk/ccc/ce102103.html.
9. http://abob.libs.uga.edu/bobk/ccc/cc111003.html. The quoted report about the work of Dr. Lonnie Thompson, of Ohio State University, can be found at http://researchnews.osu.edu/archive/quelcoro.htm.

10. Hope, *Ancient Egypt,* 221.
11. Ibid., 21.
12. Keene has three articles about these coincidences: "Cosmic Coincidences" (www.diagnosis2012.co.uk/jjk.htm); "Celestially Challenged" (www.diag nosis2012.co.uk/jjk2.htm); and "Continually Coordinating Circumstances" (www.authorsden.com/visit/viewshortstory.asp?id=11420&AuthorID=10788).

CHAPTER 12. FRINGE SCIENCE

1. www.ssrsi.org/Onsite/bobbuck.htm.
2. www.timstouse.com/EarthChanges/shiftoftheages.htm.
3. Wilcock to Stray, e-mail, June 11, 2001.
4. www.divinecosmos.com/index.php?option=com_content&task=category&sectionid=6&id=19&Itemid=36.
5. Wilcock to Stray, e-mail, June 11, 2001.
6. http://divinecosmos.com/index.php?option=com_content&task=view&id=156&Itemid=48.
7. http://divinecosmos.com/index.php?option=com_content&task=category&sectionid=6&id=20&Itemid=36.
8. Ibid.
9. 1997, updated in 1998. www.tmgnow.com/repository/global/planetophysical.html.
10. http://news.bbc.co.uk/2/hi/science/nature/3052467.stm.
11. http://divinecosmos.com/index.php?option=com_content&task=view&id=102&Itemid=36.
12. http://divinecosmos.com/index.php?option=com_content&task=view&id=47&Itemid=30.
13. www.tmgnow.com/repository/global/planetophysical1.html.
14. "Some lines of investigations on properties of self-luminous formations using a model of non-homogenous physical domains" (www.tmgnow.com/repository/solar/Dmitriev.html). See also: "Electrogravidynamic concept of tornadoes" (www.tmgnow.com/repository/planetary/tornado.html) and "Planetophysical function of vacuum domains" (www.tmgnow.com/reposi tory/planetary/pfvd.html).
15. www.tmgnow.com/repository/global/planetophysical4.html.

16. Professor James McCanney and his plasma discharge comet model, which predicts the arrival of Planet X "sometime in the next ten years"; in other words, 2002–2013; www.jmccanneyscience.com.
17. Professor Sergey Smelyakov: www.diagnosis2012.co.uk/sme3.htm.
18. www.goldennumber.net.
19. Wilcock to Stray, e-mail, May 28, 2002.
20. www.divinecosmos.com/index.php?option=com_content&task=view&id=45&Itemid=30.
21. http://cura.free.fr/xx/20smely2.html.
22. Smelyakov to Stray, e-mail, May 16, 2002.
23. Smelyakov to Stray, e-mail, May 19, 2002.
24. Smelyakov to Stray, e-mail, May 22, 2002.
25. Smelyakov to Stray, e-mail, May 23, 2002.
26. Smelyakov e-mail, June 1, 2002. I am not trying to boast that I outthought a mathematics professor, just pointing out that he agreed that I had a good point, for reasons that become clear further on in the chapter. I have more recently reread Karpenko and Smelyakov's paper and found that on pp. 31–32 they remark that in their projection of Chinese population levels, which they use to cross-reference the Auric Time Scale, the levels theoretically spiral to infinity between 2036 and 2037. They point out that their calculation of the exact date of this—April 7, 2036—has no real meaning, since their calculation was just an approximation.
27. www.divinecosmos.com/index.php?option=com_content&task=view&id=45&Itemid=30.
28. Russell, *The White Hole in Time,* 200.
29. Ibid., 202–3, and note.
30. Russell, 205.
31. Robert Anton Wilson, *Right Where You Are Sitting Now,* 29–32.
32. Robert Anton Wilson, *Trajectories Newsletter.*
33. Ibid., 17.
34. Robert Anton Wilson, *Prometheus Rising,* 240.
35. Robert Anton Wilson, *Trajectories Newsletter,* 18.
36. Atwater, *Future Memory,* 215.
37. Carey, *Starseed,* 125.
38. Ibid.
39. Carey, *Vision,* 67.

40. Carey, *Starseed,* 128.
41. Carey, *Vision,* 62.
42. Ibid., 54.
43. Carey, *The Starseed Transmissions,* 22.
44. Ibid., and also *Starseed,* 47.
45. Carey, *The Starseed Transmissions,* 22; Carey, *Vision,* 28, 67; Carey, *Starseed,* 47, 159.
46. Carey, *Starseed,* 127.

CHAPTER 13. SHAMANISM

1. McKenna, "World Youth Culture, Shamanism, Gaia, Chaos, and Plant Hallucinogens" (lecture).
2. http://web.archive.org/web/20021021082948/http://users.skynet.be/light-workers-tempel/engvisionfeb98.html.
3. Unborn Mind to Stray, e-mail, June 3, 2004.
4. Called adrenoglomerulotropine. McKenna, *True Hallucinations,* 65.
5. 6-methoxytetrahydroharman. Naranjo, "Psychotrophic properties of the harmala alkaloids," quoted in McKenna, ibid.
6. These are: tetrahydro-betacarboline; 6-methoxy-tetrahydro-betacarboline (McKenna, *The Invisible Landscape,* 1993, p. 97); and a possible third one, which has been synthesized from melatonin in vitro, 6-methoxyharmalan (Naranjo, "Psychotropic properties of the harmala alkaloids," quoted in Terence and Dennis McKenna, *The Invisible Landscape,* 65), which has also been said to be formed from serotonin. Roney-Dougal, *Where Science and Magic Meet,* p. 101, quoting McIsaac, Khairallah, and Page, "10-methoxy harmalan, a potent serotonin antagonist which affects conditioned behavior."
7. Roney-Dougal, *Where Science and Magic Meet,* 101. Pinoline is the second compound in note 6, above: 6-MeOTHBC for short.
8. R. B. Guichhait, "Biogenesis of 5-methoxy-N, N-dimethyltryptamine in human pineal gland," 187–90, quoted by Ananda at www.akasha.de/~aton/NeoDMT.html.
9. www.diagnosis2012.co.uk/dmt.htm.
10. Caroline Taylor's review of *DMT: The Spirit Molecule:* www.diagnosis2012.co.uk/dmt.htm.

11. www.diagnosis2012.co.uk/dmt.htm.
12. Roney-Dougal, *Where Science and Magic Meet,* 100.
13. Ibid., 139.
14. Ibid.
15. Ibid.
16. Devereux, *Places of Power,* 129–30.
17. Ibid., 63.
18. Roney-Dougal, *Where Science and Magic Meet,* 140–41.
19. Ibid., 140–42.
20. www.akasha.de/~aton/NeoDMT.html.
21. http://phoenix.akasha.de/~aton/Unidance.html.
22. http://phoenix.akasha.de/~aton/CV.html.
23. www.akasha.de/~aton/NeoDMT.html.
24. Ibid.
25. Callaway, "A proposed mechanism for the visions of dream sleep."
26. M. M. Airaksinen, et al., "Binding Sites for (3H) pinoline"; and M. M. Airaksinen, et al., "Structural requirements for high binding efficiency."
27. Callaway, "A proposed mechanism for the visions of dream sleep."
28. Ananda's books: *The Soma Conspiracy,* a limited edition, privately printed 642-page book, about inner and outer alchemy; and *The Rig Veda,* another limited edition, 432-page book in which Ananda reveals his discoveries of a 2012 connection in the ancient Hindu writings called the Rig Veda. These are hard to find and so is his original book, *The Unity Keys of Emmanuel,* which was to have been made available online, but the files became corrupted so it is now lost.
29. From Ananda's *Grailzine Newsletter,* September 2003.
30. www.erowid.org/archive/psychonaut/main.htm.
31. www.erowid.org/archive/psychonaut/mushrooms/mushroom3.htm.
32. www.erowid.org/archive/psychonaut/mushrooms/mushroom4.htm.
33. www.erowid.org/archive/psychonaut/mushrooms/mushroom5.htm.
34. Charlie Sabatino to Stray, e-mail, February 15, 2002.
35. Ibid.
36. Damaeus to Stray, e-mail, May 12, 2002.
37. Canul replied to my article in *Salvia Divinorum Magazine:* "Salvia Divinorum: Key to the Paradigm Shift in 2012." The reply thread follows the article, but is no longer available online.

38. www.diagnosis2012.co.uk/new5.htm#salvia.
39. Tomorrowlander to Stray, e-mail, July 7, 2003.

CHAPTER 14. OTHER ALTERED STATES

1. http://web.archive.org/web/20031012220531/http://members.aol.com/yesloveisgod/says/prophecy3.html.
2. Ibid.
3. Ibid.
4. www.inlightimes.com/archives/2-Dannion-Brinkley.htm.
5. www.inlightimes.com/archives/1999/12/dannion.htm.
6. www.gracemillennium.com/spring01/pmh.htm.
7. www.geocities.com/nulliusinverba/AR3.htm.
8. http://web.archive.org/web/20030120204023/http://www.happycampnews.com/coast2coast/2001_12_02_archives.html. See also this Wikipedia page concerning the documentation of the NDE: http://en.wikipedia.org/wiki/Pam_reynolds.
9. http://web.archive.org/web/20041205014007/http://www.soundwellness.com/ofrank/of/of2doc/ParApocalyp.html. (The text on this web page will probably be invisible until it is highlighted by right clicking and dragging the mouse across the page.)
10. www.religionen.at/irsilverb01.htm#top.
11. LaBerge, *Lucid Dreaming,* 61.
12. www.lucidity.com/LucidDreamingFAQ2.html#OBE.
13. Harary and Weintraub, *Lucid Dreams in 30 Days,* 89–90.
14. See discussion at www.robertpeterson.org/obe-vs-lucid.html.
15. www.sealifedreams.com/.
16. There is a slightly differently worded version of the same dream here: www.improverse.com/ed-articles/nick_cumbo_2003_jul_waves_sirius.htm.
17. http://web.archive.org/web/20050412204549/http://www.dreamofpeace.net/sealife/viewtopic.php?t=1175.
18. Dream date November 28, 1998. http://sstoth0.tripod.com/dreamimages/id26.html.
19. www.cropcircleresearch.com/thoth/dream_pv0001.html.
20. www.greatdreams.com.

21. http://web.archive.org/web/20070102131929/http://womenandwisdom.org/article1.shtml.
22. Karpinski, *Barefoot on Holy Ground,* 31.
23. http://web.archive.org/web/20050408024331/http://www.hindustantimes.com/news/181_72097,001100010001.htm.
24. Ibid.

CHAPTER 15. UFOS AND ETS

1. Video footage: *Messengers of Destiny*—1991–1992 footage (fifty-plus clips); *Masters of the Stars*—1992–1993 footage (fifty-five-plus clips); and *Voyagers of the Sixth Sun*—1993–1996 footage (ninety-plus clips), by Genesis 111.
2. www.diagnosis2012.co.uk/new4.htm#dresden.
3. Mike Finley's website, The Real Maya Prophecies: Astronomy in the Inscriptions and Codices, can be found at http://members.shaw.ca/mjfinley/mainmaya.html.
4. www.diagnosis2012.co.uk/new4.htm#dresden.
5. Finley to Stray, e-mail, October 15, 2002.
6. *UFO Reality,* no. 7, April–May 1997.
7. Ibid.
8. Ibid.
9. Ibid.
10. *UFO Reality,* no. 4, October–November 1996.
11. Hurtak, *The Keys of Enoch.* This book is also listed in Argüelles's bibliography, so that is possibly why some of the TUMI column sounds familiar, although by the time the magazine was published, Jon King (living in Glastonbury) would have been familiar with Argüelles's *The Mayan Factor.*
12. *UFO Reality,* no. 1, April–May 1996.
13. Ibid.
14. Ibid.
15. William and Birnes, *Unsolved Mysteries,* 234–35. See also this quote of an article by Bill Hamilton, Pam's husband, though it says the prediction was for 2011: http://web.archive.org/web/19970706153215/http://www.brotherblue.org/libers/pamsrep.htm.
16. *UFO Reality,* no. 4, October–November 1996.
17. Conversation with the author, August 2003.

18. *UFO Reality,* no. 4, October–November 1996.
19. Ibid.
20. *UFO Reality,* no. 5, December–January 1997.
21. Ibid.
22. www.karinya.com/watchers.htm.
23. Ibid.

CHAPTER 16. CROP CIRCLES

1. Thomas, *Vital Signs.*
2. Davis, *Ciphers in the Crops,* 53.
3. Robert Anton Wilson, *Trajectories Newsletter,* 17, and www.rawilson.com/trigger2.shtml.
4. *Chambers's Twentieth Century Dictionary.*
5. Séjourné, *Burning Water;* and Waters, *Mexico Mystique.*
6. Waters, *Mexico Mystique,* 180.
7. See the following video link, fast-forward to 40:20, and watch for nine minutes: http://video.google.com/videosearch?q=geoff+stray&emb=0&aq=f#.
8. Men, *Secrets of Maya Science/Religion,* 34. The symbol is now known to be an Aztec design for a cloak to be worn at the festival of lip plugs, having no overt connection to the Maya, Hunab Ku, or Galactic Center—neither visual nor astronomical. www.diagnosis2012.co.uk/new10.htm#hunab.
9. Lucy Pringle's website: http://home.clara.net/lucypringle.
10. Thomas, ed., *SC: The Bimonthly Journal of Crop Circles and Beyond.*
11. Thomas, *Vital Signs* (2002), 80.
12. www.kachina.net/~alunajoy/.
13. www.kachina.net/~alunajoy/98july.html.
14. www.swirlednews.com/article.asp?artID=110.
15. Michell, *The View Over Atlantis.*
16. Alford, *Gods of the New Millennium,* 588.
17. Ibid.
18. Alford calculated the date for the Great Flood, using Babylonian and Sumerian mythology, as 10,983 BC. He says that this is the date that precession started—in other words, Earth's angle of tilt was perpendicular to its orbit around the Sun until 10,983 BC, when a close pass of Nibiru caused an axis shift and the

subsequent flood. After this, Alford says that the *sars,* or ages of Sumeria (each ruled by a different god), changed from 3,600 years to 2,160 years, which is one zodiac age, or a twelfth of the classic precessional cycle of 25,920 years. He also points out the strange fact that 2,160, the length of a zodiacal age, is 3-6-0-0 when written in the sexagesimal Sumerian system, yet implies that Nibiru is still on a 3,600-year orbit, although he admits that it might have been forced "into a slightly different orbit" (Alford, *Gods of the New Millennium,* 569). This means that, working forward, the next pass of Nibiru is due in 3418 AD (adding a year to compensate for the missing year zero). However, if Nibiru's orbit changed to 2,160 years, as could be implied by all of Alford's information, then when would the next pass be? The year 1978 AD is the answer (adding a year to compensate for the missing year zero), and since it didn't come then, that can't be right. But Alford calculated that the next precessional age starts in 2012 (Alford, *Gods of the New Millennium,* 586–88) using the current value of precession (25,776 years, as suggested by Egyptologist Jane Sellers). So how did he work that out? One current zodiac age is a twelfth of 25,776 years, and works out as 2,148 years. But before calculating, we have to take into account the 108-year delay, or destiny, determining time (Alford, *Gods of the New Millennium,* 333, 355–56, 359, 360, 366, 368). This phrase comes from the Sumerian Lugal-e text quoted by Sitchin (*The Wars of Gods and Men,* 165) and tells us that Nergal was sent to tell Marduk that he had returned to Babylon too early, since his destiny determining time was not yet due. Alford says the discrepancy alluded to here is 108 years, because 108 years is 1.5 degrees of precessional movement (72 + 36), which he feels is the minimum time needed to detect the precession cycle (pp. 358–59). If we start at Alford's date for the Flood at 10,983 BC and subtract the 108 years, we get 10,875 BC. So six sars of 2,148 years each (12,888 years) after this corrected date for Marduk's destiny determining time brings us to 2013, except we have to add another year to compensate for the missing year zero, as in all calculations across the BC–AD divide (due to the fact that the year after Jesus's hypothetical birth was 1 AD and the year before was 1 BC—your calculator allows for a zero that isn't there, so it must be added on somewhere). In other words, using Alford's figures, we get 2014, not 2012. Oh dear, it's two years out! So how did Alan Alford get 2012? The calculation is given on p. 662, note 58: "10983 BC – 107 = 10,876 BC." He mistakenly used the number 107, when he had already stated three times that it should be 108 (Alford, 1996, pp. 359, 366, 368). Alford's second error was that he forgot about the missing year zero, so we have to add yet another year to the calculation (Alford, *Gods of the New Millennium*). On pp. 457–58, he says the thirteen-baktun cycle began in 3113 BC, which he may have used to back-calculate his original Flood date. If he did use 3113 BC to back-calculate, then all his BC dates will be Gregorian astronomical dates, so no compensation for the missing year zero is necessary, and his calculation will only be a year out.

CHAPTER 17. SECRET GOVERNMENT

1. Berlitz and Moore, *The Philadelphia Experiment,* 115–16, 165–66.
2. Ibid., 15, 176.
3. web.archive.org/web/20020202055415/http://home.inreach.com/dov/ttstrnge.htm. Gibbs now has his own website at www.hdrenterprises.net, but a much better display of his products is here: www.futurehorizons.net/time.htm. For more on do-it-yourself time machines, see this incredible page: www.keelynet.com/energy/dimshift.htm.
4. www.futurehorizons.net/time.htm.
5. www.allthingsspooky.com/letter.html (this is an alternative to the archived address at note 4).
6. www.freezone.de/english/mc/e_conv04.htm.
7. Ibid.
8. http://web.archive.org/web/20030320213027/http://2013.com/.
9. http://web.archive.org/web/20030413194122/2013.com/faq/index.html.
10. Ibid.
11. Interview with Art Bell on January 30, 1997.
12. http://mindcontrolforums.com/hambone/dame0197.html.
13. Ibid.

CHAPTER 18. NOSTRADAMUS

1. Cheetham, *The Prophecies of Nostradamus.*
2. Ibid., 417.
3. Revelation 6:13, 8:7, 11:19, 12:4, 15:21.
4. Ibid., 8:5, 8:8, 8:10–11, 9:1.
5. Ibid., 6:12, 8:5, 11:13, 16:18.

6. Quatrains 1:69, 2:15, 2:41, 2:43, 2:46, 2:62, 4:67, 5:32, 6:6.
7. Quatrains 2:18, 2:81, 2:92, 2:96, 4:100, 5:98, 6:97, 8:2, 8:10, 8:77.
8. Quatrains 1:87, 6.66, 9:31, 9:83, 10:67.
9. www.newprophecy.net/.
10. www.satansrapture.com/foundnostclock.htm.
11. http://web.archive.org/web/20030430111526/http://www.home.mastersites.com/kingofterror/index.html.
12. Dimde, *Nostradamus, Das Apokalyptische Jahrzehnt* (Nostradamus, the Apocalyptic Decade).
13. www.tmgnow.com/repository/cometary/76P_phobos1.html.
14. Ibid.
15. Ibid.
16. http://petragrail.tripod.com/codes.html.
17. http://web.archive.org/web/20021203004740/http://www.angelfire.com/wizard/magick2/2.html and http://web.archive.org/web/20031210233225/http://www.angelfire.com/rebellion/newworld_1/666.html.
18. Ibid.
19. Cheetham, *The Prophecies of Nostradamus,* 417.
20. http://web.archive.org/web/20000619101722/www.freeyellow.com/members7/howardmj/page5.html.
21. http://web.archive.org/web/20021106080154/http://www3.telus.net/themist/myrddin.htm.
22. Revelation 6:12.
23. Ibid.
24. Ibid., 6:13.
25. http://sunearth.gsfc.nasa.gov/eclipse/SEcat/SEdecade2011.html and www.eclipse.co.uk/solarspirit/solar_sequence3.htm.
26. www.nostradamus-repository.org/shipton.html.
27. http://web.archive.org/web/20021205014800/http://www.wilders.force9.co.uk/BeyondEarth/comets.htm and www.mkzdk.org/hammer/hammer.html.

CHAPTER 19. NEW AGE

1. www.aquarian-age.net.
2. Ibid.

3. From *Testimony Magazine:* www.testimony-magazine.org/back/jun2001/piperov.pdf.
4. http://ancientegypt.hypermart.net.
5. Audrey Fletcher, e-mail, July 24, 2001.
6. Vincent Bridges, e-mail, May 12, 2001.
7. See especially Stanley Messenger's articles "Lucifer and Ahriman under the Bed," www.isleofavalon.co.uk/GlastonburyArchive/messenger/sm-lucifer.html, and "Crop Circles as a Path from Space into Time (2)," www.isleofavalon.co.uk/GlastonburyArchive/messenger/sm-ccircles2.html.
8. Watson, *Lifetide,* pp. 174–77. *Hundredth monkey* is a term that caught on after Lyall Watson described some behavior changes of monkeys on a Japanese island. One female discovered that washing the sweet potatoes in the sea, to clean them and add a salty flavor, made them much more palatable. Over several months, other monkeys slowly began to copy her, until one day a critical number was reached (the hundredth monkey), and by the end of that day, all the monkeys on all the islands—without any physical communication having taken place between them—changed their behavior permanently to the washing of potatoes before eating them. However, recent reexamination of the story shows that it is unreliable. The effect took place slowly, over years, with young monkeys learning it and older ones dying off, and at least one of the monkeys is said to have swum to a neighboring island. http://en.wikipedia.org/wiki/Hundredth_Monkey.
9. Gilson to Stray, e-mail, June 1, 2001.
10. www.nvisible.com.
11. http://aa.usno.navy.mil/data/docs/EarthSeasons.php.
12. Argüelles, *The Mayan Factor,* 154.
13. Ibid., 158.
14. Interview in *UFO Reality,* no. 7, April–May 1997; and no. 8, June–July 1997.
15. Argüelles, *The Mayan Factor,* 194.
16. There are a number of examples of such errors in Argüelles's *The Mayan Factor:*

    - On pp. 45, 110, it says that the zero day of the Long Count was in 3113 BC instead of 3114 BC. The year 3113 BC is only correct if viewed as an astronomical date, implying that an extra year has been substituted for the missing year zero, to ease calculation across the BC–AD divide. Since the references at the back of *The Mayan Factor* include several books by J. Eric and

S. Thompson, one of these was probably the source. Thompson wrote these books in the 1950s and 1960s, and since then the convention for indicating astronomical dates has been updated to using a negative determinant, such that 3113 BC astronomical is now written as year – 3113. This is equivalent to 3114 BC historical, but many readers would simply have misunderstood the start-date as 3113 BC historical. This was most certainly the cause of the one-year error in Sergey Smelyakov's *The Auric Time Scale,* since Smelyakov uses Argüelles's *The Mayan Factor* as his primary source on Maya calendrics. **The error is +368 days.**

- On p. 29, we find that Good Friday in 1519 is said to be the day 1 Reed in the Aztec calendar (1 Ben in the Maya). **This is an error of +12 days**, since 1 Reed was on May 4 Julian (JD), or May 14 in the Gregorian calendar (GD), and Good Friday occurred on April 22 (JD), or May 2 (GD). Good Friday actually occurred on the day 2 Imix (Alligator), according to the True Count.
- On p. 131 is a reference to June 20, 1986, as being tzolkin day 10 Ben, whereas it was actually the day 8 Kawak in the True Count. **This is an error of +54 days.**
- On p. 131, June 20, 1986, is said to be the haab day 9 Kayeb, whereas it was actually 12 Zotz. **This is an error of –108 days.**
- On p. 131, Argüelles also says that June 20, 1986, was the Long Count date 12.18.14.18.9, whereas it was actually 12.18.13.1.19. **This is an error of +690 days.**
- On p. 131, Argüelles says that June 20, 1986, was 1,862,599 days after zero day, whereas it was actually 1,862,319 days after zero day. **This is an error of +280 days.**
- On pp. 148, 169, Argüelles says that the first day of Harmonic Convergence, August 16, 1987, was day number 1,863,022 after zero day, whereas it was actually day number 1,862,741 after zero day. **This is an error of +281 days.** Anyone using the Long Count figures or the day numbers above to calculate forward to the end of the Long Count would calculate the end point (day 13.0.0.0.0., or 1,872,000 days after zero day) to be March 16, 2012; March 15, 2012; or January 31, 2011.
- On p. 196, Argüelles writes that October 8, 1986, was the tzolkin day 1 Imix, whereas in the True Count this was the tzolkin day 12 Manik. **This is an error of +52 days.**

- On p. 196, Argüelles says that October 8, 1986, was the day 12 Zotz in the haab calendar, whereas it was actually 0 Yax. **This is an error of –110 days.**
- On pp. 161, 179, 200, Argüelles says that the final (260th) katun of the thirteen-baktun cycle will start in 1992. In Dreamspell, he later refines this to July 26, 1992. It actually started on April 5, 1993. A katun is 7,200 days—approximately 19.72 years, not twenty years. **This is an error of –254 days.**
- Tony Shearer first calculated Harmonic Convergence as being nine fifty-two-year Hell cycles after the arrival of Cortés (interpreted as the return of Quetzalcoatl by the Aztecs), which he took as April 21, 1519 (JD), the day before Good Friday. He calculated that 9 × 365.25 × 52 days (or 170,937 days) after Cortés's arrival would be August 16, 1987. However, the answer should have been April 21, 1987 JD, which is May 4, 1987, in the Gregorian calendar. But this is just the first error, since Hell cycles are Calendar Rounds, which are each fifty-two haabs in length, and a haab is a 365-day year, not the 365.25-day Julian year. This means **there was an error of +117 days** (there are 117 leap days in 52 × 9, or 468 years), and the end of the nine Hell cycles should have been 117 days earlier, which is December 25, 1986 (JD), or January 7, 1987 (GD). Shearer should have used 9 × 52 × 365 days, or 170,820 days. Instead of adjusting 117 days back, the date was adjusted 117 days forward, from April 21, 1987 (JD), to August 16, 1987, but this was taken as a Gregorian date rather than as a Julian date. This seemed to be a meaningful date, because in the correlation Shearer was using (corr. no. 584284, or one day away from the True Count), August 16, 1987, was the day 13 Ahau, which is the last day in the tzolkin. The date of August 17, 1987, was thus (in the 584284 correlation) the day 1 Imix, or the first day of the tzolkin. So Harmonic Convergence should have been on December 25, 1986 (JD), or January 7, 1987 (GD). **Total error = +221 days** or (2 × 117) – 13 (where 13 is the accumulated difference between Julian and Gregorian dates).

Some people who have tried to understand how Harmonic Convergence was figured have concluded that it must have been calculated from the day that Cortés met the Aztec emperor Moctezuma (Montezuma), since, coincidentally, this happened 117 days after April 21 (when he landed), on August 16, 1519 (JD). Exactly 468 Julian years later (9 × 52 Julian years, or 170,937 days) would be August 16, 1987. This calculation incorporates both errors of 117 days and the Gregorian and Julian errors. Argüelles used Shearer's calculation for Harmonic Convergence, but subsequently developed an Aztec/Maya-based calendar system called Dreamspell, which used the same fifty-four-day error tzolkin count featured in *The Mayan Factor,* p. 131 (see above). This system ignores the leap-year day of February 29 every four years, and the result is that the fifty-four-day error is reducing by one day every four years. In March 2004, the error was down to forty-nine days. In the Dreamspell count, Harmonic Convergence, August 16–17, 1987, happened on the days 3 Men and 4 Cib, which means that all the meaning involved in calculation of the Harmonic Convergence as 13 Ahau and 1 Imix—as the start/end of the tzolkin—is lost.

In his response to the count controversy, Argüelles says, "The Harmonic Convergence, *The Mayan Factor,* and the Dreamspell are all based on the July 26 synchronization date. It was by following the calendar based on this synchronization date that I was able to verify the Harmonic Convergence as a test of Maya time science." We can only assume that he is referring to the next July 26 after Harmonic Convergence—i.e., July 26, 1988—which was 13 Reed (Ben) in the Dreamspell. However, this was a meaningful day to the Aztecs, not the Maya. See: www.earthportals.com/Portal_Messenger/speakout.html.; also see "Investigating the Origins of Dreamspell," at www.diagnosis2012.co.uk/dspell.htm.

17. Argüelles, *The Mayan Factor,* 194.
18. www.beyond-the-illusion.com/files/Orvotron/Newsletters/92marapr.txt. Halfway down the page, repeated from a 1981 article in an Australian publication, *Flying Saucer Research Magazine.*
19. Ibid.
20. Ibid.
21. Ibid.
22. Erbe, *God I Am—From Tragic to Magic,* 225.
23. Ibid.
24. Essene and Nidle, *1994. You Are Becoming a Galactic Human,* 53.
25. See Roberto Solarion for more on this; he thinks the Sirian Nibiru will return in 2012: http://web.archive.org/web/20030811131437/http://www.apollonius.net/sirius.html.

26. www.straightdope.com/classics/a4_042.html.
27. Hurtak, *The Keys of Enoch,* 78. (Key 1-0-8; 32.)
28. Ibid., 78.
29. Ibid., 79.
30. www.cassiopaea.org.
31. http://web.archive.org/web/20031008070403/http://www.vincentbridges.com/cassiopaea.html.
32. Calleman, *The Mayan Calendar,* 207.
33. Ibid., 75.
34. www.wilsonsalmanac.com/converg.html.
35. http://edj.net/mc2012/calleman.htm.
36. Calleman, *The Mayan Calendar,* 157.
37. Ibid., 236. Calleman's reason for correlating the start point to June 17, 3115 BC, was that the Creation story of Palenque said that the solar god First Father had his seventh birthday the previous day.
38. Ibid., 206. Table 6 and fig. 27, both on p. 77 of Calleman's book *The Mayan Calendar,* each indicate a different length for the Universal Underworld (the highest level of the pyramid). Table 6 says it is 13 × 20 kin (260 days) and fig. 27 says it is 13 × 18 kin (234 days).
39. www.diagnosis2012.co.uk/jmj.htm.
40. The debate is accessible from the index page of www.diagnosis2012.co.uk, and see the readers' responses at John Major Jenkins's site, Alignment 2012.com, www.alignment2012.com/fiasco.html.
41. Scallion, *Notes from the Cosmos,* 226.
42. http://web.archive.org/web/20031207091702/http://www.citilink.com/~mjfitz/gmscallion.htm, and *End-Time Visions: The Road to Armageddon,* by Richard Abanes, 53–56.
43. Morton and Thomas, *The Mystery of the Crystal Skulls,* 143.
44. Ibid., 322–23.
45. Ibid., 349.
46. www.earthportals.com/Portal_Messenger/josevotan.html.
47. http://user.cyberlink.ch/~koenig/dplanet/staley/nema10.htm.
48. www.inventati.org/amprodias/txt/hacktivism/maat.htm.
49. www.crossroads.wild.net.au/index.html.
50. www.crossroads.wild.net.au/zuv.htm.

51. www.crossroads.wild.net.au/lib13.htm.
52. Ibid.
53. Andahadna, "Liber Pennae Praenumbra," quoted in Grant, *Outside the Circles of Time.* She is also known as Maggie Cassady and Margarete Ingalls. See www.hermetic.com/wisdom/pennae_praenumbra.html.
54. According to most Thelemites, we have been in the Aeon of Horus since 1904, when Crowley received The Book of the Law. See Nema's book *Maat Magick* (1995).
55. Grant, *Outside the Circles of Time,* 242.
56. Ibid., 185, quoting Andahadna in an unpublished paper on the mysteries of Maat.
57. Ibid.

CHAPTER 20. ANALYSIS

1. http://members.shaw.ca/mjfinley/calnote.htm.
2. www.traditionalstudies.org/journal/s_america/aztec.html.
3. I used the free Geditcom date calculator for this calculation. Download it at www.geditcom.com/DateCalc.html. Another version of this Nine Hells cycle was revealed by Patricio Dominguez at the Confederation of Indigenous Elders of America, in which he said that the Nine Hells started in 1517, when the Spanish landed in the land of the Maya, and ended in 1985. See www.diagnosis2012.co.uk/books2.htm.
4. See these sources: http://web.archive.org/web/20060106063332, http://users.pandora.be/gohiyuhi/frauds/index.htm, and http://web.archive.org/web/20020601193939/http://www.iktome.freewebsites.com/.
5. http://web.archive.org/web/20030808083901/http://www.thefourwinds.com/AProphecy.html.
6. Argüelles, *The Mayan Factor,* 161, 179, 200.
7. Len Stevens, e-mail, June 15, 2003.
8. www.alignment2012.com/mproph.htm.
9. Gilbert and Cotterell, *The Mayan Prophecies,* 249–50.
10. Retitled *The Mystery of the Great Cross of Hendaye: Alchemy and the End of Time,* in a new updated and expanded edition. Weidner and Bridges, *The Mysteries of the Great Cross of Hendaye.*

11. LaViolette, *Earth Under Fire,* 104.
12. www.etheric.com/LaViolette/Disinformation.html.
13. Paul LaViolette, e-mail, September 1, 2000.
14. http://web.archive.org/web/20021009094440/http://therealmidori.tripod.com/disaster.html.
15. www.earthchangestv.com/mitch/november_2003/05flare.htm.
16. http://news.bbc.co.uk/1/hi/sci/tech/3515788.stm.
17. http://web.archive.org/web/20050301104353/www.newscientist.com/article.ns?id=dn4321.
18. http://news.bbc.co.uk/2/hi/science/nature/3869753.stm and www.sfgate.com/cgi-bin/article.cgi?f=/c/a/2006/03/07/MNGAFHJJL91.DTL and www.stevequayle.com/News.alert/06_Cosmic/060512.solar.2012.html.
19. www.spacedaily.com/news/milkyway-03a.html, and http://spaceflightnow.com/news/n0308/10ulysses/.
20. www.spacedaily.com/news/milkyway-03b.html.

    See also www.space.com/scienceastronomy/dust_storm_030814.html.
21. Clow, *The Pleiadian Agenda,* 47.
22. See "Exploding the Myth of the Photon Belt," at http://web.archive.org/web/20020620154249/http://www.worldlightfellowship.org/classes/photon_belt.htm and www.diagnosis2012.co.uk/blt1.htm.
23. "The Physics of Crop Formations": www.diagnosis2012.co.uk/blt2.htm, best read after Nancy Talbott's "Crop Circle Analysis Shows That Most Are Not Hoaxes," at www.diagnosis2012.co.uk/blt1.htm.
24. www.diagnosis2012.co.uk/blt2.htm.
25. Ibid. The highest concentration of meteoritic dust—about one hundred times normal soil concentration—was found at a crop circle known as the H-Glaze formation, at Yatesbury, Wiltshire, in 1993. Studies on samples from this contributed to BLT's conclusions. However, a more recent scientific study by R. Ashby (January 2005), posted at www.xstreamscience.org, concludes that the formation was hoaxed by paranormal debunker Rob Irving. Irving supplied a sample of the powdered iron consisting of microscopic spheres that he claimed to have used, and this was used on crop and chalk samples that were subjected to similar conditions that existed when the pattern appeared. This produced results that were very similar to those found in the formation. However, Burke says that about twenty other formations yielded similar "meteoric dust" of at

least twenty times normal soil concentration, and these were distributed over thirteen states of the United States and five countries, and clearly had nothing to do with Irving.

26. www.tmgnow.com/repository/planetary/pfvd.html.
27. http://hessdalen.hiof.no/reports/.
28. http://dwij.org/pathfinders/linda_moulton_howe/linda_mh9.htm.
29. www.earthfiles.com/news/news.cfm?ID=164&category=Real+X-Files.
30. Bert Janssen and Janet Ossebaard (2001), www.bertjanssen.nl.
31. www.rense.com/general17/talbot.htm.
32. http://web.archive.org/web/20030923080953/http://www.newscientist.com/news/news.jsp?id=ns99994174.
33. Ibid.
34. www.guardian.co.uk/life/farout/story/0,13028,1048649,00.html.
35. www.daviddarling.info/encyclopedia/P/plasma-based_life.html.
36. www.tmgnow.com/repository/repository.html.
37. www.tmgnow.com/repository/planetary/pfvd.html.
38. http://web.archive.org/web/20010117060500/http://www.acemake.com/PaulDevereux/earthlights.html.
39. www.tmgnow.com/repository/planetary/tornado.html.
40. http://web.archive.org/web/20021210002409/http://www.timeofglobalshift.com/Science/scieblue.html.
41. www.geocities.com/ResearchTriangle/Thinktank/8863/main.html.
42. www.winstonbrill.com/bril001/html/article_index/articles/151-200/article158_body.html.
43. http://video.google.com/videoplay?docid=-7027255937915952897.

    This video shows the original levitation effects. However, some of these have been successfully duplicated by skeptics: www.trailerparkscience.com/Experiments.html.
44. www.hutchisoneffect.ca/.
45. Lecture by Terence McKenna at the University of London Student Union (ULU) on September 25, 1991: "World Youth Culture, Shamanism, Gaia, Chaos, and Plant Hallucinogens."
46. www.tmgnow.com/repository/global/planetophysical4.html.
47. http://noosphere.princeton.edu/shnoll2.html.
48. http://21stcenturysciencetech.com/articles/time.html.

49. See S. E. Shnoll, et al., 1998, "Realization of discrete states during fluctuations in macroscopic processes," in *Uspekhi Fizicheskikh Nauk* (Advances in Physical Sciences), vol. 41, no. 10, pp. 1025–35. See the article here: www.cyclesresearchinstitute.org/physics/ufn9810d.pdf. A new paper was later published, which claims to provide the proof of the hypothesis. See S. E. Shnoll, et al., 2000, "Regular variation of the fine structure of statistical distributions as a consequence of cosmophysical agents," in *Uspekhi Fizicheskikh Nauk* (Advances in Physical Sciences), vol. 43, no. 2, pp. 205–09 See the article summary here: www.iop.org/EJ/abstract/1063-7869/43/2/L13. Full text here is at www.iop.org/EJ/article/1063-7869/43/2/L13/PHU_43_2_L13.pdf?requestid=42c87lfd-394f-42ac-a729-26596e793lab. An even more recent paper finds more evidence. See M. V. Fedorov, et al., 2003, "Synchronous Changes in Dark Current Fluctuations in Two Separate Photomultipliers, in Relation to Earth Rotation in Astrophysics and Space Science, 2003," published *in Astrophyics and Space Science*, vol. 283, no. 1, pp. 3–10. Here is the preview page: www.springerlink.com/content/j43115q064560240/. Shnoll's group is based at Moscow State University.
50. http://21stcenturysciencetech.com/articles/time.html.
51. http://web.archive.org/web/20021216051917/http://www.museum.vic.gov.au/planetarium/solarsystem/sun.html., and http://umtof.umd.edu/pm/crn/carrtime.html.
52. Thanks to David Wilcock again for uncovering this Shnoll material: http://divinecosmos.com/index.php?option=com_content&task=view&id=81&Itemid=36.
53. www.tmgnow.com/repository/global/planetophysical2.html.
54. Wilcock to Stray, e-mail, May 28, 2002.
55. Stray to Wilcock, e-mail, May 28, 2002.
56. Wilcock to Stray, e-mail, May 28, 2002.
57. Ibid.
58. www.divinecosmos.com/index.php?option=com_content&task=view&id=45&Itemid=30.
59. This was actually Ra through Wilcock: www.fieldwerks.com/2003davidwilcock.htm, and http://divinecosmos.com/index.php?option=com_content&task=view&id=29&Itemid=30.

    See also comments and quotes in chapter 12.

CHAPTER 21. SYNTHESIS

1. www.erowid.org/plants/salvia/salvia_info5.shtml.
2. www.erowid.org.
3. Dr. Ross Adey of Loma Linda University Medical Center, in California (a Seventh-day Adventist Health Sciences Institution): "Cell membranes, electromagnetic fields, and intercellular communication," 1987 (no further refs. given). Gilbert and Cotterell, *The Mayan Prophecies,* p. 24. Adey is quoting Walker et al., 1983, who are also quoted by Serena Roney-Dougal here: http://paranormal.se/psi/tallkottskoerteln.html.
4. Hutchison, *Mega Brain Power.*
5. www.meta-religion.com/Psychiatry/The_Paranormal/trascendent_experiences.htm.
6. Ibid.
7. On this page (www.akasha.de/~aton/NeoDMT.html), Ananda gives the following reference: Kärkkäinen J., et al. (1988), "Urinary excretion of free bufotenin by psychiatric patients"; *Biological Psychiatry* 24:441–46. However, it seems that the patients with epilepsy whose urine showed traces of bufotenin (or alternative spelling, bufotenine) were also autistic. http://web.archive.org/web/20070606141715/http://erowid.org/chemicals/bufotenin/bufotenin_journal2.shtml.
8. Bumpus and Page, and also Evarts, referenced in Rinkel, *Chemical Concepts of Psychosis,* 58, 214.
9. Steve Scott to Geoff Stray, e-mail, July 10, 2003. See also http://web.archive.org/web/20040519221150/http://www.ianhard.2ya.com/

   In 2004, on his website, Steve's update said he didn't have TLE, but rather schizophrenia.
10. Budden, *Electric UFOs,* p. 113, and Devereux (1999), p. 7. Persinger has also demonstrated a relationship between the intensity of geomagnetic disturbance the day after birth and later visions and auditory hallucinations. K. A. Hodge and M. A. Persinger, "Quantitative increases in temporal lobe symptoms in human males are proportional to postnatal geomagnetic activity," pp. 205–8. See archive: http://web.archive.org/web/20030708015243/http://www.lycaeum.org/drugs/abstracts/L4.cgi?mode=keys&kwand=epilepsy.
11. Devereux, *Places of Power.*
12. Ibid., 207. See also Devereux, *Earthmind,* 203–25.
13. Devereux, *Places of Power,* 208.

14. http://web.archive.org/web/20051105025518/http://www.argemto.com.ar/mayas.htm. Google translation: www.google.com/intl/en_ALL/help/faq_translation.html.
15. http://web.archive.org/web/20041214035804/http://www.global-vision.org/dream/dreamch4.html.
16. http://web.archive.org/web/20021214000546/www.global-vision.org/dream/dreamint.html.
17. http://web.archive.org/web/20021213235308/www.global-vision.org/dream/dreamch1.html
18. Paraphrased.
19. http://web.archive.org/web/20030201190508/www.global-vision.org/dream/index.html.
20. http://web.archive.org/web/20031121210518/http://www.madness.crowcity.com/.
21. Rinkel, *Chemical Concepts of Psychosis,* p. 219, and Rodnight, "Urinary dimethyltryptamine and psychiatric symptomatology and classification," pp. 649–57: "[This research project] studied the excretion of dimethyltryptamine (DMT) in 122 recently admitted psychiatric patients and twenty normal Ss [subjects]. DMT was detected in the urine of 47% of those diagnosed by their psychiatrists as schizophrenic, 38% of those with other non-affective psychoses, 13% of those with affective psychoses, 19% of those with neurotic and personality disorders, and 5% of the normal Ss. Ninety-nine of the patients were interviewed in a semi-standardized fashion, and also categorized according to a variety of operational definitions of the psychoses. The operational definitions failed to reveal any group significantly more correlated with urinary DMT than did the hospital diagnosis of schizophrenia, but a discriminate function analysis of symptomatology could be used to define a group of 21 patients of whom 15 (71%) excreted detectable DMT. There was a general relationship between psychotic symptoms and urinary DMT, but specifically schizophrenic symptoms did not appear to be major determinants of DMT excretion." From http://deoxy.org/annex/daytripr.htm#6.
22. Luke 11:34–36 and Matthew 6:22.
23. Straiton, *The Celestial Ship of the North,* quoted in Jenkins, *Galactic Alignment,* 140.
24. Luke 20:35.
25. A copper bell, according to Charles Musès/Musaios, *The Lion Path.*

26. Musaios, *The Lion Path—You Can Take It With You.* Charles Musès wrote several books, but he wrote *The Lion Path* under the pseudonym of Musaios.
27. Ibid., 33–35.
28. See appendix 2: "The Astral Transition."
29. Naydler, *Temple of the Cosmos*, 201.
30. Ibid., 209.
31. Ibid., 205.
32. Ibid., 208.
33. Ibid., 285.
34. Scott, *Kundalini in the Physical World,* 77.
35. Ibid., 96.
36. Ibid.
37. Ibid. 77.
38. Arundale, *Kundalinin: An Occult Experience,* 64.
39. Ibid.
40. Scott, *Kundalini in the Physical World,* 116.
41. www.utexas.edu/courses/stross/papers/sacrum.htm.
42. Weidner and Bridges, *A Monument to the End of Time,* 61.
43. Ibid.
44. Ibid., 398.
45. http://web.archive.org/web/20030808083901/http://www.thefourwinds.com/AProphecy.html.
46. www.thefourwinds.com/shs-end-time.php.
47. http://web.archive.org/web/20030808083901/http://www.thefourwinds.com/AProphecy.html.
48. James Arévalo Merejildo (Chaski), *The Awakening of the Puma Initiation Path: Evidences of Archeo Astronomy in the Andes.* English translation by Alain Goormaghtigh (Cuzco, Peru: Chaski), 1997, quoted at http://tonebytone.com/hatunkarpay/14.shtml.
49. www.sacredmysteries.com/allVideos.htm#healingbody.
50. http://web.archive.org/web/20040813101717/sacredmysteries.com/netx4.htm.
51. http://web.archive.org/web/20030808083901/http://www.thefourwinds.com/AProphecy.html.
52. Ibid.
53. Ibid.

54. Elizabeth B. Jenkins, *Initiation,* 113.
55. Ibid., 237.
56. www.diagnosis2012.co.uk/soul.htm.
57. Elizabeth B. Jenkins, *Initiation,* 230.
58. Ibid.
59. http://web.archive.org/web/20030808083901/http://www.thefourwinds.com/AProphecy.html.
60. See Devereux, *Shamanism and the Mystery Lines,* pp. 78–83, and Aveni, *Stairways to the Stars,* pp. 147–76. A more recent study by Brian S. Bauer in *The Sacred Landscape of the Inca: The Cusco Ceque System* (1998, University of Texas Press) shows that the ceques are not perfectly straight and cannot be solely astronomical in nature.
61. Devereux, *Shamanism and the Mystery Lines,* 191.
62. www.misionrahma.com/ingles/disc_solar.htm.
63. http://web.archive.org/web/20031007034948/http://www.spiritjourneys.com/everyone/calendar/peruthomsondtl.html.
64. Wilcox, *Keepers of the Ancient Knowledge,* 22.

CHAPTER 22. PROGNOSIS

1. Luk, *Taoist Yoga.* The method described is applicable only to males.
2. The Sorcerer's Apprentice is a Leeds-based (UK) supplier of books and paraphernalia for occultists. www.sorcerers-apprentice.co.uk.
3. Luk, *Taoist Yoga,* 185.
4. Ibid., 186.
5. Ibid.
6. The kabbalistically correct spelling, according to Hurtak.
7. These aura goggles can still be obtained from the Society of Metaphysicians .www.metaphysicians.org.uk.
8. Colin Wilson, *Mysteries,* 387.
9. Ostrander and Schroeder, *Psychic Discoveries behind the Iron Curtain,* 224, 225–42; and Watson, *Supernature,* 146–47 (Coronet paperback edition).
10. Watson, *Supernature,* 147 (Coronet paperback edition).
11. Jay Kinney, "The Mysterious Revelations of Philip K. Dick." *Gnosis Magazine,* vol. 1, winter 1985, p. 7.

12. Dick, *Valis,* appendix, paragraph 25.
13. Graves, *Needles of Stone.*
14. Ibid.
15. Devereux, *Places of Power,* 87–88.
16. Graves, *Needles of Stone,* 167.
17. Devereux, *Shamanism and the Mystery Lines,* 199, 201. Devereux is quoting from *Heart of the World,* a BBC-TV documentary, December 4, 1990.
18. Devereux, *Shamanism and the Mystery Lines,* 150, 179, and Devereux, *The Long Trip,* 100–104.
19. Devereux, *The Long Trip.*
20. http://home.zonnet.nl/postbus/hats.html and www.telegraph.co.uk/news/worldnews/europe/germany/1388038/Mysterious-gold-cones-hats-of-ancient-wizards.html.
21. Ibid.
22. www.friendsofthornborough.org/about.htm.
23. BBC-2 TV program, "Time Flyers," November 4, 2003.
24. See appendix 2, "The Astral Transition."
25. Devereux, *Earth Lights,* 228–29.
26. Charles, *The Book of Enoch.*
27. Genesis 5:23. Also mentioned in the New Testament in Hebrews 11:5.
28. Heath, *Sun, Moon, and Stonehenge,* 147.
29. Enoch 1:55 (R. H. Charles/SPCK edition), 93.
30. Knight and Lomas, *Uriel's Machine,* 263–64. Enoch 14:9, 71:5 (Charles/SPCK edition), 41, 93.
31. Heath, *Sun, Moon, and Stonehenge,* 149.
32. www.mythicalireland.com/ancientsites/gaelicmap.gif.
33. Knight and Lomas, *Uriel's Machine,* 227; Knight and Lomas, *The Book of Hiram,* 197. Knight and Lomas say this double spiral is based on an experiment by artist Charles Ross, and Martin Brennan shows the same diagram in his book *The Stones of Time,* 190.
34. Enoch 39:7.
35. Ibid., 62:16.
36. Ibid., 104:2.
37. Ibid., 108:15.
38. Genesis 7:2.

39. Exodus 34:28.
40. 1 Kings 19:8.
41. Matthew 4:2.
42. O'Brien, T., *Light Years Ago,* 34–39.
43. www.lucypringle.co.uk/articles/points/.
44. Ibid.
45. www.keelynet.com/interact/Arc_7_98-12_98/00001143.htm.
46. Ostrander and Shroeder, *Psychic Discoveries,* 236–38.
47. www.keelynet.com/energy/dimshift.htm.
48. Walter Rawls's website is www.magnetlabs.com.
49. The mask was made out of a modified pair of glasses. www.keelynet.com/interact/Arc_1_98-7_98/00001120.htm.
50. www.keelynet.com/biology/biomag2.htm.
51. www.keelynet.com/energy/dimshift.htm.
52. http://divinecosmos.com/index.php?option=com_content&task=view&id=88&Itemid=36.
53. Baker, Mather, and Kennaugh, *Magnetic Bones in Human Sinuses,* referenced in Budden, *Electric UFOs,* 80.
54. Budden, *Electric UFOs.*
55. Ibid., 82.
56. Devereux, *Earthlights.*
57. Budden, *Electric UFOs,* 89.
58. Ibid., 231.
59. Devereux, *Places of Power,* 124.
60. See also http://divinecosmos.com/index.php?option=com_content&task=view&id=47&Itemid=30.
61. Séjourné, *Burning Water,* 157.
62. Butler, *Magic and the Qabalah,* 93.
63. Persinger and Lafrenière, *Space-Time Transients and Unusual Events,* 242–43.
64. Ibid., 243–44.
65. Teilhard de Chardin, *The Future of Man,* quoted in Devereux, Steele, and Kubrin, *Earthmind,* 142.
66. Alien Dreamtime—a live multimedia performance with Space Time Continuum. Recorded live on February 27, 1993, in San Francisco. Available on CD (Astralwerks/Caroline Records) and DVD (Magic Carpet Media 2003).

APPENDIX 1.
TWO ACCOUNTS OF OUT-OF-BODY EXPERIENCES

1. Brown, *Brainchild,* 1–2.
2. Terence McKenna, *True Hallucinations,* 60–61.

APPENDIX 2. THE ASTRAL TRANSITION

1. www.house-of-horus.de/lpfinal.html.
2. Fix, *Star Maps.*
3. Freke and Gandy, *The Jesus Mysteries,* 61.
4. http://web.archive.org/web/20060115134411/http://www.solsticestudios.net/santawriting.htm.
5. Jenkins, *Maya Cosmogenesis 2012,* 193.
6. Devereux, *The Long Trip,* 88.

APPENDIX 3. CROPGNOSIS

1. www.bltresearch.com/plantab.html.
2. www.bltresearch.com/otherfacts.html.
3. www.nhne.com/specialreports/srcropcircles/int.pringle.html.
4. www.gnosticchristianity.com.
5. www.gnosticchristianity.com/ch9.htm.

BIBLIOGRAPHY

Airaksinen, M. M., J. Gunther, A. Poso, J. C. Callaway, and C. Navajas. "Structural requirements for high binding affinity to the 5-HT uptake transporter for beta-carbolines." *British Journal of Pharmacology* 104 (1991): 370.

Airaksinen, M. M., J. C. Callaway, L. Rägo, P. Nykvist, E. Kari, and J. Gynther. "Binding sites for (3H) pinoline (in Melatonin and the Pineal Gland)." *Excerpta Medica.* Amsterdam: Elsevier Science Publishers B.V., 1993.

Alford, Alan. *Gods of the New Millennium: The Shattering Truth of Human Origins.* London: Hodder and Stoughton, 1996.

Allan, D. S., and J. B. Delair. *Cataclysm! Evidence of a Cosmic Catastrophe in 9,500 BC.* Santa Fe, N. Mex.: Bear & Co., 1997.

Andahadna, Soror. "Liber Pennae Praenumbra." *The Cincinnati Journal of Ceremonial Magick* 1, no. 1 (1976).

Appleby, Nigel. *Hall of the Gods: The Quest to Discover the Knowledge of the Ancients.* London: William Heinemann, 1998.

Argüelles, José. *Earth Ascending: An Illustrated Treatise on the Laws Governing Whole Systems.* Boulder, Colo.: Shambhala Publications Inc., 1984.

———. *The Mayan Factor: Path Beyond Technology.* Santa Fe, N. Mex.: Bear & Co., 1987.

———. *The Transformative Vision: Reflections on the Nature and History of Human Expression.* Berkeley: Shambhala Publications, 1975.

Arthur, James. *Mushrooms and Mankind: The Impact of Mushrooms on Human Consciousness and Religion.* San Diego, Calif.: Book Tree, 2000.

Arundale, G. S. *Kundalini: An Occult Experience.* Adyar, Madras, India: The Theosophical Publishing House, 1947.

Atwater, P. M. H. *Future Memory*. Charlottesville, Va.: Hampton Roads Publishing Company Inc., 1999.

Aveni, Anthony. *Stairway to the Stars: Skywatching in Three Great Ancient Cultures*. London: Cassell, 1997.

Azena. St. Germain: *Earth's Birth Changes*. Perth, Australia: Triad Publishers, 1991.

Baigent, Michael, and Richard Leigh. *The Dead Sea Scrolls Deception*. London: Corgi, 1991.

Baigent, Michael, Richard Leigh, and Henry Lincoln. *The Holy Blood and the Holy Grail*. London: Corgi, 1982.

Baillie, Mike. *Exodus to Arthur: Catastrophic Encounters with Comets*. London: B. T. Batsford Ltd., 1999.

Baker, Douglas. *The Opening of the Third Eye*. Wellingborough, England: The Aquarian Press, 1977.

Baker, R. R., J. Mather, and J. Kennaugh. "Magnetic Bones in Human Sinuses." *Nature* 301 (1983).

Bauval, Robert, and Adrian Gilbert. *The Orion Mystery: Unlocking the Secrets of the Pyramids*. London: Mandarin, 1995.

Bauval, Robert, and Graham Hancock. *Keeper of Genesis: A Quest for the Hidden Legacy of Mankind*. London: Mandarin, 1996.

Berlitz, Charles, and William Moore. *The Philadelphia Experiment: Project Invisibility*. London: Souvenir Press, 1979.

Bernbaum, Edwin. *The Way to Shambhala: A Search for the Mythical Kingdom Beyond the Himalayas*. Boston: Shambhala Publications Inc., 2001.

Boerman, Robert J. *Crop Circles, Gods and Their Secrets*. Brummen, The Netherlands: The Ptah Foundation, 2000.

The Book of Enoch translated by R. H. Charles. London: SPCK, 1997.

Braden, Gregg. *Awakening to Zero Point: The Collective Initiation*. Bellevue, Wash.: Radio Bookstore Press, 1993.

Brennan, Martin. *The Stones of Time: Calendars, Sundials, and Stone Chambers of Ireland*. Rochester, Vt.: Inner Traditions, 1994.

Brinkley, Dannion. *Saved by the Light: The True Story of a Man Who Died Twice and the Profound Revelations He Received*. New York: Harper Mass Market Paperbacks, 1994.

Brotherston, Gordon. *The Book of the Fourth World: Reading the Native Americas Through Their Literature*. Cambridge: Cambridge University Press, 1992.

Brown, David Jay. *Brainchild*. Las Vegas: Falcon Press, 1988.

Brunton, Paul. *A Search in Secret Egypt*. London: Rider and Co., 1969.

Budden, Albert. *Electric UFOs*. London: Blandford, 1998.

Budge, E. A. Wallis. *The Book of the Dead*. New York: Gramercy Books, 1999.

Butler, W. E. *Magic and the Qabalah*. Wellingborough, England: The Aquarian Press, 1978.

Cain, Ivan, and Dora Cain. *The Year 2012*. Frederick, Md.: PublishAmerica Inc., 2002.

Callaway, J. C. "A proposed mechanism for the visions of dream sleep." *Medical Hypotheses* 26 (1988): 119–24.

Calleman, Carl Johan. *Solving the Greatest Mystery of Our Time: The Mayan Calendar*. Coral Springs, Fla.: Garev Publishing International, 2001.

Carey, Ken. *Starseed: The Third Millennium—Living in the Posthistoric World*. San Francisco: Harper, 1991.

———. *The Starseed Transmissions: An Extra-Terrestrial Report*. Kansas City, Mo.: Uni-Sun, 1982.

———. *Vision*. Kansas City, Mo.: Uni-Sun, 1985.

Carroll, Peter. *Liber Kaos*. Boston: Red Wheel/Weiser, 1992.

Charles, R. H., trans. *The Book of Enoch*. London: SPCK, 2006.

Cheetham, Erika, ed. *The Prophecies of Nostradamus*. London: Corgi, 1991.

Clow, Barbara Hand. *Catastrophobia: The Truth behind Earth Changes and the Coming Age of Light*. Rochester, Vt.: Bear & Co., 2001.

———. *The Pleiadian Agenda: A New Cosmology for the Age of Light*. Santa Fe, N. Mex.: Bear & Co., 1995.

Coe, Michael D. *The Maya*. London: Thames and Hudson, 1966.

———. *Mexico*. London: Thames and Hudson, 1962.

Collins, Andrew. *Gateway to Atlantis: The Search for the Source of a Lost Civilisation*. London: Headline Book Publishing, 2000.

———. *Gods of Eden: Egypt's Lost Legacy and the Genesis of Civilisation*. London: Headline Book Publishing, 1998.

Cotterell, Maurice M. *Astrogenetics*. Saltash, England: Brooks Hill Robinson and Co., 1988.

———. *The Lost Tomb of Viracocha: Unlocking the Secrets of the Peruvian Pyramids*. London: Headline Book Publishing, 2001.

———. *The Supergods*. London: Thorsons, 1997.

———. *The Tutankamun Prophecies: The Sacred Secret of the Mayas, Egyptians and Freemasons*. London: Headline Book Publishing, 1999.

Crockett, Stephen E. *The Prophet Code: Precessional Encryption and the Apocalyptic Tradition*. Mount Gilead, N.C.: Aethyrea Books LLC, 2001.

Crowley, Aleister. *Magick*. London: Guild Publishing, 1986.

Davis, Beth, ed. *Ciphers in the Crops: The Fractal and Geometric Circles of 1991*. Bath, U.K.: Gateway Books, 1992.

Devereux, Paul. *Earth Lights*. Wellingborough, England: Turnstone Press, 1982.

———. *Earth Lights Revelation*. London: Blandford Press, 1989.

———. *The Long Trip: A Prehistory of Psychedelia*. London: Arkana, 1997.

———. *Places of Power: Measuring the Secret Energy of Ancient Sites*. London: Blandford, 1990.

———. *Shamanism and the Mystery Lines: Ley Lines, Spirit Paths, Shape-Shifting and Out-of-Body Travel*. Slough, England: Quantum, 1992.

Devereux, Paul, John Steele, and David Kubrin. *Earthmind: Communicating with the Living World of Gaia*. Rochester, Vt.: Destiny Books, 1982.

Dick, Philip K. *In Pursuit of Valis: Selections from the Exegesis*. Lancaster, Pa.: Underwood-Miller, 1991.

———. *Valis*. New York: Vintage Books, 1991.

Dimde, Manfred. *Nostradamus, Das Apokalyptische Jahrzehnt*. Munich: Heyne Buchen, 2001.

Drosnin, Michael. *The Bible Code*. London: Weidenfeld and Nicolson, 1997.

———. *The Bible Code 2: The Countdown*. London: Weidenfeld and Nicolson, 2002.

Erbe, Peter O. *God I Am—From Tragic to Magic*. Perth, Australia: Triad Publishers PTY Ltd., 1991.

Ereira, Alan. *The Heart of the World*. London: Jonathan Cape, 1990.

Essene, Virginia, and Sheldon Nidle. *You Are Becoming a Galactic Human*. Santa Clara, Calif.: Spiritual Education Endeavors Publishing Company, 1994.

Fix, William R. *Star Maps*. London: Octopus Books, 1979.

Freke, Timothy, and Peter Gandy. *The Jesus Mysteries*. London: Thorsons, 1999.

Frissell, Bob. *Nothing in This Book Is True, but It's Exactly How Things Are*. Berkeley: Frog Ltd., 1994.

Fulcanelli. *Le Mystère des Cathédrals*. Paris: J. Schmeit, 1957.

Furlong, David. *The Keys to the Temple*. London: Piatkus Books, 1997.

Gaspar, Dr. William A. *The Celestial Clock*. Chacon, N. Mex.: Adam and Eva Publishing, 1999.

Geryl, Patrick, and Gino Ratinckx. *The Orion Prophecy: Will the World Be Destroyed in 2012?* Kempton, Ill.: Adventures Unlimited Press, 2001.

Gilbert, Adrian. *Signs in the Sky: Prophecies for the Birth of a New Age*. London: Bantam, 2000.

Gilbert, Adrian, and Maurice M. Cotterell. *The Mayan Prophecies: Unlocking the Secrets of a Lost Civilization*. Shaftesbury, England: Element Books, 1995.

Gillette, Douglas. *The Shaman's Secret: The Lost Resurrection Teachings of the Ancient Maya*. New York: Bantam, 1997.

Godwin, Joscelyn. *Mystery Religions in the Ancient World*. London: Thames and Hudson, 1981.

Grant, Kenneth. *Outside the Circles of Time*. London: Frederick Muller Ltd., 1980.

Grasse, Ray. *Signs of the Times: Unlocking the Symbolic Language of World Events*. Charlottesville, Va.: Hampton Roads Publishing Company Inc., 2002.

Graves, Tom. *Needles of Stone*. St. Albans, Herts, U.K.: Granada, 1978.

Guichhait, R. B. "Biogenesis of 5-methoxy-*N,N*-dimethyltryptamine in human pineal gland." *Journal of Neurochemistry* 26 (1976).

Gurdjieff, G. *All and Everything (Beelzebub's Tales to His Grandson)*. London: Routledge & Kegan Paul Ltd., 1950.

Hail, Raven. *The Cherokee Sacred Calendar: A Handbook of the Ancient Native American Tradition*. Rochester, Vt.: Destiny Books, 2000.

Hancock, Graham. *Fingerprints of the Gods: A Quest for the Beginning and the End*. London: William Heinemann Ltd., 1995.

———. *Heaven's Mirror: Quest for the Lost Civilization*. London: Michael Joseph Ltd., 1998.

———. *The Mars Mystery: A Tale of the End of Two Worlds*. London: Michael Joseph Ltd., 1998.

Harary, Keith, and Pamela Weintraub. *Lucid Dreams in 30 Days: The Creative Sleep Programme*. Wellingborough, England: The Aquarian Press, 1990.

Hart, Will. *The Genesis Race: Our Extraterrestrial DNA and the True Origin of the Species*. Rochester, Vt.: Bear & Co., 2003.

Heath, Robin. *Sun, Moon, and Stonehenge: Proof of High Culture in Ancient Britain*. Cardigan, Wales: Bluestone Press, 1998.

Hodge, K. A., and M. A. Persinger. "Quantitative increases in temporal lobe symptoms in human males are proportional to postnatal geomagnetic activity: Verification by canonical correlation." *Neuroscience Letters* 125, no. 2 (1991).

Hogue, John. *Nostradamus: The Complete Prophecies*. Shaftesbury, England: Element Books Ltd., 1997.

Hope, Murry. *Ancient Egypt: The Sirius Connection*. Shaftesbury, England: Element Books, 1990.

———. *Time: The Ultimate Energy*. Shaftesbury, England: Element Books, 1991.

Howard, Kenneth W. *The Final Days of Mankind*. Middletown, Ohio: Direction Inc., 2002.

Hurtak, J. J. *The Book of Knowledge: The Keys of Enoch*. Los Gatos, Calif.: The Academy for Future Science, 1996.

Hutchison, Michael. *Mega Brain Power: Transform Your Life with Mind Machines and Brain Nutrients*. New York: Hyperion, 1994.

Jenkins, Elizabeth B. *Initiation: A Woman's Spiritual Adventure in the Heart of the Andes*. London: Piatkus, 1997.

Jenkins, John Major. *Galactic Alignment: The Transformation of Consciousness According to Mayan, Egyptian, and Vedic Traditions*. Rochester, Vt.: Bear & Co., 2002.

———. *Maya Cosmogenesis 2012: The True Meaning of the Maya Calendar End-Date*. Santa Fe, N. Mex.: Bear and Co., 1998.

———. *Tzolkin: Visionary Perspectives and Calendar Studies*. Garberville, Calif.: Borderland Sciences Research Foundation, 1994.

Johnson, Kenneth Rayner. *The Fulcanelli Phenomenon*. London: C. W. Daniel Co., Ltd., 1980.

Kalb, Ken. *The Grand Catharsis*. Santa Barbara: Calif.: Lucky Star Research Institute, 1998.

Karpinski, Gloria. *Barefoot on Holy Ground: Twelve Lessons in Spiritual Craftsmanship*. Albuquerque: Wellspring/Ballantine, 2001.

Keene, J. J. *Cosmic Locusts: The 2012 Convergence of Nostradamian, Mayan & Biblical Prophecy*. Privately published: ISBN: 0-9724582-0-4, 2002.

Knight, Christopher, and Robert Lomas. *The Book of Hiram: Freemasonry, Venus and the Secret Key to the Life of Jesus*. London: Century, 2003.

———. *Uriel's Machine: The Prehistoric Technology That Survived the Flood*. London: Century, 1999.

LaBerge, Stephen. *Lucid Dreaming*. New York: Ballantine Books, 1998.

LaPlante, Eve. *Seized*. New York: HarperCollins, 1993.

LaViolette, Paul. *Earth Under Fire: Humanity's Survival of the Apocalypse*. New York: Starlane Publications, 1997.

Lehner, Mark. *The Complete Pyramids*. London: Thames and Hudson, 1997.

Lucas, Laurence. *The 21st Century Book of the Dead*. Lechlade, England: StarDrum Books, 2002.

Luk, Charles. *Taoist Yoga*. London: Rider, 1970.

Mardyks, Raymond, and Stacia Alana-Leah. *Maya Calendar: Voice of the Galaxy*. Sedona, Ariz.: Star Heart Publications, 1999.

Masters, Robert. *The Goddess Sekhmet: The Way of the Five Bodies*. New York: Amity House, 1988.

McFadden, Steven. *Profiles in Wisdom*. Santa Fe, N. Mex.: Bear & Co., 1991.

McIsaac, W. M., P. A. Khairallah, and I. H. Page. "10-methoxy harmalan, a potent serotonin antagonist which affects conditioned behaviour." *Science* 134 (1961): 674–75.

McKenna, Terence. *True Hallucinations*. San Francisco: HarperSanFrancisco, 1993.

———. "World Youth Culture, Shamanism, Gaia, Chaos, and Plant Hallucinogens." Lecture at ULU, London. September 25, 1991.

McKenna, Terence, and Dennis McKenna. *The Invisible Landscape: Mind, Hallucinogens, and the I Ching*. New York: Seabury Press, 1975.

Men, Hunbatz. *Calendrios Mayas Y Hunab K'U*. Juarez, Mexico: Ediciones Horizonte, 1983.

———. *Secrets of Maya Science/Religion*. Santa Fe, N. Mex.: Bear & Co., 1990.

Michell, John. *The View Over Atlantis*. London: Sphere Books Ltd., 1969.

Middleton-Jones, Howard, and James Michael Wilkie. *Giza-Genesis: The Best Kept Secrets*, vol. 1. Tempe, Ariz.: Dandelion Books, 2001.

Mini, John. *Day of Destiny: Where Will You Be August 13, 1999?* Sausalito, Calif.: Trans-Hyperborean Institute of Science Publishing, 1998.

Monroe, Robert A. *Ultimate Journey*. New York: Doubleday, 1994.

Morton, Chris, and Ceri Louise Thomas. *The Mystery of the Crystal Skulls: Unlocking the Secrets of the Past, Present and Future*. London: Thorsons, 1997.

Mouravieff, Boris. *Gnosis: Study and Commentaries on the Esoteric Tradition of Eastern Orthodoxy. Book One: Exoteric Cycle*. Robertsbridge, England: Praxis Institute Press, 1989.

———. *Gnosis: Study and Commentaries on the Esoteric Tradition of Eastern Orthodoxy. Book Two: Mesoteric Cycle*. Robertsbridge, England: Praxis Institute Press, 1992.

———. *Gnosis: Study and Commentaries on the Esoteric Tradition of Eastern*

Orthodoxy. Book Three: Esoteric Cycle. Robertsbridge, England: Praxis Institute Press, 1993.

Muldoon, Sylvan, and Hereward Carrington. *The Projection of the Astral Body*. London: Rider & Co. Ltd., 1929.

Musaios (Charles Musès). *The Lion Path—You Can Take It with You: A Manual of the Short Path to Regeneration for Our Times*. Sardis, B.C., Canada: House of Horus, 1987.

Musès, Charles, and A. M. Young, eds. *Consciousness and Reality*. New York: Outerbridge and Lazard, 1972.

Naranjo, Claudio. "Psychotropic properties of the harmala alkaloids, in ethnopharmacologic search for psychoactive drugs." *Public Health Service Publication* no. 1645 (1967).

Naydler, Jeremy. *Temple of the Cosmos: The Ancient Egyptian Experience of the Sacred*. Rochester, Vt.: Inner Traditions, 1996.

Nema (Margarete Ingalls). *Maat Magick: A Guide to Self-Initiation*. Boston: Red Wheel/Weiser, 1995.

Noah, Joseph. *Future Prospects of the World According to the Bible Code*. Cameron Park, Calif.: New Paradigm Books, 2002.

North, John. *Stonehenge: Neolithic Man and the Cosmos*. London: HarperCollins, 1996.

Novak, Peter. *The Division of Consciousness: The Secret Afterlife of the Human Psyche*. Charlottesville, Va.: Hampton Roads Publishing Company Inc., 1997.

O'Brien, Christian. *The Genius of the Few*. London: Turnstone Press, 1985.

O'Brien, Shane. *The Venus Legacy: When the Moon Turns to Blood*. Telford, England: Cumorah Hill, 1998.

O'Brien, Tim. *Light Years Ago*. Dublin: The Black Cat Press, 1992.

Ostrander, Sheila, and Lynn Schroeder. *Psychic Discoveries behind the Iron Curtain*. London: Sphere Books, 1973.

Persinger, Michael A., and Gyslaine F. Lafrenière. *Space-Time Transients and Unusual Events*. Chicago: Nelson-Hall, 1977.

Petit, J. R., J. Jouzel, et al. "Climate and atmospheric history of the past 420,000 years from the Vostok ice core in Antarctica." *Nature* 399 (June 1999).

Phillips, Graham. *Act of God: Tutankhamun, Moses and the Myth of Atlantis*. London: Sidgwick and Jackson, 1998.

Poynder, Michael. *Pi in the Sky: A Revelation of the Ancient Wisdom Tradition*. London: Rider, 1992.

Ramanda, Azena. *St. Germain: Earth's Birth Changes*. Perth, Australia: Triad Publishers PTY Ltd, 1991.

Ratsch, Christian, ed. *Gateway to Inner Space: Sacred Plants, Mysticism and Psychotherapy*. Lindfield, New South Wales, Australia: Unity Press, 1990.

Rayner Johnson, Kenneth. *The Fulcanelli Phenomenon*. Jersey, Channel Islands: Neville Spearman, 1980.

Reyner, J. H. *The Gurdjieff Inheritance*. London: Turnstone Press, 1989.

Rinkel, Max, M.D., ed. *Chemical Concepts of Psychosis*. London: Peter Owen Ltd., 1960.

Rodnight, R., et al. "Urinary dimethyltryptamine and psychiatric symptomatology and classification." *Psychological Medicine* 6, no. 4 (November 1976): 649–57.

Roney-Dougal, Serena. *Where Science and Magic Meet: Exploring Our Psychic Birthright*. Frome, England: Vega, 1997.

Rudisill, Elliott. *E.Din: Land of Righteousness*. n.p.: Author House, 2001.

Russell, Peter. *The Awakening Earth: The Global Brain*. London: Ark Paperbacks, 1982.

———. *The White Hole in Time: Our Future Evolution and the Meaning of Now*. Chatham, England: Aquarian Press, 1992.

Santillana de, Giorgio, and Hertha von Dechend. *Hamlet's Mill: An Essay on Myth and the Frame of Time*. 1977–1992 editions. Boston: David R. Godine Publisher Inc., 1969–1992.

Scallion, Gordon-Michael. *Notes from the Cosmos: A Futurist's Insights into the World of Dream Prophecy and Intuition*. W. Chesterfield, N.H.: Matrix Institute Inc., 1997.

Schwaller de Lubicz, R. A. *Sacred Science: The King of Pharaonic Theocracy*. Rochester, Vt.: Inner Traditions International, 1988.

Scofield, Bruce. *Day-Signs: Native American Astrology from Ancient Mexico*. Amherst, Mass.: One Reed Publications, 1991.

Scott, Mary. *Kundalini in the Physical World*. London: Arkana, 1989.

Séjourné, Laurette. *Burning Water: Thought and Religion in Ancient Mexico*. London: Thames and Hudson, 1957.

Shearer, Tony. *Beneath the Moon and Under the Sun: A Poetic Appraisal of the Sacred Calendar and the Prophecies of Ancient Mexico*. Albuquerque: Sun Books, 1975.

———. *Lord of the Dawn: Quetzalcoatl and the Tree of Life*. Healdsburg, Calif.: Naturegraph Press, 1971.

Sitchin, Zecharia. *Genesis Revisited*. New York: Avon Books, Inc., 1990.

———. *The Lost Realms*. New York: Avon Books, Inc., 1990.

———. *The Twelfth Planet*. New York: Avon Books, Inc., 1976.

Skinner, Stephen. *Millennium Prophecies: Predictions for the Year 2000 and Beyond from the World's Greatest Seers and Mystics*. London: Virgin Books/Carlton Books, 1994.

Straiton, Valentia. *The Celestial Ship of the North*. New York: Albert and Charles Boni, 1927.

Strassman, Rick. *DMT: The Spirit Molecule—A Doctor's Revolutionary Research into the Biology of Near-Death and Mystical Experiences*. Rochester, Vt.: Park Street Press, 2001.

Stray, Geoff. *The Mayan and Other Ancient Calendars*. New York: Walker, 2007.

———. *2012 in Your Pocket*. Virginia Beach: A.R.E. Press, 2009.

Teilhard de Chardin, Pierre. *The Future of Man*. London: Fontana, 1959.

Temple, Robert. *The Sirius Mystery: New Scientific Evidence for Alien Contact 5,000 Years Ago*. London: Century, 1998.

Thomas, Andy, ed. *SC: The Bimonthly Journal of Crop Circles and Beyond* 84 (May/June 1999).

———. *Swirled Harvest: Views from the Crop Circle Frontline*. Lewes, England: Vital Signs Publishing, 2003.

———. *Vital Signs: A Complete Guide to the Crop Circle Mystery and Why It Is NOT a Hoax*. Seaford, England: Frog Books, 1998.

Timms, Moira. *Prophecies to Take You into the Twenty-First Century*. London: Thorsons, 1996.

Vallee, Jacques. *Passport to Magonia: From Folklore to Flying Saucers*. London: Tandem Publishing Ltd., 1970.

Velikovsky, Immanuel. *Worlds in Collision*. London: Sphere Books, 1972.

Waters, Frank. *Book of the Hopi*. New York: Ballantine Books, 1969.

———. *Mexico Mystique: The Coming Sixth World of Consciousness*. Athens, Ohio: University Press Books, 1975.

Watson, Lyall. *Lifetide*. Sevenoaks, England: Hodder and Stoughton, 1979.

———. *Supernature: A Natural History of the Supernatural*. Sevenoaks, England: Hodder and Stoughton, 1973–1980.

Weidner, Jay, and Vincent Bridges. *A Monument to the End of Time: Alchemy, Fulcanelli and the Great Cross*. Mount Gilead, N.C.: Aethyrea Books LLC, 1999.

———. *The Mysteries of the Great Cross of Hendaye: Alchemy and the End of Time*. Mount Gilead, N.C.: Aethyrea Books LLC, 2003.

Weldon, Randolph. *Doomsday 2012: A Survival Manual*. Indianapolis, Ind.: Cork Hill Press, 2003.

West, John Anthony. *Serpent in the Sky: The High Wisdom of Ancient Egypt*. Wheaton, Ill.: Quest Books, 1993.

Whiteside, Richard Henry. *God's New Millennium*. n.p.: God's Web Publishing Company, 2002.

Wilcox, Joan Parisi. *Keepers of the Ancient Knowledge: The Mystical World of the Q'ero Indians of Peru*. Shaftesbury, England: Element Books, 1999.

William, Harold, and Burt Birnes. *Unsolved Mysteries: The World's Most Compelling Cases of Alien Encounter*. New York: Aspect, 2000.

Wilson, Colin. *Mysteries*. London: Grafton Books, 1986.

Wilson, Robert Anton. *Prometheus Rising*. Phoenix, Ariz.: Falcon Press, 1983.

———. *Right Where You Are Sitting Now: Further Tales of the Illuminati*. n.p.: And/Or Press, 1982.

———. "Trajectories Newsletter." *The Journal of Futurism and Heresy*, no. 7 (Spring 1990).

Wilson, Terry. *The Secret History of Crop Circles: The True, Untold Story of the World's Greatest Mystery*. Paignton, England: The Centre for Crop Circle Studies, 1998.

ABOUT THE AUTHOR

Geoff Stray has been studying the meaning of the year 2012 since 1982. Around 1996 he started to compile the information into a stapled booklet called *Beyond 2012.* In September 2000, he made it available on the Internet via his website, Diagnosis 2012 (www.diagnosis2012.co.uk). Since then, Diagnosis 2012 has become renowned as the most comprehensive database available on 2012, and has gathered input from every corner of the globe ("except that globes have no corners"). This book is the project's ultimate evolution.

Geoff is also the author of *The Mayan and Other Ancient Calendars* and *2012 in Your Pocket* and was one of the contributing authors of *The Mystery of 2012.* He has given presentations across England and as far afield as Finland, the United States, and Peru, and appears in the films *2012: The Odyssey; Timewave 2013; 2012: An Awakening;* and *Time of the Sixth Sun*. In his spare time he makes footwear, rides a dirt bike, and plays the blues harmonica to passing sheep. He lives in Glastonbury, UK.

INDEX

Numbers in *italic* represent illustrations.

11:11, 280, 282
13-baktun cycle. *See* thirteen-baktun cycle
2012 Unlimited website, 37
21st Century Book of the Dead, The (Lucas), 366
5-meo-DMT, 206, 211–12, 337
5-methoxy-DMT, 203
6-MTHH, 206
80 Greatest Conspiracies of All Time, The (Whalen), 82
9/11, *63*, 65–66, 87

Act of God (Phillips), 110
acupuncture, 98, 390
acute schizophrenic break, 343–45, 397
Adam, 79–80
Adams, Cecil, 288–89
Adey, Ross, 339
Aeon of Maat, 303–4
aeons, 303–4
Africa, 39
Age of Adam, 135, 149
Age of Gemini, 279
Age of Meeting Ourselves Again, 28
Age of Pisces, 277
Age of Taurus, 279
Age of the Father, pl.4, 28
Age of the Holy Spirit, pl.4, 28
Age of the Son, pl.4
Age of the Sun, 28
AIST (National Institute of Advanced Industrial Science & Technology), 162
akh, 350
alchemy and alchemists, 140–45, 357, 370
Alcon, Andy, 76
Alcyone, 105–6, 321
Aldebaran, 152–53
Alejandro, Don, 26, 302–3
Alford, Alan, 258, 276, 278
Al-Hawali, Abd Al-Rahman, 82
aliens. *See* extraterrestrials
Allah, 82
Allan, D. S., 165, 411
Allegro, John, 411–12
Allen, Carl, 259–60
Allen, Rush, 140, 147, 318
altered states
  future life pre-call, 231–32
  lucid dreams, 225–26
  meditation, 226–27
  NDEs (near-death experiences), 218–23, *220*

OBEs, 156–58, *157*, 212, 213, 223–24, *224*, 340, 403–5, 414
past-life recall, 228–31, *229*
See also hallucinogens and hallucinogenic experiences
Ambilac website, 136, 138
American Revolution, 104
Ancient Prophecy Revealed for 1999 website, 270–71
Andahadna, Soror, 306
Anderla, George, 194
angels and archangels, 81, 115, 222–23, 384
animals, 46–47
ankh, *149*
Annunaki, 289–90
Antichrist, 270–71
apocalypse, 72
Apollo objects, 133
Appleby, Nigel, 110–12, 125–26, *126*, 127–29, 135–36, 149, 229, 231
Aquarian Age Network, 278
Aquarian Triangle, *237*, 243, 247
Aquarius, Age of, *277*, 277–79
aquastats, 355
Arabia, Gulf of, 87
archaeology, alternative
crustal displacement, 124
Dendera zodiac, 122–24, *123*
descent of the New Jerusalem cube, 145–47, *146*
eye in the pyramid, *133–34*, 133–35
Giza-Genesis theory, 138–39
Great Pyramid, 135–36
Hall of Records, 124–26, *126*
Pyramid of Kukulcan, 130–32, *131–32*
rattling of the Pleiades, 132–33
rebirth of Osiris, 140
return of the Phoenix, 126–29, *127*, *129*
secrets of alchemists, 140–45, *141*, *143–44*
signs in the sky, 147–50
star map-to-ground correlation at Giza in 2012, 136–38
Archaeology-Archaeoastronomy-Ancient Civilizations website, 136–37
Arcturus, 298
Argemoto website, 342–43
Argüelles, José, 2, 3, 13, 23, 38, 51–53, 53–54, *67*, 189, 255, 283–84, 295, 297, 303, 309, 311, 318, 337
Armstrong, Nick, 37
artifacts, 97–98
Arundale, G. S., 353–54
Asian calendars, 41–55
Aspasia, 229
asteroids, 85, 114–15
astral projection, 377–78, 380–81, 382, 407–8. *See also* OBEs
Astrogenetics (Cotterell), 92, 97–98
astrology, 49
astronomy, claims of
astronomical phenomena, 100–103
axis tilt and, 116–17
Behemoth and Leviathan, 114–15
birth and death of Venus, *107–8*, 107–10
galactic core explosion, 117–20, *118–19*
Nibiru, 110–13, *111–13*
Phobos flyby, 113–14
Venus-Pleiades conjunction, *105–6*, 105–7
Venus transit theory, 103–5
Atahualpa, 361
Athena, 107
Atlantis, 132–33, 150–51, 153, 173
atomic bombs, 3, 59
Atwater, P. M. H., 196–97, 222–23
Aubrey holes, 173

auras, 372–73, 380
Auric Time Scale, 188–92, *190*, 317–18, 332–35, *334*, 341
Auric Time Scale and the Mayan Factor, The (Smelyakov & Karpenko), 188
Avebury, 388
Aveni, Anthony, 9–10
Awakening Earth, The (Russell), 105, 193, 198–99
Awakening to Zero Point (Braden), 155–56
Axayacatl, 18
axis tilt, 116–17, 164, 164–65, 168, 168–69
ayahuasca, 204, 206
Aztec Pizza crop circle, 255–57
Aztecs, 21–23, 32, 52–53, 172, 310–11
Aztec Sunstone, pl.1, pl.3, *18–19*, 18–21, *21*, 247, 255–56, 311

ba, 350
Bache, Christopher, 223
Bailey, Alice, 395
Baillie, Mike, 174, 177
Baker, Douglas, 46
baktuns, 134, 158–59, 254–55, 257, 291, 293, 294–95. *See also* thirteen-baktun cycle
balls of light (BOLs), 326–27, 382
Barbury Castle, 242–47
bardo body, 144, 150
Barefoot on Holy Ground (Karpinski), 227
Barrios, Carlos, 26–27
Barrios, Gerardo, 27
Bascule, Orryelle, 304–5
Bauval, Robert, 118–19, 128, 135, 149, 410
Behemoth, 114–15
Bell, Alexander Graham, 105
Bell, Art, 223, 263
Beltz, Ellin, 158–59
Beneath the Moon and Under the Sun (Shearer), 51
Benest, Daniel, 33
Bentov, Itzhak, 144
Berlitz, Charles, 259–60, 272
betacarbolines, 209
Bible, 72–74, 75–77, 87, 347, 347–48, 417. *See also* New Testament; Old Testament
Bible Code, 84–88
Bible Code, The (Drosnin), 84–86
bicycle accident, 211–14
blind springs, 355
BLT Research (and website), 322–26, 415–16
Blue Star Kachina, pl.5, 30–33, 298
Boerman, Robert, 257–58, 278
Boerner, Brendan, 68
Book of the Hopi (Waters), 31, 32
Bosman, Ananda, 211–14, 366–67
Bower, Doug, 242–43
Boylan, Richard, 241
Braden, Gregg, 119, 155–56, 158–59, 298, 339–40, 370, 397
Brasseur de Bourbourg, Charles Étienne, 107–8
Brennan, Martin, 98
Bridges, Vincent, 141–42, 143–44, 150, 278, 280, 318, 358, 396
Brinkley, Dannion, 219–22, 337
Bronze Age, 142
Brotherston, Gordon, 172
Broughton, Glenn, 324–25
Brown, David Jay, 403
Buck, Bob, 178–79
Budden, Albert, 392
Buddha, 50
Buddhism, 50
bufotenine, 203
Burke, John, 322–23, 415

Cain, Ivan and Dora, 70
cairns, *99*
Cairo, *137–38*, 137–38
Calendar Round, 21–23, 102, 132, 133
calendars. *See* specific calendars
Calendar Stone. *See* Aztec Sunstone
Callaway, Carl, 337
Callaway, J. C., 213
Calleman, Carl Johan, 103, 104–5, 291–97
Cameron, Edward, 260, 262
Canada, 88
Carey, Ken, 197–99, 290, 394
Carn Ingli, 393
Carrington rotation, 331, 367–68
Carroll, Peter, 304
Cassiopaeans, 289–90
Cassiopaea website, 289–90
Catastrophobia (Clow), 230, 280, 411
cathedrals, 140
Cathie, Bruce, 179
Cayce, Edgar, 125, 128, 179, 230
Celestial Clock, The (Gaspar), 164–65, 316
celestial gates, 43–46, *44*
Celestial Ship of the North, The (Straiton), 347–48
Center of the Sun website, 253
Cetus, 249
chakras, *25*, 25, 30, *46*, 46, 208–11, *211*, 306, 352–53
Chambers of the First Creation, *125*, 125–26
channelers and channeling, 179–80, 197, 286–87
Chaos Magick, 304–5
Chatelain, Maurice, 179
Cherokee calendar, 36–37, *37*
Cherokee Nation, 35–36
Cherokee Sacred Calendar, The (Hail), 35
Chilam Balam of Chumayel, 132
China, 87
Chinese calendar, 47
Chorley, Dave, 243
Christ. *See* Jesus
Christians and Christianity, 70–78, 82, 192, 386, 408, 417–20
Christmas, 412–13
cleansing, 366–67
climate, 163
Clow, Barbara Hand, 113, 229, 230, 280, 288, 320–21, 411
Codex Magliabecchiano, 20, *21*, 247
Coe, Michael, 32, 124, 198
Collins, Andrew, 124–25, 132
Comas Sola, José, 285
comets, 31, 85–86, 132, 138, 172–74, 177, 183, 266–67, 269–70
communication, 104–5
compassion, 144–45
Complete Pyramids, The (Lehner), 139
Complexity Consciousness, Law of, 192–94
compression, 194
Computer Revolution, 195
concrescence, 59, 67
Conrow, Edgar, 347–48
Contact video (Janssen), 327
Coptic Church, 142
Coricancha, 367–69
coronal mass ejection (CME), pl.8, 94–95
Cortés, Hernan, 22–23, 52, 53, 104
cosmic dust, 318–20. *See also* galactic core explosion
Cosmic Locusts (Keene), 77, 177
Cotterell, Maurice, 91, 108–9, 111, 165, 169, 248, 309, 316–18, 339
Craig, John, 237–38
Crane, Ian, 147
Creation Day, 17, 78
critical mass, 192–94, 198

Crockett, Stephen E., 78
Crocodile Tree, 15, *17. See also* Milky Way
crop circles, 236
analysis of, 322–26, *323*
Aztec Pizza, 255–57
Barbury Castle, 242–47
calendar references of, 247–48
endocrine systems and, 389–90
Gnostic perspective of, 415–20
Maya glyphs, 252–55
Nibiru and, 257–58
pictured, pl.17–20, *243–45, 250–54, 256, 389–90*
quincunxes, 246–47
sacred sites and, 388–90
as solar predictors, 248–49
as Venus transit predictor, 250–52
Crop Circles, Gods and Their Secrets (Boerman), 257–58
Crossing points, 36–37
Cross of Hendaye, pl.16, 140–44, *141, 143–44*, 146, 318–19, 357–59, 364, 396
crustal displacement, 124, 127
crystal skulls, 299–303, *300*, 313–14
Cuauhtitlan Annals, 20
Cuba, 132
cube of space, 145–47, *146*
Cumbo, Nick, 226
Cuzco, 144, 358, 367
CyberSky software, 105–6

Dajjal, 81–82
Dames, Ed, 263
Daniel (prophet), 73–74
Daniel, Book of, 72–74, 75–77
Davis, Albert Roy, 391
Day of Destiny (Mini), 313
Day of Judgement, 75
Day of Wrath, The (Al-Hawali), 82
day-signs, 10, *11*, 35–36
Day-Signs (Scofield), 10, 35
Dean, Robert O. (Bob), 239–40
Dechend, Hertha von, 44–45, 123–24, 147–48, 410
Decker, Jerry, 390, 391–92
Declaration of Independence, 133–34
Dee, Sir John, 145
Deen, Glen, 81, 114, 269
Delair, J. B., 165, 411
Dendera, 113
Dendera, Temple of, 122–23, *123*, 153–54
Dendera zodiac, 122–24, *123*, 153–54
Deravy, Echan, 237–38
Devereux, Paul, 210, 327–28, 341–42, 368, 375, 377–78, 382, 393, 413–14
Diagnosis 2012 website, 3, 116, 152
Diamond Body, 358, 366–67
Diary of a Psychonaut (Copehead), 214–15
Dick, Philip K., 25, 133, 373, 374–75
Dimde, Manfred, 268
Dionysus, 409
Dire Gnosis project, 5
Divine Cosmos, The (Wilcock), 180, 181
Divinecosmos website, 180
Djed, 142, 147, 410–11
Dmitriev, Alexey N., 180–83, 321–22, 325, 328–29, 331, 332, 396, 415
DMT, 204, *205*, 205–8, *207*, 211, 336, 337, 339, 397
DMT: The Spirit Molecule (Strassman), 206–7
DNA, 24, 96
Dogon, 33–34, 39, 84
Doomsday 1999 AD (Berlitz), 272
Doomsday 2012 (Weldon), 228–29
Douglas, Karen, 255
dowsing, 97–98, 353–54
DreamImages website, 226
Dreampeace website, 226

Dreamspell, 295, 311
Dreamspell Time Shift, 27
Dresden Codex, 10, 12, 92–93, 233, 234, 250
Drosnin, Michael, 84–86
Duvent, J. L., 33

Earth, 96, 97–98, 110–11, 250
 ice cycles of, 164–70
 solar activity and, 163
 speed of core, 158–59
 See also axis tilt; critical mass; magnetic fields
Earth Ascending (Argüelles), 3, 67
Earth Changes Report (Scallion), 297–99
Earth kundalini, 353–57
Earth Lights (Devereux), 375, 382
earthquakes, 77, 88, 91, 96, 266–67
Earth's Birth Changes (Ramanda), 287
Earth Under Fire (LaViolette), 117–18, 120–21
ecdysone, 406–7
Echo Wheel, 37
eclipses, 49, 75, 95–96, 100, 233, 268
Ecliptic Pole, *16*, 16–17
ecstasy, 353
Edfu texts, 125
E.Din Land of Righteousness (Rudisill), 114–15, 116, 316
egg of creation, 125
Egypt and Egyptians, 149, 348–49, 349–52, 356–57, 407–8, 413–14
Egyptian Book of the Celestial Cow, 174
Egyptian Book of the Dead, 150–51, 154
El, 114–15
Elder Gods, 306
Elders, Lee and Britt, 233, 313
Eldridge, USS, 259–60, *260*
electromagnetic hypersensitivity (EMH), 392–93, 399
electromagnetic pulse (EMP wave), 120–21, *121*, 319
elements, 47
elves, 325
Emergence, 28–29
Emmanuel, 212
Encke, 173
endocrine glands, 91, 208–11, 317, 389–90
end of days, 73–74
end point, 158–59
energy body, 372–73, 391–92. *See also* chakras
Enoch, Book of, 110, 114, 289, 290, 382–86, 383, 384, 386
Epic of Gilgamesh, The, 171
equinoxes, 42, 97, 130, 148
Erbe, Peter O., 286
ESP (extrasensory perception), 331–32
Espinosa, Don Andreas, 363
Eta Carinae, 187
Ethiopian calendar, 83–84
Eve, 80
Exodus, 110–11, 228
Exodus, Book of, 87
Exodus 2006 website, 87–88
Exodus to Arthur-Catastrophic Encounters with Comets (Baillie), 174
extraterrestrials, 234–40, 239–40, 286–87
Eye in the Pyramid, *133–34*, 133–35
Eye of Horus, 119, 175–76, *175–76*
Eye of Ra, 112–13, 119, *175–76*, 175–76
Eye of the Centaur (Clow), 229
Ezekiel, Book of, 78

Fatima, 177
Feast of Trumpets, 75
feng shui, 98, 355
Feynman, Richard, 328
Fibonacci, Leonardo, 187–89

Fibonacci series, *187*, 187–88
field-effect energy reversal, 17
Fifth Sun, 247, 395
Fifth World, 27, 28, 34, 88, 104, 315
fig tree parable, 74–75, 150
Fingerprints of the Gods (Hancock), 124
Finley, Mike, 2, 234
First Father, One Hunahpu, *19*
First World, 28
Fletcher, Audrey, 279–80
floods and flooding, 91, 124, 132–33, 171–72, 386
Four Horsemen of the Apocalypse (Ward and Goodwin), 268–69
Fourth World, 27, 28–29, 34, 88, 315
Four Winds website, 361
Fractal Time, 57, 195
Fractal Time website, 57
fractal wave, 59
Franklin, Ben, 67
Franklin Timewave, 66–69, *68–69*
Freke, Timothy, 408
Frissell, Bob, 261–62
Fulcanelli, 140–45, 318–19
Fulcanelli Phenomenon, The (Johnson), 140–41
Furlong, David, 174–75
future life pre-call, 231–32
Future Memory (Atwater), 196–97
Future Prospects of the World According to the Bible Code (Noah), 87

Gabriel, Archangel, 81
Gaia, 105, 196
Gaia-The Trajectories of Her Evolution (Wilson), 195
galactic alignment, 15–18, *16*, 43–46, 127, 330–32, 396–97
Galactic Alignment (Jenkins, J. M.), 43, 347–48, 356
Galactic Anticenter, 45
Galactic Calendar, 115
Galactic Center, 15–17, 43, 64–65, 101, 115, 118, 177, 321
  crop circles and, 247
  EMP waves and, 120–21
galactic core explosion, 117–20, *118–19*, 144, 318–20, 396
galactic equator, 16–17
galactic synchronization, 283–84
Galactic Tsunami (Souers), 120
Gandy, Peter, 408
Gaspar, William A., 164–65, 316, 340, 386, 388
Gateway to Atlantis (Collins), 132
Geb, 175
Gemini, 44–45
Genesis, Book of, 78
Genesis Race, The (Hart), 103–4
genetic memory, 25
Geolab, 160
geomagnetic fields, 209–10, 341–43
geomagnetic reversal, 158–59, 181
geopsyche, 397–99
Gerloff, Sabine, 380–81
Geryl, Patrick, 129, 150–54, 316
ghosts, 210
Gibbs, Steven, 260–61
Gilbert, Adrian, 91, 108–9, 111, 118–19, 135, 147, 149–50, 150, 309, 316–18, 339
Gilson, Robert, 281–82, 337
Giza-Genesis: Best Kept Secrets (Middleton-Jones & Wilkie), 138–39, 318
Giza-Genesis theory, 138–39
Giza Plateau, 118–19, 124–26, 139, 140
Giza pyramids, 125–26, *129*, 129, 152–53
glaciation, 166–67. *See also* ice cycles
Glatzmaier, Gary, 159–60

Glickman, Michael, 248
global warming, 168–69
Gnostics and Gnosticism, 142, 346–48, 357, 374–75, 415–20
God, 78, 134
God I Am-From Tragic to Magic (Erbe), 286
Gods of Eden, The (Collins), 124–25
Gods of the New Millennium (Alford), 258, 276
GodsWeb website, 71
Godwin, Joscelyn, 44
Gold Age, 142
Golden Gate, 44–45, *45*, 148–49
golden hats, pl.22, 379–81
Golden Mean, 98, 187
golden ratio, 188–89
Gonzales, Patricia, 26
Goodman, Joseph, 12
Goodman correlation (GMT), 124
Goodwin, Gary, 268–69
government, secret, 259–63
Grand Catharsis, The (Kalb), 219, 337
Grant, Kenneth, 306
Grasse, Ray, 55
Graves, Tom, 375–76, 390
Great Conjunction, 279–81
Great Cycle. *See* thirteen-baktun cycle
Great Day of Purification, 31–32
Great Dreams website, 226
Great Pyramid, 83, 128, 135–36, 139, 156
Great Return, 142
Great Seal of America, *133–34*, 133–35
Great White Brotherhood, 286–87
Greeks, 73
Gregorian calendar, 1–3, 11–13, 51–52, 79
Grist, Brian, 246
Grondine, E. P., 174
Guardian (newspaper), 327–28
Guénon, René, 43
Gunn, Bert, 311
Gurdjieff, G., *42*, 148

haabs, 10–11, 12, *12*, 21–23, 53, 102–3, 133, 134, 251, 254, 303, 311
Hail, Raven, 35–36
Hale-Bopp, 31, 138, 190, 266
Hall of Records, 124–26, *126*, 128, 140, 318
Hall of the Gods (Appleby), 110, 125, 136, 316
hallucinogens and hallucinogenic experiences, 5, 50, 57, *67*, 75–77, 94–95, 203–8, 214–15, 336–40, 378, 379, 413–14
Hamilton, Pam, 238
Hamlet's Mill (de Santillana and von Dechend), 44–45, 113, 123–24, 147–48, 410
hanaq pacha, 28
Hancock, Graham, 118–19, 124, 128, 135, *346*, 347
Hannibal, 274
Harary, Keith, 225
Harding, Jan, 301, 381, 388
harmine, 336
Harmonic Convergence, 23, 38, 52, 156, 286–87, 303, 311–12
Harner, Michael, 314
Hart, Will, 103–4
HaShem, 80
Hatun Karpay, 363, 366
Hawara, 153
Healing the Luminous Body: The Way of the Shaman (video: Weidner), 360
Heath, Robin, 383–84
Heaven and Hell cycles, 38, 51–53
Heaven's Mirror (Hancock), *346*
Hell cycles. *See* Heaven and Hell cycles
Hemi-Sync, 404

Hendaye, 140–44
Hercules, Mike, 403–4
Hermetic Order of the Golden Dawn, The, 145
Hernandez, Juan Martinez, 12
Hessdalen valley, 326–28
Hesse, Paul Otto, 285, 289
higher selves, 17
Hiroshima, 59, 69, 84
Hoagland, Richard, 179
Hogue, John, 266
holocaust, 84
holographic energy field, 366
holons, 212–13
Homo luminous, 360
Homo spiritus, 96
Hope, Murry, 32, 112, 175
Hopi, 28–30, *29–30*, 30–33, 88, 305
Horus, 123
Howe, Linda Moulton, 179, 326
Huang Ti, *63*, 65–66, 69
Huayta, Willaru, 27
Hucha Mikhuy, 366
Huitzilpochtli. *See* Tonatiuh
Hulot, Gautier, 1590160
human spiritual embryogenesis, 17
Hunab Ku, 283. *See also* Galactic Center
Hunahpu, *18*
Hunter, Keith, 116–17
hurricanes, 77
Hurtak, J. J., 135, 289
Hussein, Saddam, 69
Hutchinson, John, 329–30
Hutchinson Effect, 329–30
Hutchison, Michael, 156
Hyades, 140, 152–53
Hyperdimensional Resonator, 260–61

ice age, 96, *118*, 124, 161–63
ice cores, 161–62
ice cycles, 164–70, *166–69*
I Ching, 2–3, 56–59, *60–61*, 98, 318
ida, 208
Illuminati, 134
Incas, 367–69
Industrial Revolution, 195
information doubling, 194–97, *196*
Initiation: a Woman's Spiritual Adventure in he Heart of the Andes (Jenkins), 364
Inka, 360–61, 362
Inka mallku, 364–65
InterUniversals, 212
Invisible Landscape, The, 2
Invisible Landscape, The (McKenna), 56, 336–37
Iron Age, 142
Isis, 123, 357
Islam, 81–82, 142
Israel, 150
Izapa, Mexico, 1

Jadczyk, Laura Knight, 290
Jaguar, 20
Jaguar-toad, *19*
Janssen, Bert, 327
Japan, 87
Jaykar, Smita, 231–32
Jenkins, Elizabeth, 359, 362–63, 364–66
Jenkins, John Major, 9–10, 20, 29, 332, 396–97, 411
  on end point of thirteen-baktun cycle, 148, 247–48
  on kundalini, 356
  on opening of celestial gates, 45–46
  on Pyramid of Kukulcan, 130–32
  reputation of, 15
  on sunspot cycles, 91–92
  on tzitzimime, 20
  on Vedic yuga system, 43

Jerusalem, 74, 82
Jesus, 13, 72, 73, 83–84, 174, 409
Jesus Mysteries, The (Freke & Gandy), 408
jewelry, 97–98
Jews and Judaism, 78–81, 110, 271
Joel, Book of, 177
Johnson, Kenneth Rayner, 140–41
Journal of Astronomy and Astrophysics, The, 33–34
Jubilee calendar, 116
Judgement Day, 71
Jung, Carl, 3, 326, 344
Jupiter, 9, 77, 183, 230
Jupiter-Saturn conjunction cycle, 49

kachinas, 30–33
Kalachakra prophecy, 47, 50–54
Kalb, Ken, 219, 337, 345
Kali Yuga, 41–43, *42*
Kalki, 54–55
Karpenko, Yuri, 188
Karpinski, Gloria, 226–27
katuns, 133
kay pacha, 28
Keeleynet website, 390
Keene, Joel J., 77, 177
Keeper of Genesis (Bauval and Hancock), 128
Keepers of the Ancient Knowledge (Wilcox), 27–28, 356
Kelley, Royce, 66, 69
Kerridge, David, 160
Keys of Enoch, The (Hurtak), 135, 283, 289
Keys to the Temple (Furlong), 172
Khafre Pyramid, 139
Khufu complex, 139
Kilner, Walter J., 372–73
King, Jon, 235–36, 243, 283–84, 337
King Wen sequence, *58*, 58–59, 66
kivas, *29–30*, 29–30
Knight, Christopher, 384–85
Knights Templar, 142, 384–85
Kogi people, 377
Kriyananda, Goswami, 55
Kuauhtlinxan, Ehekateoti, 311–12
Kukulcan. *See* Quetzacoatl
kundalini, 25, 46, 222, 305, 352–53, 353–57
Kundalini - An Occult Experience (Arundale), 353
Kyoto Protocol, 224, 345, 396

LaBerge, Stephen, 225
Labyrinth (lost Egyptian astronomy complex), 153, 154
Lafrenière, Gyslaine, 327–28, 397–99
Laing, R.D., 344
LaPlante, Eva, 340–41
Lash, John, 122–24, 154
Last Time of Osiris, 128–29
laughing gas, 214–15
LaViolette, Paul, 115, 117–18, 119, 120, 135, 144, 161–62, 177, 318–19, 321–22, 358, 396
Left Behind (video), 71
Legend (Rohl), 171
Legend of the Suns, 22
Lehner, Mark, 139, 153
Lemesurier, Peter, 274–75
Leo, 128
Levengood, William C., 322–23
Leviathan, 114–15
Lewis, Bill, 210
Ley Hunters Companion, The (Devereux), 382
ley lines, 355, 377, 382, 414
Liber Kaos, 304
Lion Path, The (Musès), 406
liquid lightbody, 366
Lockyer, Sir Norman, 356

Lomas, Robert, 384–85
Long Count, 11–12, 133–34, 174, 311
  Africa and, 39
  correlations with Gregorian calendar, 1–3
  crop circles and, 254
  extended version of, 291–97
Long Trip (The): A Prehistory of Psychedelia (Devereux), 414
Lord of the Dawn, 22–23
Lord of the Dawn (Shearer), 51
Lost Science of the Stone Age: Sacred Energy and the I Ching (Poynder), 98
Lounsbury, Floyd, 12
Lounsbury correlation, 124, 295
Lovecraft, H.P., 306
Lucas, Laurence, 366
lucid dreams, 225–26
Lucifer's Hammer (Niven and Pournelle), 71–72, 275
Luke, Gospel of, 74–75, 347–48
Lu K'uan Yu, 370
lunar calendars, 2–3
lunar nodes, 49

Maat Magick, 303–4
machine elves, 204, 330, 337, 405
Mack, John, 208
Mackey, S.A., 154
Macrobius, 43–44
Magic, Music, Myth rock opera, 70
magick, 303–5
magic squares, 66–67, *67*, 318
magnetic fields, 92–93, *93*, 96, 97–98, 178–79, 181, 298, 339–40
magnetic reversals, 93, 127–28, *155*, 155–58, *157*
  reversal of, 159–60, 393–94
magnetism, pineal gland and, 391–93
magnetized band of plasma, 183–88, *184–85*, 321–22, 332, 394, 396
major electrical event (MEE), 393
Mandelbrot, Benoit, 246
Mandelbrot set, 246
Mandeville, Michael, 126–27, 135–36
Maoris, 40, 315, 369
Mardyks, Raymond, 100–103, 105, 133, 134–35
Mark, Book of, 71–72
Mars, 9, 81, 110–11, 182–83, 265, 298–99
Mars Code, 270
Masonic calendar, 83
Masons, 83, 135, 352, 384–85
Matthew, Book of, 71–72, 347–48
Maunder Minimum, 104, 181
Maussan, Jaime, 233
Maya, 1, 32, 52, 78, 88, 317, 356–57
  collapse of, 97–98
  creation date of, 12
  galactic alignment and, 15–18
  precession of the solstices, 148
  Sirius and, 134–35
  Venus and, 104, 250–51
Maya, The (Coe), 124
Maya calendar systems, 156, 247–48
  Calendar Round, 21–23
  Calleman "solution" and, 291–97
  crop glyphs and, 252–55
  galactic alignment and, 15–18, *16–17*
  reliability of prophecies, 310–15
  See also Aztec Sunstone; haabs; Long Count; thirteen-baktun cycle; tzolkin
Maya Cosmogenesis 2012 (Jenkins), 1, 15, 20, 411
Mayan Factor, The (Argüelles), 2, 3, 13, 51, 189, 255, 283–84, 303, 309, 337
Mayan Prophecies, The (Gilbert & Cotterell), 91, 91–93, 97–98, 107, 126, 169, 309, 316, 339
Mazatecs, 215
McClellan, Michael, 267

McCormac, Gerry, 174
McCracken, Andy, 87–88
McFadden, Steven, 26–27, 31–32
McKenna, Dennis, 2, 56–59, 204, 336–37
McKenna, Terence, 2, 56–59, 63–65, 194, 204, 214, 246, 329–30, 336–37, 399, 404–5
Meaden, Terence, 323–26
meditation, 226–27, 394
melatonin, 212, 339, 389
Melchizedek, Drunvalo, 370
Men, Hunbatz, 23–24, 247, 321
Menghin, Wilfred, 380
Mercury, 67, 77, 137, 183
Merejildo, James Arévalo, 360
Mer-Ka-Ba, *155*, 156, 370–71
Mesopotamia, 173
Messenger, Stanley, 281
metal, 88
metatonia, 394
meteors, 114–15
Metonic cycle, 379
Mevryl, Paul, 141
Mexico Mystique (Waters), 32, 124, 247, 336
Meyer, Peter, 57, 59, 61, 65–66, 67–68
Michael, Archangel, 81, 115
Middleton-Jones, Howard, 136, 136–37, 138–39, 318
Milankovitch theory, 163, *166–69*, 166–70
Milky Way, 15, 118, 331, 413
Millennium Group, The, 268–69
Miller, Stephen, 275
Mini, John, 313
Mir space station, 31
Missing Earth formation, 250–51
Mitchell-Hedges, Anna, 299–300
Mitchell-Hedges, Frederick, 299–300
Mithras, 409
Monroe, Bob, 156–58, *157*
Montauk Project, 259–62
Monument to the End of Time, A (Weidner & Bridges), 141–42, *146*, 147, 318, 357
Moon, 49, 77, 96, 147, 181, 249, 355
Moore, William, 259–60
Mormons, 37
Morton, Chris, 109, 300–301, 313–14, 316
Motoyama, Hiroshi, 96
Mountain of the Ark (Rohl), 171
Mouravieff, Boris, 365
Mouth of the Crocodile, 15. *See also* Milky Way
Musès, Charles, 349, 351, 357, 367–68, 406–7
Musgrave, Cassandra, 87, 218–19, 350, 367, 404–5
mushrooms, 203–6, 214–15, 412–13
Mutwa, Credo, 38–39
Myrddin's Warning webpage, 273–74
Mysteries of the Great Cross of Hendaye, The (Weidner & Bridges), 357
Mystery of the Cathedrals (Fulcanelli), 140
Mystery of the Crystal Skulls (Morton & Thomas), 109, 300–301, 316
mystery religions, 408–9. *See also* specific religions
Mystery Religions in the Ancient World (Godwin), 44

Nag Hammadi, 346, 374
naked singularity, 178–79
nandana, 49
Narmer Palette, 279, 280
natural disasters, 20, 27, 85, 91, 96, 114–15, 124, 132–33, 181, 232, 266–67
Naydler, Jeremy, 349–52
NDEs (near-death experiences), 213, 218–23, *220*

Nebuchadnezzar, 72–74, *73*
Needles of Stone (Graves), 375–76
Nephilim, 290
Neptune, 183
Neugebauer, Gerry, 112
New Age, 276, 314–15
New Fire ceremony, 21, 22, 29–30, 132
Newgrange, 98, *385*, 385–86, 386–88, *387*
New Gravity, The, 178–79
Newitt, Larry, 160
New Jerusalem, 145–47, 150
New Prophecy Almanacs, 267
New Scientist (journal), 327
New Testament, 264, 382
Newton, Kondaveti, 232
New World Order and Antichrist: Years 2002-2012 website, 270–71
Nexus Magazine, 284–86
Nibiru, 24–25, 110–13, *111–13*, 114–15, 137–38, 229–30, 257–58, 276, 288
Nidle, Sheldon, 287
Nine Hells, 302–3, 311–12
Niroma, Timo, 173
Nitsch, Grandmother Twylah, 38
Niven, Larry, 275
Njau, E.C., 170
NOAA Palaeoclimatology Program website, 167
Noah, Joseph, 87
Nommo, *33*, 33–34
nonmolecular body, 349–52, 373
Noone, Richard, 165
noosphere, 193, 197, 213–14, 318, 325–26, 345
Nostradamus, 230, 264–75
Nostradamus: The Apocalyptic Decade, The Crucial Years Until 2012 (Dimde), 268
Nostradamus: The Complete Prophecies (Hogue), 266
Notes from the Cosmos (Scallion), 32–33, 298
Nothing in This Book is True, but it's Exactly How Things Are (Frissell), 261–62
Novelty Theory, 57, 195
nuclear installations, 231
Nunez del Prado, Juan, 359, 361, 362–63
nust'a, 364–65
Nut, 175
Nyingmapa, 50

OBEs, 156–58, *157*, 212, 213, 223–24, *224*, 340, 403–5, 414
obliquity, 116–17, 165, 168–69, 386, 388
O'Brien, Christian, 229
O'Brien, Joy, 229
O'Callaghan, Michael, 333–34, 397
Oh Shinnah Fastwolf, 31–32
Old Testament, 84–86, 138–39, 171–72
Oljato, 173
Ollin, 20
Olmecs, 302
Omega Manu, 213–14
Omega Point, 192–93, 212
One Hunahpu, 356–57
Order of Sion, 142
Order of the Temple of the East, 305–6
orgone, 375–76
Orion, *119*, 128, 147, 152–53, *153*
Orion, Age of, 149
Orion Mystery, The (Bauval & Gilbert), 108–9, 148, 149, 410
Orion Prophecy, The (Geryl & Ratinckx), 150–54, *151–53*, 316
Orion's Belt, 125, *126*, *127*, *129*, 147, 148, 149, 301, 381
Osiris, 108–9, 115, 140, *147*, *149*, 348–49, 408–9, 410
out-of-body experiences. *See* OBEs

Pachakuteq, 360–61
pachakuti, 28, 359–60, 363, 369
Page, Steve, 324–25
Papa, 40
Paris Papyrus, 144–45
Parker spiral, *186*, 186, 332
Passport to Magonia, 325
past-life recall, 228–31, *229*
Paxon, Gregory, 231
Paytiti, 365, 369
Peacock, Kev, pl.10, 95–96, 319, 397
Pegasus, Square of, 123–24
period of justification, 281–82
Perrault, Bruce, 261
Perry, John Weir, 344–45
Persinger, Michael, 253, 327–28, 341, 368, 397–99
Pettigrew, Mike, 220
Phantasm 2012 AD/ Phantasm's End (film), 70–71
Pharaohs and Kings: A Biblical Quest (Rohl), 171
phi, *187*, 187–88
Philadelphia Experiment, The (Berlitz and Moore), 259
Philippines, 87
Phillips, Graham, 110, 229
Phobos, pl.12, 81, 87, 113–14, 268–70, 298–99
Phoebe, 34
Phoenix, return of, 126–29, *127*, *129*
Phoenix Cycle, 135
photon belt, 284–89, *285*, 311, 320–22
Pickover, Clifford, 340
Pi in the Sky (Poynder), 97–98, 316
Pike, Bishop Jim, 412
pineal gland, 91, 134, *205*, 205–8, *207*, 209, 306, 339–40, 345, 347–48, 386, 391–93
pingala, 208
pinoline, 206, 337, 397
Piora Oscillation, 172–73
Pisces, 123
Places of Power (Devereux), 375–76, 393
Planetophysical State of the Earth and Life (Dmitriev), 180–81
Planet X, 187
plasma
  body and, 373
  crop circles and, 323–26
  thermal plasmas at Hessdalen, 326–28
plasma, magnetized bands of, 183–88, *184–85*, 332, 394, 396
Platonic (Great) Year, 42–43
Pleiades and Pleiadians, 24–25, 25–26, 101, 102, 131, 132–33, 285–86, 288–89, 290, 301
Pleiadian Agenda, The (Clow), 288, 320–21
Pluto, 183
polarity, 155
pole shifts, 116–17, 154, 158–59, 221–22, 298–99, 320, 331
Poleshifts: Theosophy & Science Contrasted (Pratt) website, 154
Popol vuh, 20, 172
Pournelle, Jerry, 275
Poynder, Michael, 316
prana, 353–54
Pratt, David, 154
precession of the equinoxes, *18*, 45–46, 49, 88, 124, 142, 148, 167–68, 276, 289, 321, 396, 410–12. *See also* Long Count
preparation, 393–96
Pringle, Lucy, 389
Profiles in Wisdom (McFadden), 31–32
Prometheus Rising (Wilson), 195–96
prophecies, 24–40
  preparation for, 393–96
  reliability of, 310–15

Prophet Code, The (Crockett), 78
psychic abilities, 331–32. *See also* telepathy
Pueblos, 34
Pulstar, 403–4
Pyramid of Kukulcan, pl.14, 22, 29, 102, 130–32, *131–32*, 321
Pyramid Texts, 407–8
Pythagoras, 176
Pythagorean comma, 176

Q'ero, 27–28, 359, 368
Qoriwaman, Don Benito, 363
Qoyllur Riti, pl.21, 364–65
Quantum Awakening, 197–99, 394
Quechua people, 359
Quetzacoatl (Kukulcan), 22–23, 25–26, 52, 107–8, *130*, 149, 174. *See also* Pyramid of Kukulcan
Quetza-Sha, 25–26
quincunx, 246–47

Ra, 179–80, 410
Rabin, Yitzhak, 84–85
Rain, 20
Rainbow Warriors, 305
Ramanda, Azena, 287
Rangi, 40
Rapture, The, 71
Rasool, Jazz, 96, 250
Ratinckx, Gino, 129, 150–54, 316
Rawls, Walter, 391
Real Maya Prophecies website, 234
Realm Border Wave, 289–90
Regulus, 76
Reich, Wilhelm, 375–76
Reichenbach, Baron Karl Von, 372
Reiser, Oliver, 45
remote viewing, 262–63
Return of the Phoenix, The (Mandeville), 126–27, 136
Revelations, Book of, 71–72, 75–77, 86, 87, 115, 145, 266–67
Reynolds, Pam, 223
Right Where You Are Sitting Now (Wilson), 194
Rig Veda, The (Bosman), *213*
RISA website, 54–55
Roberts, Paul, 159–60
Rodriguez, Roberto, 26
Rohl, David, 171–72
Roney-Dougal, Serena, 208–10, 337, 397
Rosh Hashanah, 79–81
Rosicrucians, 142
Rudisill, Elliott, 114–15, 116, 316
Rudra Cakrin, 50, 54. *See also* Kalki
Rueckert, Carla, 180
Russell, Peter, 105, 193–94, 198–99
Russia, 87

Sabatino, Charlie, 215–16
Sabians, 148
sacred sites, 24, 341–42, 355, 368, 375–77, 394. *See also* specific sites
Sagittarius, 44–45, 49, 115, *119*, 137
Sahagún, Bernardino, 203
sahu body, 150, 350
saiwas, 356
Salem, Kenneth G., 178–79
Salvia divinorum, 215–17, *216*, 338–40
samadhi, 353
Sanduloviciu, Mircea, 327
Santillana de, Giorgio, 44–45, 123–24, 147–48, 410
Saquasohuh, 31, 32
Sargon of Akkad, 69
Sarmoung brotherhood, 148
satellite pyramids, 136–37
Saturn, 183
Satyananda, Swami Saraswati, 208–9
Saved by the Light (Brinkley), 220

Scallion, Gordon Michael, 32–33, 297–99
schizophrenia, 343–45, 397
Schnoll, Simon, 330–31
Schoch, Robert, 128
Schumann resonance, 96, 156
Schwaller de Lubicz, R.A., 122–23
science, fringe, 194–97
  Auric Time Scale, 188–92
  critical mass, 192–94
  frequency changes, 178–79
  magnetized band of plasma, 183–88, *184–85*
  Quantum Awakening, 197–99
  solar system changes and, 180–84
  thirteen-baktun cycle and, 179–80
science of immortality, 346
Science of Oneness, The (Wilcock), 180
Science of Spirit, The (Turner), 223
Scofield, Bruce, 10, 35
Scorpio, *119*
Scott, Mary, 352–53, 353–54, 377
Scott, Steve, 341
Scully, Nicki, 147
Second Coming, 71, 74, 75, 262
Second World, 28
Secret History of Crop Circles, The (Wilson), 246–47
Secret of the Andes (Brother Phillip), 368–69
Secrets of Maya Science/Religion (Men), 23–24
seed-people, 27
Séjourné, Laurette, 395
Sekhmet, 112–13, 175
Seneca, 38
serpents, 130–31, 132
Seven Macaw, *18*, 411
Sewell, Helen, 279–80
sexual energy, 348, 350, 375–76
Shakti, 352–53
Shamanism and the Mystery Lines (Devereux), 382, 414
shamans and shamanism, 203–5, 310–15, 344–45, 348–49, 359, 377–78, 407–8
Shambhala, 50, 53–54
Sharma, Mukul, 162
Shavuot, 80–81
Shaytan, 81
Shearer, Tony, 14, 22–23, 37, 38, 51, 303
Sheliak, John, 61, 68
Sheliak wave, *64*
Shift of the Ages, The (Wilcock), 179–80, 316
Shining Ones, 98
Shipton, Mother, 272–73, *273*
Shiva, 352–53
Shnoll, Simon, 367, 396–97
Shoemaker-Levy, 183
Shongo, Moses, 38, 314–15
Short Count, 133
Siberia, 412–13
Signs in the Sky (Gilbert), 147–50, 316
Silver Age, 142
Silverbird, J. Reuben, 224
Silver Gate, 44–45, *45*, 148–49, 150
Sirius, 32–33, 33–34, 46, 112–13, 122–23, 134–35, 147, 175–76
Sirius Mystery, The (Temple), 33, 141, 276
Sitchin, Zecharia, 110, 137–38, 229, 257–58
Sixth Generation, 27
Sixth Sun, 104, 233–34, 315
Skyboom website, 55
SkyGlobe software, 108–9, 129, 135, 151–52
Slosman, Albert, 153–54
Smelyakov, Sergey, 188–92, 317–18, 332–35
Smith, Joseph, 37

Snake Handler, *119*, 122
Solara, 282
solar activity, 104, 162–63, 180–83, *182*, 249, 316–17, 319–20
solar eclipse, 75, 100, 233–34, 273
solar flares, pl.7, pl.9, 95, 141, 269, 298, 337
solar maximum, 94–95, 95–96, 319–20
solar megacycle, 95, 181, 316–17, 414
solar system activity, 95–96
solstices, 43–46, 49, 97, 98, 100, 102, 148, 342
Solving the Greatest Mystery of Our Time (Calleman), 297
soma, 412
Soma Conspiracy, The (Bosman), *213*
Soma: Divine Mushroom of Immortality (Wasson), 412
Sonic Resonator, 260–61
Space Daily website, 320
Space-Time Transients and Unusual Events (Persinger & Lafrenière), 397–99
Speaking Wind, 34
speed of light, 178–79
Sphinx, 128
spine, 356–57
spirit molecule. *See* DMT
Spotiswoode, James, 331
spring equinox, 130–31. *See also* Aquarius, Age of
sprites, 325
Stanford, Ray, 177
Starr, Jelaila, 113
Starseed Transmissions, The (Carey), 197
Steiner, Rudolph, 281
Stevens, Len, 315
Stonehenge, 173, 386
Stones of Time, The (Brennan), 98
Straight Dope website, 288
Straiton, Valentia, 347–48
Strand, Erling, 326
Strassman, Rick, 206–8, 213, 337, 397
Stray, Geoff, 20–21, 68, 105–6, 117
Streiber, Whitley, 393
Stross, Brian, 356–57
Sukkot, 77
Sullivan, Jack, 248–49, 250–51
Sullivan, William, 28
Sumerians, 24, 110
summer solstice, 148, 149. *See also* solstices
Sun, 16, 17, 49, 96, 137, 250
  activity level of, 95, 320
  Sirius and, 134
Sun, Moon and Stonehenge (Heath), 383–84
Sun disk, *368*, 368–69
Sun kundalini, 353–57
Sun of Movement, 20, 247
Sun-Pleiades zenith conjunction, 131, 132
Suns (Aztec and Maya eras), 92–93, 172, 247
Suns (Aztec gods), 18–20, *19*, 22
sunspot cycles, 91–93, 95–96, 104, 150–51, 179–80, 316
  artifacts and, 97–98
  change in solar maximum, 94–95
Supernature (Watson), 373
sushumna, 208
Sussex Circular (journal), 248–49, 250–51
Swiftdeer, Harley, 302, 314
Swirled News website, 249
synchronicity, 3
Syrian Rue, 214–15

Talbott, Nancy, 322–23, 327
Talisman - The Sacred Cities and the Secret Faith (Hancock and Bauval), 135
Tamil calendar, 49–50
Taoist yoga, 370–72, *371–72*
Taripay Pacha, 28, 359–60, 363

tarot, pl.5, *76*, 119, 142
Taurids, 173
Taurus, *119*
Taylor, Caroline, 206–8
Taylor, Leon, 66
Teilhard de Chardin, Pierre, 192–93, 325–26, 345, 398–99, 419
telepathy, 96, 105, 210–11, 237–38, 376–77, 388–89, 397
Temple, Robert, 33–34, 141, 276
Temple Mount, 74, 75
Temple of the Cosmos (Naydler), 349–50
temporal lobe epilepsy (TLE), 340–43, 397
Tenochtitlan, 20, 313
Teordorani, Massimo, 326
Teotihuacan, 300–301
Teotitlan calendar, 53
Teotitlan Calendar Round, 23
Teshuvah, 80
Testimony (magazine), 278–79
Test of Time, A (Rohl), 171
Thelemites, 303–4
theory filtration, 316–18
Thera, 229
thermal plasmas, 326–28
third eye. *See* pineal gland
Third World, 28–29
thirteen-baktun cycle, 1–2, 11, 13–15
  climate cycles and, 164–65
  crop circles and, 247–48
  double catastrophe and, 143–44
  EMP waves and, 120–21
  end date of, 32, 198–99, 247–48
  extraterrestrials and, 240
  flooding and, 172
  fractal waves and, 61–63
  galactic synchronization and, 283–84
  Great Seal pyramid and, 133–35
  Mercury magic square and, 67
  obliquity cycle and, 386, 388
  phi and, 189–92
  rotation of Earth and, 158–59
  significance of, 111
  Suns and, 247
  sunspot cycles and, 91–93, *93*
  Venus and, 103–4, 109–10
  Wilcock on, 179–80
Thirteen Heavens, 302–3
Thomas, Andy, 249
Thomas, Ceri Louise, 109, 300–301, 313–14, 316
Thompson, Eric, 12, 295
Thompson, Lonnie, 174
Thornborough Henge, 381–82, 388
Thoth, 84, 175
thunderstorms, 375
Tibet, 50
Tibetan calendar, 47–48, 50–54, 53
Tikal Calendar Round, 23
Timbavati, 39
Time of the Warning, 302
time travel, 260–62
Timewave, the, 57–65, *60–64*
Timewave Zero, 57, 59, 194, 198, 204, 337–38
Time Wave Zero 2012 Internet discussion group, 65
Timms, Moira, 31, 142
Tishya constellation, 54–55
Tlakaelel, 310–11
toad venom, 203
Toltecs, 29, 32, 130
Tonatiuh, 20
Torah, 78
tornadoes, 328–29
Toutatis, pl.11, 273–74
track lines, 355
Transcendent Experience and Temporal Lobe Epilepsy (Pickover), 340

Transformative Vision, The (Arguelles), 51
Tree of Knowledge, 347, 411
Tree of Life, 142, 145, *346*, 347, 411
tribal prophecies, 24–40, 310–15, 393–96
Tribulation, 75, 76
Troxell, Dan, 36–37
True Count, 295
tryptamines, 203, *205*, 341
tryptophan, 212, *213*
T Tauri stars, 119–20
Tully, Australia, 242
Turner, O. Frank, 223
Twelfth Planet, The (Sitchin), 110, 112–13
Typhon, 112–13, *113*, 306
Typhonian OTO, 306
tzitzimime, 20, *21*
Tzolkin, 9, 13–15
tzolkin, *14*, 21, 97, 291–97, 295–97, 318
Tzoltze ek', 24–25

uayeb, 10
UFO Magazine (UK), 238
UFO Reality (magazine), 235, 237
UFOs, 98–99, *99*, 212, 233–34, 283, 326, 337, 397
  alien telepathy, 237–38
  assessment of by military, 239–40
  contact with extraterrestrials, 235–36
Uinals, 291–97
Ultimate Secret of the Mayan Calendar, The (Wilcock), 188, 191–92
Unborn Mind, 204–5
Underwood, Guy, 353–54
Unites States, 82, 87, 133–35
Urcos, 364–67
Uriel, 384
Usoskin, Ilya, 320
Ussher, James, 79

vacuum domains, 185, 328–29
Vallee, Jacques, 325
Van Impe, Jack, 71
Vedas, 41, 43–46, 213–14
Vedic time cycles, 41–43, 88
Velikovsky, Immanuel, 107, 108, 110, 228, 231
Venus, 9, 21–22, 76, 100–102, 102–3, 137, 181, 229–30, 233–34
  birth and death of, *107–8*, 107–10
  crops circles as transit predictor, 250–52
  Orion and, 149
  thirteen-baktun cycle and, 103–4, 109–10
Venus loop, pl.13, 129, 149, *151–52*, 151–52
Venus Round, 22, 250–51, 254
Venus transit, 100, 100–102, *101*, 103–5
vernal equinox, 148
Vigay, Paul, 226
Villoldo, Alberto, 314, 359, 359–60, 363
Virgin Mary, 177
Virgo, 77, 123–24
Vishnu, 54–55
Vital Signs (Thomas), 249
volcanos and volcanic activity, 96, 110–11, 172–73, 181
Vostok (Antarctica) ice cores, *161*, 161–62, 169–70

Walker, Dean, 225
Ward, Ray, 268–69
Wasson, Gordon, 412
Water, 20
water lines, 353–54
Waters, Frank, 30, 31, 124, 198, 247, 336
Watkins, Mathew, 61, 66, 69
Watson, Lyall, 373
Way to Shambhala, The (Bernbaum), 54–55

weather
need for control of, 378–82, *380*
sacred sites and, 375–76
See also natural disasters
Weidner, Jay, 141–42, 143–44, 150, 278, 280, 318, 358, 360, 396
Weldon, Randolph, 228–31
West, John Anthony, 128
Whalen, John, 82
Wheeler, John, 328
When the Earth Nearly Died (Allan & Delair), 165, 411
Where Science and Magic Meet (Roney-Dougal), 208
White, Paul, 135, 136
White Feather, 31
Whitehead, Alfred North, 57, 59
White Hole in Time, The (Russell), 193–94
Whiteside, Richard Henry, 71
Wilcock, David, 179–80, 181, 186, 188, 191–92, 316, 329, 332, 396
Wilcox, Joan Parisi, 27–28, 356, 359–60, 362–63, 369
Wilkie, James, 136, 138–39, 140, 318
Williamson, George Hunt, 369
Wilson, Robert Anton, 194–96, 403–4
Wilson, Terry, 246–47
Wind, 20
winter solstice, 17–18, *19*, 49, 98, 116, 148, 150, 167, 279–80, 342
Wiracocha Foundation website, 366
Withbroe, George, 94
Wooley, Sir Leonard, 171
Worlds in Collision (Velikovsky), 107
World War II, *64*, 69
World War III, 87
Worp, James van der, 113–14, 269
Wright, Paul, 137
Wúwuchim ceremony, 31

X-Files, The, 240–41
Xibalba, *19*
X-ray flares, 94–95

Yahowah, 37
Yaxk'in, Aluna Joy, 253
Year 2012, The (Cain), 70
Yom Kippur, 75, 80–81
York, Andrew, 327–28
You Are Becoming a Galactic Human (Nidle), 287
yugas, 41–42
Yukteswar, Sri Swami, 43, 46

Zep Tepi, 128–29
zero point, 328–29
Zeus, 107, *113*
zodiac, 119, 122–24, 137, 153–54, 277, *285*
Zuidema, Tom, 367
Zulus, 38–39
Zyzygyz, pl.6, 65–69

BOOKS OF RELATED INTEREST

Maya Cosmogenesis 2012
The True Meaning of the Maya Calendar End-Date
by John Major Jenkins

Galactic Alignment
The Transformation of Consciousness According to Mayan, Egyptian, and Vedic Traditions
by John Major Jenkins

The Mayan Calendar and the Transformation of Consciousness
by Carl Johan Calleman, Ph.D.

The Mayan Code
Time Acceleration and Awakening the World Mind
by Barbara Hand Clow

Catastrophobia
The Truth Behind Earth Changes
by Barbara Hand Clow

The Mayan Factor
Path Beyond Technology
by José Argüelles

2012 and the Galactic Center
The Return of the Great Mother
by Christine R. Page, M.D.

Return of the Children of Light
Incan and Mayan Prophecies for a New World
by Judith Bluestone Polich

Inner Traditions • Bear & Company
P.O. Box 388
Rochester, VT 05767
1-800-246-8648
www.InnerTraditions.com

Or contact your local bookseller